1st International Conference on Natural Resources and Environmental Conservation (ICNREC2018)

Impact of Oil Palm Plantation on Physical and Chemical Environment, Biodiversity and Local Socio-Economic Diversity of the Surrounding Communities

IOP Conference Series: Earth and Environmental Science
Volume 336

Bogor, Indonesia
23 October 2018

ISBN: 978-1-5108-9668-0
ISSN: 1755-1307

Printed from e-media with permission by:

Curran Associates, Inc.
57 Morehouse Lane
Red Hook, NY 12571

Some format issues inherent in the e-media version may also appear in this print version.

Institute of Physics
Dirac House, Temple Back
Bristol BS1 6BE UK

Phone: 44 1 17 929 7481
Fax: 44 1 17 920 0979

techtracking@iop.org

Additional copies of this publication are available from:

Curran Associates, Inc.
57 Morehouse Lane
Red Hook, NY 12571 USA
Phone: 845-758-0400
Fax: 845-758-2633
Email: curran@proceedings.com
Web: www.proceedings.com

TABLE OF CONTENTS

IMPACT OF OIL PALM PLANTATION ON SOCIAL AND ECONOMIC

IMPACT OF OIL PALM PLANTATION ON PHYSICAL AND CHEMICAL ENVIRONMENT

IMPACT OF OIL PALM PLANTATION ON BIODIVERSITY

IOP Conf. Series: Earth and Environmental Science **336** (2019) 011001 doi:10.1088/1755-1315/336/1/011001

PREFACE

The positive and significative contribution of oil palm plantations to the development of the Indonesian economy (*through increasing foreign exchange earnings, employment and other multiplier effects*) has drawn many negative accusations in environmental aspects (*as causes of deforestation, degradation of biodiversity, fires, sources of emissions and other*) from certain parties, so that it becomes a polemic in the public space. In connection with this and encouraged by the enthusiasm to achieve "sustainable and environmentally friendly oil palm plantation management", Faculty of Forestry IPB in collaboration with the Center for Nature Study, Advocacy and Conservation (PUSAKA KALAM) and CIRAD France consider it necessary to invite experts/specialist from various parts the world and the archipelago and are multi-disciplinary to discuss objectively and scientifically about the "impact of Indonesian oil palm plantations on the physical/chemical components, biodiversity and socio-economic diversity of the surrounding communities".

The First International Conference on Natural Resources and Environmental Conservation (ICNREC) which was opened by the Deputy for Food and Agriculture Coordination of the Coordinating Ministry for Economic Affairs RI featured 2 keynote speakers; 5 prominent speakers and 41 paper presenters. This proceeding consists 33 papers that have been through peer review process. Based on the conference topic, there are 15 papers relating to socio-economic and cultural aspects of local communities, 8 papers relating to physical/chemical aspects, and 10 papers related to biodiversity. It is expected that the results of this international conference will be able to minimize the diversity of opinions among experts/researchers, central/regional government officials, NGOs and stakeholders related to plantations and the palm oil industry.

In particular, we would like to express our gratitude and appreciation to the Center for Nature Study, Advocacy and Conservation (PUSAKA KALAM), the French Agriculture Research Center for International Development (CIRAD), Institut Français Indonesia (IFI), Coordinating Ministry of Economic Affairs of Indonesia, Oil Palm Plantation Fund Management Agency (BPDPKS), Indonesian Palm Oil Association (GAPKI), PT. Musim Mas and PT. Astra Agro Lestari for their financial support and good cooperation so that this conference can be carried out as expected. We also convey our gratitude and appreciation to the speakers, the team of reviewers, the organizing committee and all the participants of The First ICNREC. Like Indonesia's old proverbs saying that "there is no ivory that is not cracked", we apologize if there is a lack of perfection in the implementation of The First ICNREC and the preparation of the ICNREC proceedings.

Best Regards,

Prof. Dr. Ir. Yanto Santosa, DEA
Chairman of Committee of ICNREC 2018

IOP Conf. Series: Earth and Environmental Science **336** (2019) 011001 doi:10.1088/1755-1315/336/1/011001

Organized by:

Bogor Agricultural University (IPB)

Co-organized by:

Center for Nature Study, Advocacy and Conservation (PUSAKA KALAM)

French Agriculture Research Center for International Development (CIRAD)

Sponsored by:

Indonesia Oil Palm Plantation Fund Management Agency (BPDPKS)

Coordinating Ministry of Economic Affair Indonesia

Institut Francais Indonesia (IFI)

Indonesian Palm Oil Association (GAPKI)

PT. Musim Mas

PT. Astra Agro Lestari

IOP Conf. Series: Earth and Environmental Science **336** (2019) 011001 doi:10.1088/1755-1315/336/1/011001

CONFERENCE PHOTOGRAPHS

ICNREC
Impact of Oil Palm Plantation on Physical and Chemical Environment, Biodiversity and Local Socio-Economic

ICNREC

ICNREC

Peer review statement

All papers published in this volume of *IOP Conference Series: Earth and Environmental Science* have been peer reviewed through processes administered by the proceedings Editors. Reviews were conducted by expert referees to the professional and scientific standards expected of a proceedings journal published by IOP Publishing.

IOP Conf. Series: Earth and Environmental Science **336** (2019) 012001 doi:10.1088/1755-1315/336/1/012001

Oil palm agroforestry: an alternative to enhance farmers' livelihood resilience

Budiadi[1], Ari Susanti[1]*, Hero Marhaento[1], M Ali Imron[1], Dwiko B Permadi[1], Hermudananto[1]

[1]Universitas Gadjah Mada, Faculty of Forestry, Jl. Agro No. 1 Bulaksumur 55281 Yogyakarta

*E-mail: arisusanti@ugm.ac.id

Abstract. Oil palm has been rapidly expanding in Indonesia and becoming an important export commodity for the country. However, oil palm production has been promoted only as a monoculture, and this has led to adverse impacts on natural and social systems. At the same time, the demand for oil palm products is likely to increase, and therefore there is an urgency to find alternatives for more sustainable oil palm production. We argue that the monoculture system is not necessarily to be the only way to produce oil palm and this article aims at analysing alternatives for oil palm production systems beyond monoculture with agroforestry as the main focus. We visited oil palm agroforestry practiced by smallholders in Jambi and Central Kalimantan and asked questions about their motivation in adopting agroforestry, farm management, value chain and institutions. We found that smallholders have various motivations in adopting oil palm agroforestry but improving their household resilience becomes the most prominent reason. The adoption of perennial species, oil palm and other crops in the farms may improve the diversity of the farms and thus the diversity of household income. However, smallholders encounter problems on productivity because of the competitions in spaces. Lack in the knowledge of species selection, quality of planting materials and spacing arrangement have influenced oil palm productivity. In addition, market access also becomes a crucial issue when their farms are located in remote areas. There is currently no institutions, both government and non-government which promote and facilitate the adoption of oil palm agroforestry although this system has been practiced especially by smallholders. We suggest that oil palm agroforestry system needs to be further investigated not only regarding financial benefits for the smallholders but also its social acceptability and barriers of adoption as well as its potentials in contributing to the improvement of ecosystem functions.

1. Introduction

Oil palm has been seen as the green gold as it contributes to national and regional income generation, job creation and economic growth and its development has been supported by national and regional regulations [1], [2]. It has been rapidly expanding and becoming an important export commodity for the country. In 2016, Indonesia produced more than 31 million tons of palm oil from its 11 million ha oil palm plantations where around 40% of it is managed by smallholders. Around 24 million tons of oil palm was exported with total export value amounted USD 16.27 billion. The whole palm oil production contributed around 13.52% of national GDP [3].

Most of the oil palm plantations in Indonesia are cultivated as a monoculture. The establishment of these monoculture oil palm plantations has involved the conversion of significant size of land [4] which includes forest lands [5], agricultural lands [6] and peatlands [7], [8]. These land conversions have led to deforestation [5], habitat fragmentation [9], [10], biodiversity loss [11], [12], [13], disturbance on hydrological system [14], land subsidence [15], [16], recurrent fire events and increased GHG emissions [17], [18].

Regions that highly depend on oil palm tend to experience boom and bust economic growth [19] instead of long-term sustainable growth because their growth is subsidized by the exploitation of lands and natural resources [20]. The conversion of diverse land uses into monoculture oil palm plantations has also transformed people's livelihood and potentially threatened food security [21], [22], [23]. The monoculture practice has created a monoculture livelihood strategy that highly depends on merely oil palm [19] and entails social and economic risks [24]. Forest dwellers have potentially found difficulties to survive because of forest loss [25] as well as have been challenged by increasing competition for lands and the conflicts embedded in it [26].

At the same time, the demand for oil palm products is likely to increase. The Coordinating Ministry for Economic Affairs of the Republic of Indonesia estimated that the growth of demand for oil palm products would continue to grow at about 5% annually [27]. In addition, two major sustainability certifications of oil palm, Roundtable on Sustainable Palm Oil (RSPO) and Indonesia Sustainable Palm Oil (ISPO), also require improvement in the performance of production unit for more sustainable production [28], [29]. Developing and implementing an action plan that supports increasing sustainable oil palm production as well as maintaining and preserving biodiversity are necessary for smallholders under the ISPO principle and criteria 4 and 3.3 for smallholders, respectively. Additionally, long-term livelihood benefits by regularly reviewing the performance of the production unit become requirements for RSPO smallholder standard in criteria 6.3. Those conditions encourage plantation managers to find an alternative oil palm management system. For those reasons, there is an urgency to find alternatives for more sustainable oil palm production to minimize those aforementioned adverse impacts on natural and social systems by promoting more sustainable production of existing oil palm plantations without expanding its area.

In this article, we argue that the monoculture system is not necessarily to be the only way to produce oil palm. Mixed cropping between oil palm and other crops [30], [31] or livestock [32] has been practiced in limited scales. Experiments indicated that the biodiversity enrichment in oil palm plantations with multi-purposes trees species (MPTS) could contribute to the improvements of ecological and social functions of the landscapes [24]. The land sharing in this agroecosystem matrix of agroforestry seems to be a more realistic to promotes the multiple uses of landscapes [33] than land sparing with intensified agriculture [34]. Therefore, this article aims at analysing alternatives for oil palm production systems beyond monoculture with agroforestry practiced by smallholders as the main focus.

2. Methods and analysis

As the oil palm agroforestry has not been widely adopted, the selection of the cases is mainly based on the oil palm agroforestry practice. The selection of the cases was done through a series of focused group discussions (FGD) with key informants. This includes government officials, NGOs and village leaders. We visited oil palm agroforestry practiced by smallholders in Jambi and Central Kalimantan (two cases). Sumatra is an old oil palm region that hosts the oldest commercial oil palm plantation in Indonesia [35] while Kalimantan becomes a relatively new frontier for oil palm expansion in Indonesia [36], [37]. We asked questions about their motivation in adopting agroforestry, farm management, value chain and institutions. This includes the problems they have perceived in managing oil palm agroforestry. The interviews were conducted in a semi-structured nature [38].

The results of the interviews have been employed to describe the existing types of oil palm agroforestry practiced by farmers, the motivation of adopting oil palm agroforestry, the problems related to its management, institution and market of the oil palm fresh bunch fruits.

This article proceeds in three sections. The first section links our field investigation with theoretical consideration of agroforestry, livelihood and conservation. The second section presents examples of agroforestry practices from Jambi and Central Kalimantan that includes the motivation in adopting oil palm agroforestry, farm management and barriers in adopting oil palm agroforestry. The third section offers discussion and conclusion.

3. Agroforestry, livelihood and conservation

Lundgren [39] defines agroforestry as a collective name for land use systems and technologies in which woody perennials (trees, shrubs, palms, bamboos, etc.) are deliberately combined on the same management unit with herbaceous crops and animals, either in some form of spatial arrangement or temporal sequence. In agroforestry systems, there are both ecological and economic interactions among the different components. Therefore, agroforestry has been considered to contribute to improving the livelihood and ecological resilience [40].

Agroforestry could be established through the integration of trees into farming systems or integration of farmers into forests [41]. The adoption of trees species into agricultural system could improve rural livelihood in the tropics because the function of tree products is ubiquitous in rural livelihoods such as in construction, fencing, furniture, foods, medicines, fibres, fuels and livestock feed, and their cultural value [42]. This way, agroforestry offers an opportunity to diversify sources of household income and more sustainable rural development [41].

In addition, the current expansion of crops such as oil palm in the tropics has replaced a significant size of tropical forests. This has led to habitat fragmentation [12], biodiversity loss [11] increased GHG emissions [17] and land-related conflicts [43], [26]. Agroforestry system has been widely recognized of having potential in contributing to climate change mitigation because of its capacity in carbon storage and sequestration [44]. In addition, the integration of farmers into forest could become an alternative to resolve or mitigate tenure problems in relation to agricultural expansion into forest areas [41].

Moreover, debates on land sharing and land sparing also suggest that the future conservation should be integrated with the agro-ecological matrix [34]. This will become an alternative to land sparing, and intensified agriculture becomes the prominent conservation practices [45] and has led to adverse impacts on livelihood and ecological resilience [40]. Agroforestry might become one of the realistic ways to sustain forest resources in the current situation where the pressure to forestlands has been continuously increasing [46] driven by the process of globalization of space [47].

However, market of commodities is very crucial in influencing farmers´ decision for the adoption of agroforestry system [48]. In other words, the agroforestry landscape should be productive and could offer meaningful financial benefits, social and environmental impact for the households so that farmers will maintain and improve the diversity and the quality of trees species in their farm [49]. Institutional interventions could facilitate to improve the process in farmers decision-making [50] such as in species selection, farm management and marketing of agroforestry products.

4. Practices of oil palm agroforestry

We identified three types of oil palm agroforestry managed by households in Jambi and Central Kalimantan based on the agroecosystem matrix combination in the farm: (a) mixed cropping between oil palm and *Shorea spp*, (b) mixed cropping between oil palm and sengon (*Falcataria molucana*), and (c) mixed cropping between oil palm, jelutung (*Dyera costulata*), rubber and other food crops. This section analyses each type of oil palm agroforestry. The summary of each agroforestry system visited is summarized below.

IOP Conf. Series: Earth and Environmental Science **336** (2019) 012001 doi:10.1088/1755-1315/336/1/012001

Table 1. Summary of each agroforestry type

Information	Oil palm & *Shorea spp*	Oil palm & sengon	Oil palm, jelutung, rubber, and other food crops
Location	Kuamang Kuning, Jambi	Sei Gohong, Central Kalimantan	Sei Gohong, Central Kalimantan
Year of adoption	2000	2016	2008
Area (ha)	2	2	4
Motivation	Timber scarcity for construction and housing & improve timber self-sufficiency	Decrease of rubber latex price & development plan of new sengon mill	Volatility of commodity prices & improve income stability
Land status	Privately owned	Privately owned	Privately owned
Spacing strategy	(9x9) meter for OP trees and (3x3) meter for *Shorea spp*	(9x8) meter with two rows of sengon in between OP rows	irregular pattern with 80 OP trees/ ha
Harvesting of OP FFB	700 kg/ two weeks	-	400 kg/two weeks

4.1. Oil palm and Shorea spp agroforestry

Mixed cropping between oil palm and *Shorea spp* was practiced in SP A, an oil palm transmigration village in Kuamang Kuning, Jambi. The adoption of *Shorea spp* in oil palm plantations was stimulated by the perceived timber scarcity in the region. The community perceived this timber scarcity because they could not easily find timbers for construction and housing in the region and its price has been steadily increasing. The adoption of *Shorea spp* was expected to increase timber self-sufficiency, especially for construction and housing. The main current source of household income has been oil palm as their main job is oil palm farmer. The farmer could harvest around 700 kg every two weeks. The farmer joined the oil palm production system through the transmigration programme as a supported smallholder.

This agroforestry farm is located in adjacent to the house. The parcel size is 2 ha, and it was acquired through transmigration programme. The spacing of oil palm trees is (9x9) meter for oil palm trees and (3x3) meter for *Shorea spp* trees. The integration of *Shorea spp* occurred when the oil palm trees are already 15 years old. This way, the seedlings of *Shorea spp* could get sufficient shading from the oil palm, which is necessary for the juvenile stage of *Shorea spp*. In addition, after 15 years, oil palm trees have reached their peak production phase and the integration of *Shorea spp* should minimize the nutrient and sunlight competition with oil palm trees. The horizontal projection of the farm is depicted in figure 1.

Figure 1. Horizontal projection of oil palm and *Shorea* agroforestry

Because he is a supported smallholder, the farmer gained support from the oil palm nucleus company in establishing oil palm plantation. This includes the certified planting materials and in-house training to maintain his oil palm farm. The *Shorea spp* planting materials were also gained from the field research station of Faculty of Forestry, Universitas Gadjah Mada in Jambi which is located near the village. The field research station was a research center for tropical rain forest rehabilitation, joint cooperation research between Universitas Gadjah Mada and *Kansai Electrical Engineering Center* (KEEC) Japan. The farmer who managed this agroforestry farm is a member of a farmer group which has collaboration with the nucleus oil palm company under the nucleus estate smallholder (NES) scheme. Therefore, the harvest of oil palm fresh fruit bunches (FFB) has been sold directly to the nucleus company.

4.2. Oil palm and sengon agroforestry

Mixed cropping between oil palm and sengon was practiced in Sei Gohong, Bukit Batu near Palangkaraya, Central Kalimantan. The adoption of sengon in oil palm farm was started two years ago and was stimulated by the continuous decrease of rubber latex price and the rumour about the development plan of the new sengon mill in Pulau Pisang, Central Kalimantan which is hoped to elevate the price of sengon timber. The current income of the household is sourced mainly from rubber and oil palm which have been managed in other parcels. The future incomes generated from sengon timber and oil palm become the household strategy to cope with the loss of income caused by a continuous decrease in rubber latex price.

This oil palm and sengon agroforestry cover 2 ha on a privately owned parcel. This agroforestry parcel is not located surrounding the house such as home garden. The distance of this parcel from the house is around 0.5 km. With this distance, it is handy to have oil palm and perennial trees such as sengon because they do not demand daily maintenance. The spacing for oil palm trees is (9x8) meter. This spacing allows the incorporation of sengon trees in between the oil palm trees. In this case, there are two rows of sengon trees in between oil palm rows. The canopy configuration of the farm is depicted in figure 2.

The oil palm planting materials were obtained from the local market without information on quality assurance or certificate. The farmers do not have access to the knowledge about good quality planting material because the agricultural extension or other sources or agricultural information current do not exist in the region. Besides, the price of oil palm certified planting materials is relatively expensive and most of the smallholders cannot afford it. Similar case also occurred to the sengon planting materials, and even worse because the young sengon trees were destroyed by a group of wild deer. The farmer also uses fertilizer mainly manure and pesticides especially in the establishment stage of the farm establishment.

During the field investigation they were more or less two years old, and therefore oil palm trees were not yet producing fruits, and the sengon trees are expected to be mature at around six years old. So far, smallholders sell their rubber latex and oil palm FFB through middlemen who play roles as collectors. These middlemen collect the rubber latex of oil palm FFB from the farm or along the roads and transport the rubber latex or oil palm FFB. Often, these middlemen also become the investors for the new oil palm plantation. This occurred mainly when the smallholders have limited access to formal financial schemes. The future harvest of oil palm agroforestry and the sengon timber will likely experience similar path of the value chain.

The farmer we have visited has been a member of a farmer group. However, it seems that there are not many activities organized by the farmer group and becoming a member of a farmer group is more administrative than a real need.

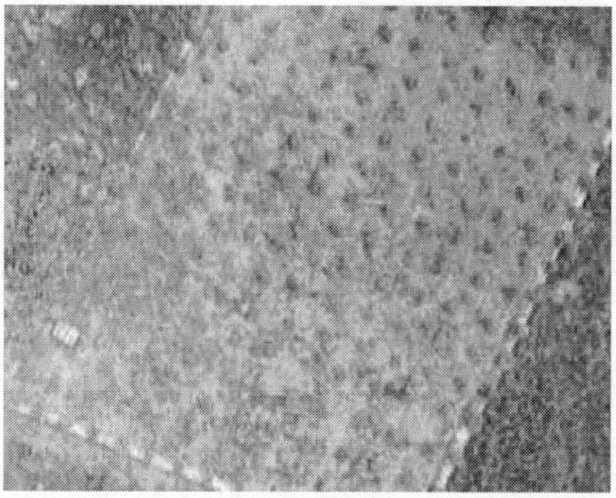

Figure 2. Vertical projection of oil palm and sengon agroforestry

4.3. Oil palm, jelutung, rubber, and other food crops agroforestry

In Sei Gohong we also found agroforestry practice which combines oil palm, rubber, jelutung (*Dyera costulata*), petai (*Parkia speciosa*), and pineapple (*Ananas muricata*). The spacing between oil palm, rubber jelutung and petai trees does not follow certain distance, and therefore the spatial distribution of the trees has an irregular pattern. This agroforestry covers around 4 ha private land which located surrounding his house. Although it the size of his farm is relatively large, he manages to do all the work by himself with the help of the family members.

This farmer has been managing agroforestry farm for ten years, but initially, he mixed rubber and jelutung and pineapple. Oil palm trees came later in his farm. It started when the rubber trees were around three years old. Therefore the average age of the rubber trees is around ten years while the age of oil palm trees is ranging from five to seven years. The introduction of oil palm trees into his farm was stimulated by the volatility of commodity prices in the global market and managing only one type of commodity in the farm was perceived as a high risk in term of household income stability.

With this combination of several crops in his farm, he could manage to diversify his household income sources and improve the household resilience. He applies certain yield regulation so that he can manage daily income from various harvests of his farm. This was done by dividing his farm into several blocks of management to regulate the harvests of rubber, oil palm, jelutung and pineapple. He manages to harvest around 400 kg oil palm FFB every week from different blocks alternately. He sells this oil palm FFB for IDR 800 per kg to a middleman who comes to his farm every week.

Figure 3. Vertical projection of oil palm, rubber, jelutung and other food crops agroforestry

5. Discussion

From the agroforestry practices aforementioned above, we learned that smallholders have various motivations in adopting oil palm into their agroforestry farm, but the financial benefits in improving their household resilience become the most prominent reason. The price volatility of commodities in the global markets, the decrease of rubber latex price and the increase of oil palm have been stimulating the adoption of oil palm agroforestry. The adoption of perennial species such as *Shorea spp*, sengon, jelutung, rubber, and petai as well as other crops such as cocoa, pineapple and seasonal vegetables in the farms may improve the diversity of the farms. This way, the farmers could diversify their sources of income for the household. It likely the market of commodities play important roles in determining the farmers' decision in adopting certain species [48] and increase the adaptability of agroforestry.

However, smallholders encounter problems related to farm management. First, the species selection is mainly based on the market instead of the biophysical suitability or combination compatibility. For example, when the price of oil palm increases farmers will likely adopt oil palm and tend to plant the trees in an irregular pattern which could lead to space competitions. This space competition could lead to severe adverse impacts on productivity when the species also compete for the same nutrients. In particular, soil type plays important roles in determining oil palm productivity. A study suggested that oil palm trees planted in mineral soil performed better than those in peat soil [51].

Moreover, farmers also tend to use uncertified or unknown quality of planting materials for perennial plants, oil palm, fruits and vegetables which could also lead to low productivity. This is mainly because farmers do not have access to information about good quality planting materials and farmers do not have resources to access certified planting materials because they are relatively more expensive. In the case of oil palm, farmers often use planting materials which were produced from "brondolan" or oil palm fruits that break away from oil palm FFB. Therefore, the quality of planting materials produced from these fruits is unknown which could lead to low productivity of oil palm. It is quite different with oil palm supported smallholders which gained on the job training from the nucleus companies [35]. For those reasons, it is important to consider the species selection, spacing arrangement and the quality of planting materials to maximize association, minimize competition and improve the productivity of the farm.

The market also becomes a crucial issue especially because agricultural products often experience a very long value chain before they could reach the end consumer. This way, the farmers will gain a very limited margin because they are in the low-end position of the whole production chain. For example, most farmers sell their oil palm FFB through middlemen instead of directly to the oil palm mills. This mainly the case of independent smallholders who have no contract with nucleus companies or oil palm mills, their farms are located relatively far from oil palm mills with poor transportation infrastructures. These situations have led to high dependency on middlemen who become the only connection between farmers and oil palm mills [19].

In addition, access to land resources for smallholders also becoming a crucial issue in the regions because lands become scarce. This is mainly because most lands are already occupied by large-scale licensing for various land-based activities [36]. During the fieldwork, we also observed potential land conflicts in the regions especially related to oil palm expansion into state forest areas. Agroforestry could contribute to solving these tenure problems in the forest areas by integrating farmers into forests which includes the granting of use rights of the forest lands [41]. This will require strong institutions both in the grass root and forest management levels.

There is currently no institutions, both government and non-government which promote and facilitate the adoption of oil palm agroforestry although this system has been practiced especially by smallholders. Farmer groups and extension workers could help in solving asymmetric information such as the quality of planting materials as well as promoting oil palm agroforestry as an alternative to its monoculture intensified management.

We also noticed that the adoption of oil palm in agroforestry had mainly occurred during the oil palm boom. Although we are optimistic that oil palm agroforestry could contribute to improving household livelihood resilience and more sustainable oil palm production, it was not clear whether oil palm agroforestry will become the permanent land use-cover, or it will be a transient land use/cover before the farmers slowly convert their farms into monoculture oil palm plantations or other types of crops which have more lucrative financial benefits.

6. Conclusion

In this article, we argue that oil palm agroforestry could be an alternative to enhance farmers' livelihood resilience. We found at two provinces in Indonesia namely Central Kalimantan and Jambi that the most prominent reason of smallholders to adopt oil palm agroforestry in their farms is to improve household resilience especially in terms of income. However, smallholders often encounter problems related to farm management such as low productivity due to poor species selection that are combined in agroforestry. The species selection is mainly based on the market instead of the biophysical suitability or combination compatibility. Farmers also tend to use uncertified or unknown quality of planting materials for perennial plants, oil palm, fruits and vegetables. In addition, farmers gain a very limited margin because they are in the low-end position of the whole production chain. These situations have led to high dependency on middlemen who become the only connection between farmers and oil palm mills. Access to land resources becomes a crucial issue for smallholders because most lands are already occupied by large-scale licensing for various land-based activities. This situation triggers smallholders to occupy state forest areas, which later could become law violation issues. Finally, we suggest that oil palm agroforestry system needs to be further investigated not only regarding financial benefits for the smallholders but also its social acceptability and barriers of adoption as well as its potentials in contributing to the improvement of ecosystem functions.

Acknowledgement

We thank to Yayasan KEHATI and JAVLEC for providing support for our study. Our special recognition to Irfan Bahtiar, Hery Santoso, Diah Y Suradiredja and Anton Sanjaya for intensive discussion. Earlier workshop to discuss the concept was participated by various stakeholders such as Faculty of Forestry UGM, Faculty of Agriculture UGM, INSTIPER, IRE, AURIGA, and Ministry of Environment and Forestry of Republic of Indonesia.

References

[1] Setiawan, E.N., Maryudi, A., Purwanto, R.H., Lele, G. (2016) Opposing interests in the legalization of non-procedural forest conversion to oil palm in Central Kalimantan, Indonesia. Land Use Policy **58**, 472-481

[2] Susanti, A., Maryudi, A. (2016) Development narratives, notions of forest crisis, and boom of oil palm plantations in Indonesia. Forest Policy and Economics **73**, 130-139

[3] BPS, (2016) Statistik kelapa sawit Indonesia. Badan Pusat Statistik, Jakarta, Indonesia.

[4] Wicke, B., Sikkema, R., Dornburg, V., Faaij, A. (2011) Exploring land use changes and the role of palm oil production in Indonesia and Malaysia. Land Use Policy **28**, 193-206

[5] Margono, B.A., Potapov, P.V., Turubanova, S., Stolle, F., Hansen, M.C. (2014) Primary forest cover loss in Indonesia over 2000-2012. Nature Climate Change **4**, 730-735

[6] Koh, L.P., Ghazoul, G. (2010) Spatially explicit scenario analysis for reconciling agricultural expansion, forest protection, and carbon conservation in Indonesia. Proceedings of the National Academy of Sciences **107**, 11140-11144

[7] Koh, L.P.e.a. (2011) Remotely sensed evidence of tropical peatland conversion to oil palm. Proceedings of the National Academy of Sciences

[8] Miettinen, J., Hooijer, A., Tollenaar, D., Page, S., Malins, C., Vernimmen, R., Shi, C., Liew, S.C. (2012) Historical analysis and projection of oil palm plantation expansion on peatland in Southeast Asia. ICCT White Paper 17

[9] Sodhi, N.S., Koh, L.P., Brook, B.W., Ng, P.K. (2004) Southeast Asian biodiversity: an impending disaster. Trends in Ecology & Evolution **19**, 654-660

[10] Sodhi, N.S., Posa, M.R.C., Lee, T.M., Bickford, D., Koh, L.P., Brook, B.W. (2010) The state and conservation of Southeast Asian biodiversity. Biodiversity and Conservation **19**, 317-328

[11] Fitzherbert, E.B., Struebig, M.J., Morel, A., Danielsen, F., Brühl, C.A., Donald, P.F., Phalan, B. (2008) How will oil palm expansion affect biodiversity? Trends in Ecology & Evolution **23**, 538-545

[12] Koh, L.P. (2007) Potential habitat and biodiversity losses from intensified biodiesel feedstock production. Conservation Biology **21**, 1373-1375

[13] Koh, L.P., Wilcove, D.S. (2008) Is oil palm agriculture really destroying tropical biodiversity? Conservation Letters **1**, 60-64

[14] Mustafa, Y., Amin, M., Lee, T., Shariff, A. (2012) Evaluation of land development impact on a tropical watershed hydrology using remote sensing and GIS. Journal of spatial hydrology **5**

[15] Hooijer, A., Page, S., Jauhiainen, J., Lee, W., Lu, X., Idris, A., Anshari, G. (2012) Subsidence and carbon loss in drained tropical peatlands. Biogeosciences **9**, 1053

[16] Saputra, E., Hartmann, T., Zoomers, A., Spit, T. (2017) Fighting the Ignorance: Public Authorities' and Land Users' Responses to Land Subsidence in Indonesia. American Journal of Climate Change **6**, 1-21

[17] Fargione, J., Hill, J., Tilman, D., Polasky, S., Hawthorne, P. (2008) Land clearing and the biofuel carbon debt. Science **319**, 1235-1238

[18] Ravindranath, N., Mauvie, R., Fargione, J., Canadell, J., Berndes, G., Woods, J., Watson, H., Sathaye, J., (2009) Greenhouse gas implications of land use change and land conversion to biofuel crops. Cornell University Library's Initiatives in Publishing (CIP)

[19] Susanti, A., (2016) Oil palm expansion in Indonesia: serving people, planet and profit? Eburon Academic Publishers

[20] Barbier, E. (2007) Frontiers and sustainable economic development. Environmental and Resource Economics **37**, 271-295

[21] Byerlee, D., Falcon, W.P., Naylor, R.L. (2016) The Tropical Oil Crop Revolution: Food, Feed, Fuel, and Forests. Oxford University Press

[22] Naylor, R.L., Liska, A.J., Burke, M.B., Falcon, W.P., Gaskell, J.C., Rozelle, S.D., Cassman, K.G. (2007) The ripple effects: biofuel, food security, and the environment. Environment **49**, 30-43; 30

[23] Zilberman, D., Hochman, G., Rajagopal, D., Sexton, S., Timilsina, G. (2012) The impact of biofuels on commodity food prices: Assessment of findings. American Journal of Agricultural Economics **95**, 275-281

[24] Gérard, A., Wollni, M., Hölscher, D., Irawan, B., Sundawati, L., Teuscher, M., Kreft, H. (2017) Oil-palm yields in diversified plantations: Initial results from a biodiversity enrichment experiment in Sumatra, Indonesia. Agriculture, Ecosystems & Environment **240**, 253-260

[25] Colchester, M., et al. (2006) Justice in the forest: rural livelihood and forest law enforcement. Center for International Forestry Research (CIFOR), Bogor, Indonesia

[26] KPA (2015) Reforma Agraria dan Penyelesaian Konflik Agraria Disandera Birokrasi. Jakarta: Konsorsium Pembaharuan Agraria

[27] CMEA, (2011) Masterplan Percepatan dan Perluasan Pembangunan Ekonomi Indonesia 2011-2025 (MP3EI). Kementerian koordinator Bidang Perekonomian Republik Indonesia (CMEA), Jakarta, Indonesia

[28] ISPO. 2015. Permentan No.11/Permentan/OT.140/3/2015 Lampiran VI Prinsip dan Kriteria Perkebunan Kelapa Sawit Berkelanjutan Indonesia (Indonesian Sustainable Palm Oil/ISPO) untuk Usaha Kebun Swadaya. Kementerian Pertanian Republik Indonesia. Jakarta

[29] RSPO. 2018. RSPO Standard for Smallholder Certification, Support and Inclusion, Document for Public Consultation, June 4, 2018

[30] Ismail, S., Khasim, N., Raja Omar, R. (2009) Double-row avenue system for crop integration with oil palm. MPOB Information Series **465**, 1-4

[31] Khasim, N., Omar, R.Z.R., Ismail, S., Omar, W. (2009) Integration of tongkat Ali with oil palm. MPOB Information Series
[32] Tohiran, K.A., Omar, R.Z.R., Ismail, S., Ghani, E., (2007) Integration of Boer Goats in Oil palm
[33] Wilson, G.A. (2007) Multifunctional agriculture: a transition theory perspective. Cabi
[34] Perfecto, I., Vandermeer, J. (2010) The agroecological matrix as alternative to the land-sparing/agriculture intensification model. Proceedings of the National Academy of Sciences, 200905455
[35] Budidarsono, S., Susanti, A., Zoomers, E. (2013) Oil palm plantations in Indonesia: the implications for migration, settlement/resettlement and local economic development
[36] Susanti, A., Budidarsono, S. (2014) Land governance and oil palm development: examples from Riau province, Indonesia. The global land grab: beyond the hype. Zed Books, London.
[37] Van der Laan, C., (2016) Tropical landscapes in transition?: Widespread land-use change and measures to maintain forests, carbon stocks and biodiversity in North and East Kalimantan, Indonesia. Utrecht University
[38] Neuman, W.L. (2013) Social research methods: Qualitative and quantitative approaches. Pearson education
[39] Lundgren, B. (1982) Agroforestry Systems. Journal of Plantation Crops **1**, 3-6
[40] Tscharntke, T., Clough, Y., Bhagwat, S.A., Buchori, D., Faust, H., Hertel, D., Hölscher, D., Juhrbandt, J., Kessler, M., Perfecto, I. (2011) Multifunctional shade-tree management in tropical agroforestry landscapes–a review. Journal of Applied Ecology **48**, 619-629
[41] Raintree, J.B. (1986) Agroforestry pathways: land tenure, shifting cultivation and sustainable agriculture. Unasylva **38**, 2-15
[42] Dawson, I.K., Leakey, R., Clement, C.R., Weber, J.C., Cornelius, J.P., Roshetko, J.M., Vinceti, B., Kalinganire, A., Tchoundjeu, Z., Masters, E. (2014) The management of tree genetic resources and the livelihoods of rural communities in the tropics: non-timber forest products, smallholder agroforestry practices and tree commodity crops. Forest Ecology and Management **333**, 9-21
[43] Colchester, M., Jiwan N, A.S.M.F.A.Y.S.A.P.H. (2006) Lahan yang dijanjikan: Minyak Sawit dan Pembebasan Tanah di Indonesia: Implikasi terhadap Masyarakat Lokal dan Masyarakat Adat. Forest Peoples Programme, Perkumpulan Sawit Watch, HuMA dan the World Agroforestry Centre
[44] Albrecht, A., Kandji, S.T. (2003) Carbon sequestration in tropical agroforestry systems. Agriculture, Ecosystems & Environment **99**, 15-27
[45] Perfecto, I., Vandermeer, J. (2008) Biodiversity conservation in tropical agroecosystems: a new conservation paradigm. Annals of the New York Academy of Sciences **1134**, 173-200
[46] Raintree, J., Lundgren, B., (1985) Agroforestry potentials for biomass production in integrated land use systems, Symposium on Biomass Energy Systems: Building Blocks for Sustainable Agriculture
[47] Zoomers, A. (2010) Globalisation and the foreignisation of space: seven processes driving the current global land grab. Journal of Peasant Studies **37**, 429-447
[48] Leakey, R.R., Tchoundjeu, Z., Schreckenberg, K., Shackleton, S.E., Shackleton, C.M. (2005) Agroforestry tree products (AFTPs): targeting poverty reduction and enhanced livelihoods. International Journal of Agricultural Sustainability **3**, 1-23
[49] Dawson, I.K., Lengkeek, A., Weber, J.C., Jamnadass, R. (2009) Managing genetic variation in tropical trees: linking knowledge with action in agroforestry ecosystems for improved conservation and enhanced livelihoods. Biodiversity and Conservation **18**, 969
[50] Regmi, B.N., (2003) Contribution of agroforestry for rural livelihoods: A case of Dhading District, Nepal, International Conference on Rural Livelihoods, Forests and Biodiversity, pp. 19-23
[51] Budidarsono, S., Rahmanulloh, A., Sofiyuddin, M. (2012) Economics assessment of palm oil

production. Technical Brief

IOP Conf. Series: Earth and Environmental Science **336** (2019) 012002 doi:10.1088/1755-1315/336/1/012002

The impacts of oil palm plantations on local and migrant smallholders' incomes

Diana Chalil[1], Riantri Barus[1], Zulkifli Alamsyah[2], Jullimursyida[3], Mawardati[3], Isfenti Sadalia[1]

[1]University of Sumatera Utara
[2]Jambi University
[3]University of Malikussaleh

*E-mail: chalildiana@gmail.com

Abstract. In Indonesia, oil palm plantations have long affected income levels positively. Furthermore, Indonesia has developed oil palm smallholdings in 1980 to improve rural incomes. The success of this program has attracted other farmers and migrants to further developed oil palm plantations. Those who collaborated with companies were known as scheme smallholdings, while the others are known as independent smallholdings. In general, the lack of professional assistance has caused independent smallholdings to perform lower than the scheme ones. In addition, on average, the migrants also performed better as they were familiar with the collective actions of group activity management. Using as many as 210 and 219 scheme and independent smallholders from the Provinces of North Sumatra, South Sumatra, Jambi and Riau, such assumption were tested using the ANOVA compare means test of the four groups, namely schemed local and migrant, and independent local and migrant. Comparisons between the smallholders' incomes and the Regional Minimum Wages of each province were analysed. The results showed that oil palm plantations have significantly improved the smallholders' incomes in all provinces, although the impacts were observed to be higher for the scheme-migrant smallholders, not only concerning income, but also for productivity and selling price. However, this was likely due to the difference between scheme and independent rather than locals and migrants. This was partly explained due to the collective actions of all migrant-schemed smallholders in both the input purchases and the output sales and around 50% of the local-schemed, while almost all independents did it individually.

1. Background

In 1980, Indonesia utilized transmigration programs to spread regional development. People from Java were sent to Sumatra and Kalimantan to start agribusiness. Initially, food crops and horticulture were chosen as agribusiness commodities. However, the results were not as expected and oil palm plantations were then proposed as an alternative. A number of empirical studies show the substantial benefits of the high returns of oil palm smallholdings [1], [2], [3], [4]. A decade later, these transmigration regions appear to be more developed with a higher income per capita than the non-oil palm plantation centres [5]. Since then, transmigration has continued, even after the government program ended, which was shown by the increased in spontaneous transmigration. The latter refers to all kinds of migrations that are not supported by the government transmigration scheme.

The development of oil palm business creates new jobs and raises income not just for big companies, therefore oil palm agribusiness is often considered as inclusive agriculture, although the totality in all of the inclusive components still needs further studies [8]. The massive population inflow requires additional employment opportunities [6], however, compared to seasonal crops such as food crops and horticulture, oil palm plantations require relatively little maintenance and fewer workers. Although per capita incomes of the oil palm plantation centres are higher, their unemployment rates also tend to be higher [5]. Reviews and studies show that development has both disadvantages and advantages for the local populations. Some locals do not easily adapt to the oil palm plantation agribusiness, hence tend to be excluded from the industry. Stakeholders of the oil palm agribusiness need to have good vertical and horizontal coordination. Oil palm production is not directly consumed; therefore, smallholder farmers need to be integrated into the processor/mills and to do so, smallholders need to have a minimum production scale. The average smallholding size is around 2-4 ha, while the minimum land size required to fulfil the minimum requirement as a supplier is generally between 60-80 ha. Otherwise, the smallholders need to sell their Fresh Fruit Bunches (FFB) through traders with lower prices, higher quantities and price fluctuations [9]. These justify the needs for collective actions among the smallholders, which have been shown among the transmigrants once involved in the program. This is also supported by their Javanese culture that in general has a patron-client relationship, whereas in contrast, the local Sumatrans tend to be more individualistic and relatively difficult to coordinate. Such conditions might lead to a decrease in local smallholder incomes. Unlike other topics related to the development of oil palm smallholdings, studies on both local and migrant smallholder incomes are relatively limited [7]. Therefore, this study is conducted to examine the impacts of oil palm plantations on the local and migrant smallholder household incomes.

2. Methods

This study was conducted in four oil palm centre provinces, namely North Sumatra, South Sumatra, Riau and Jambi. Data were collected using convenient cluster sampling from 233 local smallholders and 155 ex-transmigrants, covering both the independent and scheme smallholders. Details of the sample distribution are as follows.

Table 1. Sample Distribution Based on Location

Province	Schemed	Independent	Total
North Sumatra	89	124	**213**
South Sumatra	38	28	**66**
Riau	27	23	**50**
Jambi	39	22	**61**
Total	**193**	**197**	**390**

The average monthly income of each group was calculated by the difference between the smallholders' total revenues and total costs. The total revenue accommodated the impact of low and high harvest seasons by weighting the first 3 months of each of the seasons and considered the remaining 6 months as normal. The impact of oil palm plantations on the local and migrant smallholder incomes are descriptively analysed by comparing it with the Regional Minimum Wage (*Upah Minimum Provinsi/UMP*) of each province. The differences of income, land size, productivity and smallholders' selling price between these groups were then further analysed using Analysis of Variance (ANOVA), using the formula:

$$V.R. = \frac{Among\ groups\ means\ square}{Within\ groups\ means\ square}$$

Following [10], 4 variables were included, namely migrant status, land size, oil palm cultivation work experience and FFB price. Author [10] found that the migrant status did not show clear trends in predicting gross monthly income, unlike the other variables. However, they also found that scheme smallholders received higher incomes due to higher selling prices. Therefore, this study differentiated the cluster between migrants and locals, as well as between scheme and independent smallholders. The average values of land size, productivity, selling price and income of the local and ex-transmigrant smallholders were compared as these were assumed to be influenced by the differences of their collective actions and networking.

3. Results and Discussion

Data collected from convenient sampling showed that transmigrant respondents could be found in South Sumatra and Riau, while the local respondents were from North Sumatra, Jambi and Riau. Composition details are as follows.

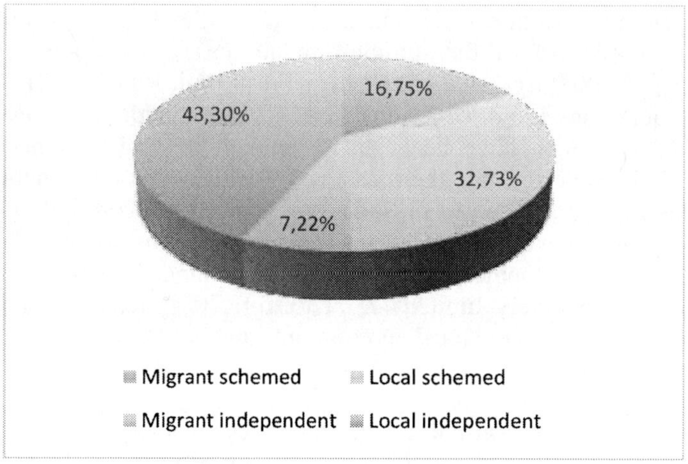

Figure 1. Distribution of Samples

The incomes of all smallholder categories in all provinces were higher than the UMP, showing that oil palm plantations have brought positive impacts on both the migrants and the locals. The ratio between smallholders incomes and the UMP could reach as high as 3.66, but even the lowest was still slightly higher than 1 (Table 2). Out of all the smallholders, the independents received the lowest incomes in all provinces as shown below in Table 2.

Table 2. Comparison of Scheme and Independent Smallholders Incomes

Types of smallholders	UMP (IDR/month)	Land size (ha)	Income (IDR/month)	Income/UMP
Migrant schemed				
Riau	2,460,000	2.19	9,005,687	3.66
Sumsel	2,595,994	2.13	6,977,039	2.69
Local schemed				
Sumut	2,100,000	2.87	6,856,494	3.26
Jambi	2,234,000	3.76	7,421,441	3.32
Migrant independent	2,595,994	3.21	6,238,340	2.40
Local independent				
Sumut	2,100,000	2.74	4,482,719	2.13
Riau	2,460,000	2.09	2,592,571	1.05
Jambi	2,234,000	2.30	4,403,107	1.97

The transmigrant programs in South Sumatra and Riau were started in 1990 and 1980, respectively. Either the state or private companies have provided training for the smallholders that participated in the transmigration programs. Following this, the farmers might become independent smallholders but with proper training experiences. As a result, both scheme and independent migrants had relatively

IOP Conf. Series: Earth and Environmental Science **336** (2019) 012002 doi:10.1088/1755-1315/336/1/012002

higher percentages of trained smallholders than the locals did. On the contrary, most of the second generation or spontaneous migrant smallholders did not receive any training. Therefore, currently 55% and 54% of the scheme and independent migrants had no proper training experiences as portrayed in Figure 2.

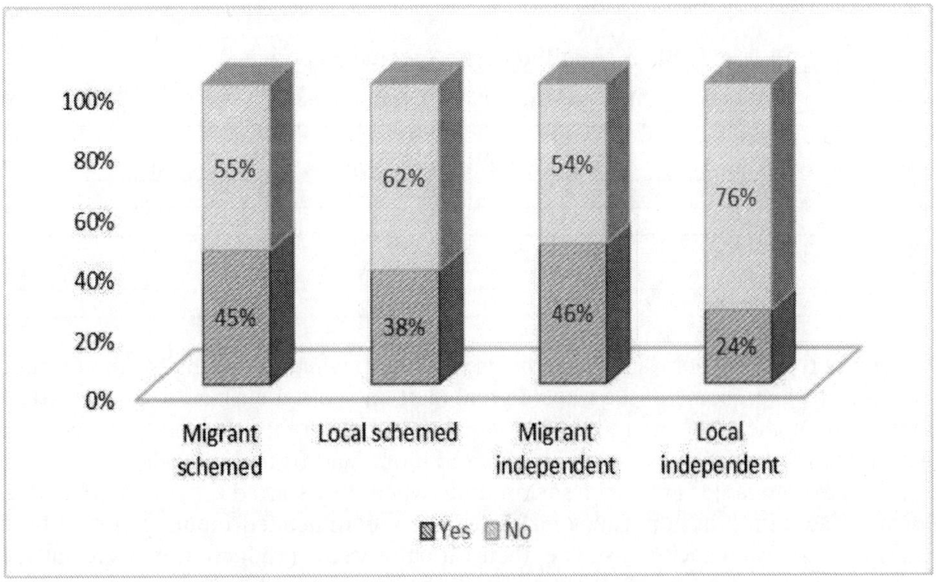

Figure 2. Smallholder Training Experiences

Similarly, both scheme and independent local smallholders also had low level of formal education, where only 40% of the scheme and 34% of the independent smallholders, had completed their senior high school, while the migrants showed higher proportions, 52% for the scheme and 63% for the independent migrants (Figure 3).

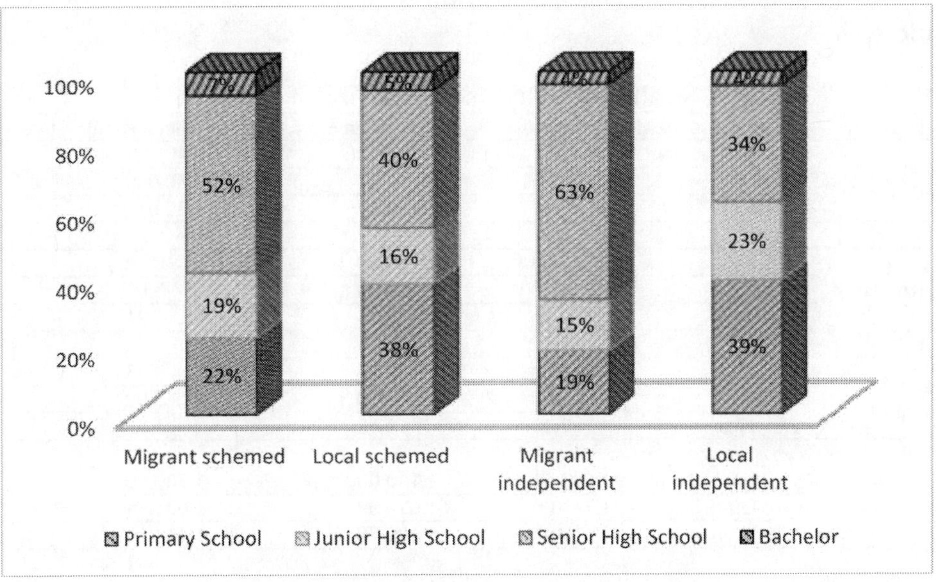

Figure 3. Smallholders' Level of Education

The data showed that on the average, migrants and local smallholders were in their 40's (Table 3), with local smallholders slightly older with a broader range of age. However, the migrant smallholders had more experiences in running oil palm plantations than the local smallholders. Both had wide ranges of experiences; some still very new and have not even started their first harvest, while some already manage their 2nd oil palm trees' cycle. Details are as follows.

Table 3. Smallholders' Age and Experience

		Migrant schemed	**Local schemed**	**Migrant independent**	**Local independent**
Age (year)	Min	29	22	29	21
	Max	68	76	63	84
	Average	45	47	46	48
Experience (year)	Min	4	2	4	3
	Max	30	37	29	56
	Average	22	17	23	16

Transmigrants received their land from the government, which was 2 ha for the farming area and 1 ha for housing. However, some of the transmigrants left and sold their land to those who stayed, which caused variations in the total land sizes belonged to the migrants. In addition, most transmigrants enjoyed sufficient incomes to save up and purchased more land from the locals.

State or private companies assisted transmigrants when they started the oil palm business as plasma smallholders. In addition, these plasmas also used the recommended inputs supplied by their partner companies (nucleus). With such assistance, their oil palm trees' productivities were satisfactory. After the nucleus-plasma program ended, some were still engaged with big companies as scheme smallholders, while others not. Some participated in both, as they had sufficient land to be developed into scheme and independent smallholdings. Transmigrant partnerships started from land clearing, selecting seeds to be harvested and selling the FFB. Differently, the local partnership in Jambi started after the smallholders had independently cultivated their oil palm trees. Thus, many still used non-certified seeds. Therefore, migrant-scheme smallholders had the highest productivity and selling prices, consequently enjoyed the highest incomes (Table 4). The same intensive partnership model also appeared in the local partnership found in North Sumatra Province. Combined with high land suitability, the local-scheme smallholders in North Sumatra also enjoyed high productivities and selling prices.

Table 4. Smallholders' Income, Land Size and Productivity

		Migrant-scheme	Local-scheme	Migrant-independent	Local-independent
Income (IDR/ha/month)	Min	458.751	138.750	240.139	134.900
	Max	7.466.406	4.686.057	3.640.000	4.058.417
	Average	3.575.904	2.177.864	1.882.681	1.603.947
Land size (ha)	Min	1,00	0,60	0,50	0,30
	Max	6,00	48,00	8,00	22,00
	Average	2,16	3,15	3,21	2,60
Productivity (ha/month)	Min	1,00	0,25	0,33	0,30
	Max	4,00	3,00	3,00	4,00
	Average	2,40	1,56	1,53	1,52
Price (IDR/kg)	Min	1.509,91	1.150,00	1.200,00	1.150,00
	Max	1.763,09	1.670,00	1.621,38	1.665,00
	Average	1.611,97	1.409,20	1.414,16	1.435,42

The scheme smallholdings attained higher productivities than the independents for both migrants and local smallholders. Although insignificant, the local-schemed received lower prices than the independent smallholders did. This might partly be explained by the composition of the independent respondents who were mostly from North Sumatra, which in general received a higher price than other provinces. Such condition was supported by the results of the ANOVA Test as follows.

Table 5. ANOVA Test results

		Sig			
		Migrant-scheme	Local-scheme	Migrant-independent	Local-independent
Income (IDR/ha/month)	Migrant-scheme		,000	,000	,000
	Local-scheme			,641	,000
	Migrant-independent				,667
Land size (ha)	Migrant-scheme		,099	,382	,641
	Local-scheme			1,000	,426
	Migrant-independent				,792
Productivity (ha/month)	Migrant-scheme		,000	,000	,000
	Local-scheme			,915	,962
	Migrant-independent				,781
Price (IDR/kg)	Migrant-scheme		,000	,000	,000
	Local-scheme			,996	,156
	Migrant-independent				,762

Table 5 shows that both the migrant and local smallholders' performances significantly differed only between the migrant-scheme and local-scheme, but not between independents of both classifications for all variables, *i.e.*, income levels, land size, productivity and selling prices. The migrant-scheme performances differed from the other groups as they have been trained by companies for years and have enjoyed high FFB productivities and qualities, whereas the local-scheme smallholders' partnership did not start from the beginning, thus most were still using uncertified seeds and could not achieve optimum productivities and qualities. Therefore, the local-scheme smallholders' productivities and selling prices were no different from those of the local independents.

4. Concluding Comments and Implementations

The oil palm plantations in all four smallholdings' centres appeared to obtain higher incomes than the UMP, indicating that not only migrant smallholders enjoyed the positive impacts of oil plantations, but so do the independent ones. The impacts were not significantly different between the migrants and the local smallholders, the same for the scheme smallholders. Although the average scheme smallholders' land sizes were not significantly different from other groups, their incomes, productivities and selling prices did. Therefore, it could be concluded that the positive impacts of income levels from oil palm plantations development programs were not only enjoyed by the migrants but also by the local smallholders. However, to reach optimum performance, the smallholders required assistance from the exceptional oil palm companies.

Acknowledgements
Funded by the Ministry of Research, Technology and Higher Education of The Republic of Indonesia in accordance with *Hibah Strategis Nasional Konsorsium* Research Contract 2018 No 170/UN5.2.31/PPM/ KP-RDPM/2018 dated 16 February 2018.

References

[1] Rist L, Feintrenie L, Levang P. 2010. The livelihood impacts of oil palm smallholders in Indonesia, Biodiversity and Conservation **19**(4): 1009-1024
[2] Duryat, Raflegan S, Cannon MP. 2013. Dynamic of plantation of oil palm smallholdings in Riau Province, Sumatra, Indonesia. Jurnal Sylva Lestari **1**(1): 93-100
[3] Kawanichi M, Mimura N. 2013. Rice farmers' response to climate and socio-economic impacts: a case study in North Sumatra, Indonesia. Journal of Agricultural Meteorology **69** (1): 9-22

[4] Euler M. 2015. Oil palm expansion among Indonesian smallholders adoption, welfare implications and agronomic challenges [Dissertation]. Universitait Gottingen, Germany.

[5] Tim Kajian, 2015. Dampak Sosial Ekonomi Pengusahaan Kelapa Sawit terhadap Kesejahteraan Petani dan Kemajuan Wilayah, Badan Pengelola Dana KelapaSawit

[6] Budidarsono S, Susanti A, Zoomers A. 2013. Oil Palm Plantations in Indonesia: The Implications for Migration, Settlement/Resettlement and Local Economic Development. Biofuels - Economy, Environment and Sustainability, Chapter 6, INTECH: 173 – 193

[7] Kouzumi Y. 2016. Migration and Its impact in Riau Province, Indonesia: An analysis of population census data and topographical maps. Journal of Asian Network for GIS-based historical studies **4**: 3-10

[8] Chalil D, Barus R. 2018. Partnership Models for Inclusive Oil Palm Smallholdings. Paper presented in CSSPO International Conference 2018: Towards Inclusive & Sustainable Agriculture. Sarawak, 9-11 July 2018

[9] Chalil D, Barus R. 2018. Risk analysis for sustainability of oil palm smallholdings. Paper presented in CSSPO International Conference 2018: Towards Inclusive & Sustainable Agriculture. Sarawak, 9-11 July 2018

[10] Lee J, Ghazoul J, Obidzinski K, Koh L. 2014. Oil palm smallholder yields and incomes constrained by harvesting practices and type of smallholder management in Indonesia. Agronomy for Sustainable Development. Springer Verlag/EDP Sciences/INRA **34** (2): 501-513

IOP Conf. Series: Earth and Environmental Science **336** (2019) 012003 doi:10.1088/1755-1315/336/1/012003

The impact of replanting oil palm plantations on the farming income of the Sungai Bahar community in Muaro Jambi Regency

Ernawati HD[1]*, Suandi[1], Mirawati Yanita[1], and Nina Qoirina[2]

[1]Lecturer at Faculty of Agriculture, Jambi University, Jambi Province
[2]Agribusiness Study Program Faculty of Agriculture, Jambi University

*E-mail: ernahamid@unja.ac.id

Abstract. The research objectives are: (1) to know the description of oil palm farming income in replanting to underplanting and conventional method (2) to know the level of oil palm farmer income in replanting to underplanting and conventional method, (3) to know the ratio of oil palm farming income in replanting to underplanting and conventional method. The research conducted in Sungai Bahar Subdistrict: Suka Makmur, Mekarsari Makmur and Marga Mulya Village. The study area is chosen purposively. The study conducted three calendar months. Respondents were 60 farmers consisting of 30 farmers applied the conventional method, and 30 farmers applied underplanting method. Data were analysed by descriptive and t-test statistics. The results showed that there are differences in the number of plants in both replantings that affect production and income. The oil palm farming income who applied conventional method is higher than the underplanting method. The analysis shows that the oil palm farming income through conventional method is significantly different from the income of underplanting method.

1. Introduction

The plantation sub-sector has a huge opportunity to become a mainstay of exports. Development in the plantation sector has directed to further accelerate the rate of growth of production from both large plantations, private and state plantations to support the industrial development, as well as increasing the utilization and sustainability of natural resources (SDA) in the form of land and water. The role of the plantation sector is so great for increasing the income of the farmers and providing raw materials for domestic industries and as a source of foreign exchange.

Oil palm plantations in Sungai Bahar District opened in 1983 by PTPN VI. The establishment of PTPN VI in Sungai Bahar District in the background behind the Central Government Program for the development of areas outside Java through the transmigration program. Then the transmigration program integrated with the development program of oil palm plantations with a partnership pattern between large plantations and smallholder plantations. The partnership pattern is known as the PIR-Plasma Pattern. In the partnership program between large plantations and the transmigration program, the plasma was a participant in the transmigration program so that the pattern of PTPN VI Sungai Bahar's Oil Palm Plantation Program called the PIR-Trans Pattern [1].

During the partnership process, PTPN VI was entirely responsible for all facilities and infrastructure in the development of oil palm plantations, ranging from fertilizing, gardening and road maintenance, while the Village Unit Cooperative (KUD) acts as a deposit of production to the company during the partnership process. From the results of the sale of the product, each farmer deducted for a credit instalment of 30% or as much as the instalment stated in the Debt Recognition Letter (SPH). The company after receiving a list of name of farmers, who sell products from KUD, make payments to KUD by 70% or according to the list of farmers' liability payments, and 30% paid to the Bank as instalments [2].

Apart from the pattern of partnership carried out by PTPN VI with farmers in Sungai Bahar Subdistrict in 2002, farmers began to cultivate or develop their oil palm plantations privately, and farmers who separated from the partnership pattern given the term Ex-PIR Plasma PTPN VI. Ex-PIR Plasma is a term made as a differentiator between purely independent palm oil farmers and transmigration oil palm farmers who have worked or collaborated with PTPN VI. Given the long-term planting of oil palm in Sungai Bahar Subdistrict, which began in 1983, indicates that oil palm plants located in Sungai Bahar Subdistrict must be immediately rejuvenated due to the declining ability of plants to produce fresh fruit bunches (FFB).

Sungai Bahar Subdistrict consists of 11 villages, which out of the 11 villages there were only three villages that had carried out the rejuvenation of oil palm, namely Suka Makmur Village, Mekarsari Makmur Village and Marga Mulya Village. The decision of the farmers to rejuvenate oil palm is due to a large number of old plants that cause a decrease in production and crop productivity so that there is a decrease in farmers' income. The area of the old palm oil plantations of Ex-PIR Plasma in Sungai Bahar Subdistrict is 7,260 Ha or 75.9% of the total area of oil palm in the Sungai Bahar District.

Rejuvenation is the replacement of unproductive old plants with new crops that are more productive and profitable. In the cultivation of oil palm plants, there are applied several alternative rejuvenation models including conventional rejuvenation, underplanting rejuvenation, and intercropping rejuvenation. People's palm oil worship needs to consider the sustainability of the farmer's income during rejuvenation. The selection of appropriate rejuvenation techniques needed, to provide alternative income for farmers during the revival period [3].

The productivity standard that used as a reference for the rejuvenation period is around 12 tons of FFB/ Ha/Year. In addition to productivity, crop effectiveness and plant density are other considerations in determining the rejuvenation period. The effectiveness of harvest will be low if the height of the oil palm tree has exceeded 12 meters. In addition, rejuvenation needs to be done if the plant density is <80 trees/ha [1].

Conventional rejuvenation technique is a developed rejuvenation technique through the development of oil palm rejuvenation models/demonstration plots. It was carried out starting in 2011 in Mekar Sari Makmur Village, Marga Mulya Village and in 2012 in Suka Makmur Village, Sungai Bahar Subdistrict, Muaro Jambi Regency, through APBN and BI funds in collaboration with the Plantation Office of Jambi Province (http://rosenmanmanihuruk.blogspot.co.id). Conventional rejuvenation carried out by farmers from three villages totalling 30 people through government programs has now entered the age of 4 to 5 years and has produced fresh outer bunches (FFB).

The selection of oil palm rejuvenation techniques in Sungai Bahar Subdistrict will affect the size of the production costs that farmers will spend on their oil palm plantations. The low production costs will have a significant effect on the obtained farmer income.

Based on the description above, this research is aimed to, 1) To find out the description of oil palm farming from each rejuvenation technique in Sungai Bahar Subdistrict; 2) To find out the level of oil palm farming income from each rejuvenation technique in Sungai Bahar Subdistrict; and 3) To compare the level of oil palm farming income from each rejuvenation technique in Sungai Bahar Subdistrict.

2. Material and methods

This research was carried out in Sungai Bahar Subdistrict namely in Suka Makmur Village, Mekarsari Makmur Village and Marga Mulya Village. The choice of this location purposively made, with the consideration that in this area, the farmers have been rejuvenating oil palm using conventional rejuvenation techniques and underplanting. The number of farmers as many as 60 farmers, namely 30 farmers who apply conventional techniques and 30 farmers who apply the underplanting technique. A sampling of conventional techniques used census and underplanting techniques used Snowball Sampling techniques.

Data analysis used in this study is descriptive analysis and differential analysis. Descriptive data analysis is an analytical technique used to analyse data by describing or describing the collected data soberly without any intention to generalize from the results of the research. For the first purpose, the analysis used is income analysis consisting of an analysis of the acceptance of oil palm farming. Inferential data analysis is a statistic used to perform data analysis by making conclusions, which are generally applicable. Characteristic of inferential data analysis is the use of specific statistical formulas, and then the calculations have done the results are what will become the basis for making generalizations that come from the sample for the population.

2.1. Farming income

To calculate the oil palm farming income the formula used:

$$Pd = TR - TC$$

Where:

Pd = Revenue (Rp)
TR = Total Receipt (Rp)
TC = Total Expenditures (Rp)

2.2. Significant Testing (t-test)

Significance test (t-test) to see the significance performed of the influence of independent variables on unbound variables individually and considers other variables constant. To test whether there are differences in oil palm farming income in conventional rejuvenation techniques and done with the statistical hypothesis formula of underplanting testing:

H_0: $\mu_1 = \mu_2$
H_1: $\mu_1 > \mu_2$

Hypothesis testing using t-test two separated variance independent samples [4] formulated as follows:

$$t_{hit} = \frac{X_1 - X_2}{\sqrt{\frac{S_1^2}{n_1} + \frac{S_2^2}{n_2}}}$$

Information:

n_1 = Number of respondents of conventional oil palm rejuvenation farmers
n_2 = the number of respondents to the rejuvenation of oil palm underplanting
X_1 = Average income of conventional palm oil rejuvenation farmers
X_2 = Average farmers rejuvenation of oil palm underplanting rejuvenation
S_1^2 = Variety of average oil palm rejuvenation income conventional
S_2^2 = Variety of average income for oil palm rejuvenation farmers underplanting

The test conclusion did by comparing the t_{count} value with the t_{table} value as follows:

a. If $t_{count} > t_{table}$, then H_0 is rejected, and H_1 is proven right which means that conventional palm oil rejuvenation techniques provide higher income compared to and underplanting rejuvenation techniques.

b. If $t_{count} \leq t_{table}$, then H_0 cannot be rejected, H_1 is not proven right which means that conventional oil palm rejuvenation techniques do not provide higher income compared to underplanting rejuvenation techniques.

3. Results and discussion

3.1. Characteristics of respondents

The number of respondents in the study area was 60 farmers, namely 30 farmers who applied conventional techniques and 30 farmers who applied the underplanting technique. The average age of farmers in the study area shows that farmers who apply conventional techniques and underplanting are mostly not in productive age in farming, with the level of pursued formal education mostly at the primary school level. All oil palm farmers have experience working on two years. The average number of dependents of farmer families is 1-2 people. The area of oil palm farming in the study area is quite varied. However, the average is less than 2 hectares, which is 1.90 hectares from what set by the government through the transmigration program for each farmer who applies conventional rejuvenation techniques and underplanting. Land ownership status of farmers who apply conventional techniques and the common underplanting technique are as owners.

3.2. Overview of oil palm farming in conventional rejuvenation techniques and underplanting in research areas

Sungai Bahar Subdistrict is one of the Districts producing palm oil in Muaro Jambi District. Sungai Bahar Subdistrict has been working on rejuvenated palm oil starting in 1983. Sungai Bahar Subdistrict cultivates oil palm by applying two rejuvenation techniques namely conventional techniques and underplanting. Conventional oil palm rejuvenation techniques carried out in 2011 and 2012 while underplanting rejuvenation techniques implemented from 2008 to present.

The number of farmers who apply conventional rejuvenation techniques and underplanting in this study amounted to 30 farmers. In cultivating oil palm farming, farmers are more effective in using underplanting techniques compared to conventional petroleum techniques. This is due to the ability of farmers to rejuvenate oil palm in their gardens. The main difference from conventional rejuvenation techniques and underplanting is in the process of planting and planting oil palm, but there are several other things that distinguish conventional rejuvenation techniques and underplanting in the study area and can be seen in the following description.

3.3. Land processing

The land is a factor that influences the level of production. It said that the land area has a positive effect on production. The ownership status of farmers' land in the research area, namely the status of the owner with the land area per farmer in each rejuvenation technique is approximately two ha. In 2011, 20 farmers carried out the rejuvenation with conventional techniques in Mekarsari Makmur Village and Marga Mulya Village with 40 Ha of total land area.

The process of planting old plants and processing land takes approximately 10 days. Land processing begins with the overthrow of old trees, the making of planting holes and initial fertilization using RP (Rock Phosphate). In 2012, the application of conventional rejuvenation techniques carried out again in Suka Makmur Village with 10 farmers with a total land area of 20 Ha. The process of land clearing and land processing in Suka Makmur Village is not much different from the farmers who did the rejuvenation with previous conventional techniques. The 30 farmers who applied the underplanting technique in this study on average did not process their land as done with conventional techniques. The process of planting young plants in the underplanting technique directly planted between the old plants that still exist.

3.4. Oil palm planting

Oil palm planting begins with the preparation of plant seeds. Superior and quality seeds are one of the keys to success in the cultivation of oil palm plants. The use of good seeds expected to get good results later. Seedlings used from the two rejuvenation techniques at the research location are types of marihat obtained from PTPN VI. Plant spacing used in both rejuvenation techniques in the study area is 8 x 9 m. There was no difference in the distance of planting in both techniques because each farmer followed the recommendations from PTPN VI, while the underplanting technique followed the planting distance of the old plants, which cultivated by PTPN VI.

In the process of planting oil palm, farmers who apply conventional techniques do the erection first on their land in order to straighten the direction of the line and the spacing determined before the making and planting take place. While on the land of farmers who apply the underplanting technique, farmers do not do the erection because farmers plant between old plants that are still in production. The process of piling up to planting in conventional techniques takes approximately 26 days per 20 ha.

3.5. Care

Plant maintenance carried out with the aim that plants can grow well and are able to provide good production as well. The treatment of oil palm plants is done after the young plants are planted in the field, which first done by farmers in conventional techniques that are fertilizing plants after the age of 3 weeks to 1 month is planted. This fertilization aims to stimulate growth and to get maximum yield. Fertilization at the research location in each technique is carried out twice in one year, and in underplanting techniques, there are fertilizers too once a year. Fertilizers used by farmers in the study area include NPK Phonska, Urea, KCL, Sp-36 and Dolomite.

Treatment or spraying is done twice a year. Herbicides used by farmers in the research area are Round-Up and Gramaxon. Furthermore, pruning or pruning done to stimulate oil palm plants to quickly flower and bear fruit. The process of pruning palm oil midribs on conventional techniques carried out at each harvest takes place because, considering the small oil palm plantations, the process of processing is not too difficult to do. While underplanting rejuvenation techniques farmers do pruning fronds once a year. This midrib trimming process can take approximately 3 to 4 days.

Harvesting is the process of taking TBS (Fresh Fruit Bunches) from oil palm plants. The harvesting process of both rejuvenation techniques is carried out 24 times a year. Palm kelp plants start production from the age of 3 to 4 years. During the process of oil palm plantation maintenance, farmers in conventional techniques use more labour in the family than outside the family because small plants and uniform plants make it easier for farmers to carry out maintenance, whereas in the underplanting technique farmers use more outside the family labour because of the not uniform and difficult plant maintenance due to older plants that are getting taller and harder during the harvesting process.

3.6. Production

Production is an activity that converts the input into output. In economic theory, a producer or entrepreneur must decide: (1) how much output that must be produced, and (2) how and in combination how the factors of production (input) are used. Production according to technical function is the integration of various resources used, such as land, labour, capital and management resources into the output, while production according to economic functions where the production process includes elements of transportation, trading costs, and other costs. Research on oil palm farming in the study area which is meant by production is the production of oil palm which is obtained from the oil palm farming products, namely in the form of fresh fruit bunches (FFB). Field data obtained by the average production obtained by farmers in 2016 is 18,168 tons/ha/year in conventional techniques and 16,274 tons/ha/year in underplanting techniques (table 1).

The production produced by farmers who rejuvenated conventional techniques in 2011 and 2012 was not much different because the amount of product produced was strongly influenced by the number of plants and the treatment carried out by farmers in each month such as fertilizer administration, weed eradication and midrib trimming. While farmers who do rejuvenation with production yield underplanting techniques are not so maximal because there is no variety in the age of the plants on the land because they are dominated by two plants, young plants and old plants with an average number of plants per hectare of 260 and 166 plants with a different age range.

Plant age that is not uniform in the underplanting technique causes the growth of young plants have to be damaged so that the development of young plants is not good due to the struggle for nutrients and not optimal lighting due to the protection of old plants that still exist. Production produced by young plants in underplanting techniques is less than 1 Ton/ha, the amount of production received by farmers is still dominated by old plants that are still in their fields. We can see performance differences in oil palm farming in conventional rejuvenation and underplanting techniques in table 1.

Table 1. Comparison of Oil Palm Farming Performance in Conventional Rejuvenation Techniques and Underplanting in 2016 Research Areas

Performance	Conventional Technique	Underplanting Technique
Plant Age	4 – 5 Years	4 – 5 Years
Plant distance	8 x 9 m	8 x 9 m
Seed types	Marihat	Marihat dan Others
Number of plants	260	260 dan 166
Fertilization	Twice per year	Once – twice per year
Treatment	Twice per year	Twice per year
Midrib Pruning	24 times per year	Once per year
Harvesting	24 times per year	24 times per year
Production	18,168 Ton/Ha/Year	16,274 Ton/Ha/Year
Price	Rp. 1.621/Kg	Rp. 1.707/Kg

3.7. Cost analysis, revenue and income of oil palm farming

Conversion of forests into oil palm plantations on peatlands increases BD or soil density in the peat surface layer. Increased BD indicates that there has been compression of peat pores in oil palm plantations, peat soil was more compact, and the capacity of peat capillary will also increase. Therefore, the movement of water from the grown water table to the surface layer increases; it can maintain soil moisture on the surface layer of the peat, so maintained the sustainability of peat. Compaction of peat soils can support the peat soil to reduce oblique plant, reduce soil nutrient leaching, reduce CO_2 emissions and reduce peat risk to fires because of smaller peat soil pores but increase crop yields.

Farming income is the difference between total revenue and total costs incurred. The income earned by farmers in this study is the amount of palm oil production multiplied by the price then reduced by the number of costs incurred during the production process. We can see the amount of income of the respondent farmers in oil palm farming in conventional rejuvenation and underplanting techniques in the study area in table 2.

Table 2. Analysis of Oil Palm Farming Revenues in Conventional Rejuvenation Techniques and Underplanting in 2017 Research Areas

Description	Conventional		Underplanting	
	Rp/Farmer	Rp/Ha	Rp/Farmer	Rp/Ha
A. Acceptance				
1. Production (Kg)	34.520	18.168	30.920	16.274
2. Price (Rp)	1.621	1.621	1.707	1.707
Total Revenue (Rp)	**55.956.920**	**29.450.328**	**52.780.440**	**27.779.718**
B. Fees Paid (Rp)				
1. Fertilize	7.807.000	4.108.947	8.142.000	4.285.263
2. Medicines	598.333	314.912	626.667	329.825
3.TKLK	1.219.000	641.579	2.646.333	1.392.807
4. Other Fees	1.250.000	657.895	1.640.000	863.158
Total Cost	**10.874.333**	**5.723.333**	**13.626.667**	**7.171.929**

Description	Conventional		Underplanting	
	Rp/Farmer	Rp/Ha	Rp/Farmer	Rp/Ha
C. Unpaid Fees (Rp)				
1. Depreciation	396.265	208.561	415.088	218.467
Total Cost	**396.265**	**208.561**	**415.088**	**218.467**
D. Total Cost (B+C) (Rp)	**11.270.598**	**5.931.894**	**14.041.755**	**7.390.396**
E. Income (A- D) (Rp)	**44.686.321**	**23.519.117**	**39.267.953**	**20.667.343**

Table 2 shows that the average income of farmers who apply conventional techniques is Rp. 55,956,920/Farmer/Year and income of Rp. 44,686,321/Farmer/Year. While the average income of farmers who apply the underplanting technique is Rp. 52,780,440/Hectare and income are Rp. 39,267,953/Farmer/Year. The size of the income earned by farmers greatly influenced by the size of the use of production costs incurred.

The costs used in the study are the total costs calculated in one year in oil palm farming activities that cover fertilizer costs, medical expenses, labour costs in the family, labour costs outside the family, depreciation costs and deduction costs. The amount of costs incurred in the underplanting technique is due to the uneven age of the plant and the difficulty in performing maintenance due to the age of different commodities in one field.

3.8. Comparison of oil palm farming revenues in conventional rejuvenation and underplanting techniques

To compare oil palm farming income in conventional rejuvenation techniques and underplanting, a different test analysis was carried out on average with computer assistance, namely the SPSS (statistical package for social science) program at a 95% confidence level (t-table $\alpha = 5\%$). Test results are provided in table 3.

Table 3. Test the Difference between the Average Palm Oil Farming Revenue in Conventional Rejuvenation Techniques and Underplanting in the 2017 Research Area

No.	Description	Conventional Technique		Underplanting Technique
1.	Average Revenue (Rp)	44.686.321		39.267.953
2.	Sig. (2 tailed)		0.000	
3.	t-count		6.158	

Based on table 3 it can be seen that the results of the data analysis using SPSS with the results of the test shows that the t-count is 9.404 and the t-table value at α = 5% is 2.001. So that a decision obtained that refuse, Ho accepts H_1. This means that the income received by farmers who apply conventional rejuvenation techniques is higher than the income obtained by farmers who apply underplanting rejuvenation techniques.

The difference in income from the two techniques strongly influenced by the amount of income received by farmers. Acceptance of the conventional technique is higher than underplanting technique; this is because the age and number of plants in the oil palm fields used two different techniques. The difference in age and number of plants dramatically influences the farmer's amount obtained of production. In addition, the costs incurred by farmers in the underplanting technique are also greater compared to conventional techniques because on the land underplanting techniques there are oil palm plants with different ages so that the cost of maintenance is greater.

In line with the above decision, income in conventional techniques is higher than underplanting techniques. This is supported by research [5] said that from the point of view of conventional rejuvenation farmers are considered more effective and efficient. Farmers will not be difficult in management and supervision because there is relatively the same age of plants so it will optimize the use of production tools and streamline operational costs. Farmers who apply the underplanting technique consider the financial aspects so that the underplanting rejuvenation is more in line with their socio-economic conditions. Farmers still have rejuvenation the income to finance their lives while some oil palm plants. Nevertheless, both rejuvenation techniques still provide financial benefits for each farmer.

In line with the results of the study above [6] in his research results said that the recommended technique of replanting is intercropping with food crops during the vegetative period of oil palm plants. This model still provides income for farmers during the TBM period. Simultaneous planting will produce plants and with consistent growth, facilitate maintenance and harvesting. In connection with the above results [7] explained that in his research there was a very significant difference between the income of the group of palm oil farmers and the group of plasma oil palm farmers where the income of the group of oil palm farmers was greater than the income of the plasma oil palm group.

4. Conclusion

Oil palm farming in conventional rejuvenation and underplanting techniques in 2016 in Sungai Bahar District has a difference in the yield of FFB produced. This is because the number of plants and the age uniformity of the plant affects the amount of products produced and obtained the farmer's income.

The income of oil palm farmers in the study area who apply conventional rejuvenation techniques is higher than the income from underplanting techniques that is Rp. 44,686,321/Farmer/Year and Rp. 39,267,953 with a difference of 13.8 percent.

Evidenced by the t-test of two oil palm farming income, made a decision that the income of conventional rejuvenation techniques was significantly higher than the income from underplanting rejuvenation techniques.

References

[1] My story 2015. Model of oil palm rejuvenation in Jambi Province. http://rosenmanmanihuruk.blogspot.co.id/2015/02/model-peremajaan-kelapa-sawit-di.html (Accessed March 24, 2017)

[2] Ernawati, HD. 2011. *Implementasi Kemitraan Agribisnis Kelapa Sawit di Provinsi Jambi*. http://ebookbrowse.com/ernawati-hd-implementasi-kemitraan-agribisnis-kelapa-sawit-docx-d449860203 (Accessed on March 24 2017)

[3] Edy, S.S, et al. 2008. Rejuvenation of Palm Oil Underplanting System. CV Mitra Karya. Field. Medan

[4] Walpole, RE. 1995. *Pengantar Statistika Edisi Ke-3.* Gramedia Pustaka Utama. Jakarta

[5] Eti, S. 2014. Comparative Analysis of Alternative Conventional Palm Oil Rejuvenation Models with Underplanting of Smallholder Nucleus Plantation Patterns (PIR) in Sei Lambu Makmur Village, Tapung District, Kampar District. Essay. Faculty of Agriculture. Riau University

[6] Lamtiur, PM, Sakti H and Shorea K. 2015. Analysis of the Rejuvenation Model of Oil Palm Plantations in Plasma Patterns in Meranti Village, Pangkalan Kuras Subdistrict, Pelalawan Regency, Riau Province. Journal. Faculty of Agriculture. Riau University

[7] I Wayan, M. 2013. Comparative Analysis of Oil Palm Farming Income in Ribs and Plasma Groups in Gunungsari Village, Pasangkayu District, North Mamuju Regency. Journal. Faculty of Agriculture. Tadaluko University

IOP Conf. Series: Earth and Environmental Science **336** (2019) 012004 doi:10.1088/1755-1315/336/1/012004

Impacts of oil palm plantation on the livelihood of the local communities and the conservation area

Sudarsono Soedomo

Department of Forest Management, Bogor Agricultural University, Darmaga, Bogor, Indonesia

*E-mail: ssoedomo@gmail.com

Abstract. Oil palm plantations have created so many controversies, particularly their impacts on the environment and the livelihood of the local communities. The controversies are usually raised by non-governmental organizations working on environmental and social-related issues. This paper focuses on the socio-economic impacts of plantation development. To observe what is going on, case studies were conducted in two villages strongly influenced by the development of oil palm plantation. The research finds out that those controversies are generally exaggerated. In general, economically the local communities benefit from the development of oil palm plantations. It does not mean that the negative socio-economic impacts are absent, but those impacts are manageable and have been addressed seriously, particularly when the plantation involves a member of Roundtable on Sustainable Palm Oil (RSPO). The real issue is not yes or no towards oil palm plantation, but rather how to develop oil palm plantation justly.

1. Introduction

Many negative issues have been associated with the development of oil palm plantation, particularly environmental and social-economic issues. The main environmental issues are deforestation and CO2 emission [1], while the main social-economic issues are land-related conflict and the loss of traditional livelihoods. According to [2], oil palm has not been a major driver of deforestation. Much of the forest in Jambi was cleared more than 20 years ago, and rubber was an established cash crop long before the oil palm boom started. However, oil palm growth occurs in locations with ongoing logging activities, so indirect effects on deforestation are possible. The average opposite position was taken by [3]. This paper focuses on socio-economic issues.

If the ownership ratio between corporations and smallholders is used to measure inequality and just level, then forestry businesses on the state lands are the most unequal and unjust since corporations own nearly 100% of those businesses. Meanwhile, nearly 40% of 11-12 million ha of the oil plantations in Indonesia is owned by smallholders. In addition, the communities living within or near forests are generally the most deprived communities. The question is why people all over the world are more concerned with inequality and unjust in the oil palm industry rather than in the forestry industry? Is there any sort of hypocrisy here?

In general, the productivity of smallholder oil palm plantation is lower than that of the corporate oil palm plantation. However, with a strong institutional arrangement, [4] found that by collective action smallholder oil palm production system outperformed its nucleus estate plantation, which is the corporate plantation. Its institutional setup allowed farmers to combine the advantages of small-holder and plantation agriculture by capitalizing on collective action.

Institutionally, social issues and impacts related to oil palm plantations have been better addressed. RSPO requires all of its members to conduct social impact assessment (SIA) and free prior informed consent (FPIC). RSPO has principles and criteria that must be fulfilled by each member. There are eight principles in the RSPO, in which RSPO members are required to undertake a social impact assessment as outlined in principle six on "Responsibility to workers, individuals and communities affected by plantation and palm oil mills." That way the company must carry out SIA activities as one of the requirements to obtain or retain RSPO certification.

As suggested by [5], many smallholders have benefited substantially from the higher returns to land and labour afforded by oil palm, but district authorities and smallholder cooperatives play critical roles in the realization of benefits. Conflicts between communities and companies have resulted almost entirely from the lack of transparency, absence of FPIC and unequal benefit sharing, and have been exacerbated by the absence of clear land rights.

2. Objective

The objective of this paper is to know direct and indirect socio-economic impacts of oil palm plantation through case studies. Although this paper was built by case studies in two villages, which are Teluk Pulai Village in Kalimantan Tengah and Sungai Melayu in Kalimantan Barat, the two companies involved are also operating in many other places with more or less similar standard operating procedures, especially in dealing with social issues. Indeed, adjustments to meet local specific needs have been made whenever necessary.

3. Methods

Studies were conducted in two villages by interviewing several farmers and the leaders of the villages. The two villages are Teluk Pulai and Sungai Melayu. The Teluk Pulai Village is located in the District of Kota Waringin Barat - Province of Central Kalimantan and the Sungai Melayu Village is in the District of Ketapang - Province of West Kalimantan. In-depth interview were conducted to gain deep insights into the respondent's experience prior to and during the partnership with the companies. Respondents included village leaders (village heads and village secretaries), cooperative managers, community leaders, and oil palm farmers. Confirmation, especially for the case of Teluk Pulai Village, interviews with TNTP managers were also carried out. Questions were raised mainly to motivate the respondents to respond in-depth.

The two villages have established a partnership with the oil palm companies, which are PT. Andalan Sukses Makmur (PT. ASMR) with the Teluk Pulai Village and PT. Inti Sawit Lestari (PT. ISL) with the Sungai Malayu Village. The fundamental key of the partnership is nucleus-plasma relation, where the company is the nucleus, and the community of the villages is the plasma.

Before visiting the villages, we collected information regarding land conflict and any negative view associated with the oil palm plantation nearby. Helped by two assistants, this research was done by holding focus group discussions and reviewing the reports and documents.

Having arrived in Teluk Pulai village, we met the village leader to express our intention for visiting the village. In essence, this was the first person interviewed in the village. This subject was then asked for the names of other people who could provide more information regarding the research topic. Certainly, it was not a probabilistic sampling. I stopped asking for a new respondent when the previous stories provided had reached a strong convergence or consistency. The same process was repeated in Sungai Melayu Village. There were 11 respondents in Teluk Pulai Village and 16 respondents in Sungai Melayu Village. The main issues explored are (1) what are the roles that the companies have done that affect their lives, (2) what is the impact of oil palm development on their livelihoods, and (3) how do they view the issues of land conflicts.

The data gathered from the respondents were examined to see their consistencies and coherence. The data were then arranged to construct more meaningful insights and on the basis of those insights the conclusions are derived.

4. Results

4.1 Teluk Pulai Village

4.1.1 The main issues

The Teluk Pulai Village is an isolated village because there is no road connecting this village and other villages. The only way to reach this village is through water transportation. The village had existed away before the designation of the Tanjung Puting National Park (TNTP), even before the independence of Indonesia. For a living, they practice shifting cultivation and collect forest products from the forests surrounding their village. The community was isolated and lived in very poor condition. In the first step of state forest land establishment, the village was included as a part of state forest land.[1] The people of Teluk Pulai Village felt that there was no economic prospect to live in the village. To express how poor the condition it was, any credit application made by anyone from the village always ended up with a rejection. The community as a whole was marked by having an extremely high credit risk. Some households migrated to neighbouring cities, such as Pangkalan Bun. The population of Teluk Pulai Village dropped. The local community calls this period as a darkness age to illustrate that there was no future to live in the village.

In an amazing moment, the Head of the TNTP visited the field, in which Teluk Pulai Village was still included, and he realized the situation; the land is not suitable to be included as a part of a national park. The head agreed to exclude Teluk Pulai Village from the TNTP. Legal processes of separation were immediately initiated. Finally, the existence of Teluk Pulai Village has been acknowledged legally by the government. After the acknowledgement, the leaders of the village searched for an investor of oil palm plantation. Even though the land available in the village is less than the smallest economical size for a unit of oil palm plantation, in fact, there is an enterprise that is willing to develop oil palm plantation in the village, PT. ASMR. It is quite reasonable when the community also supports the presence of the company in its village as suggested by [4].

Three parties, which are the community, the TNTP, and the company, who have different objectives can work together. The TNPT awarded the community a planting tool to help the community to utilize the land with less labour.

4.1.2 Role of the corporation

The company developed oil palm plantation for its own and the community. The available land has been split into two parts, for the community and the company. In return to the land handed over to the company, the company agreed to develop oil palm plantation for the community for free. Moreover, the management of all oil palm plantations is orchestrated by the company. The company purchases the FFBs produced by the community's plantations at a reasonable price. Well-designed contracts are very critical here [6]. The net profit, however, is given to the community and the community itself conducts the distribution of the profit among the community members.

The company offers employment to the community members who are within the product range of age, which is 18 to 55 years. Usually, they are employed in the community's plantations because of the distance from the settlement. Although in essence, those people work in their own plantation, they ought to follow the direction given by the company and receive a regular wage.

In addition, the company also developed a communal farm field to grow annual plants, such as rice, corn, and vegetables. The farm field is controlled and managed fully by the community. The household who is interested in using the land can send a request to the community leader. Usually, the household will be given one to two ha. So far, very few households have been interested in using the land due to available labour force in the households.

[1] There are four steps to fully establish a state forest land: designation, boundary arrangement, mapping, determination

4.1.3 The social-economic impacts

The livelihoods of the local community have been improved and diversified. Although the main activities are still related to the land, they employ better technologies leading to higher productivity. More importantly, these activities can be conducted near their settlement; working on their own land, community land, or working for PT. ASMR as an employee. In effect, the average income becomes higher and more stable. These improved livelihoods have interesting implication for the utilization of the TNTP and the education of a young generation.

In relation to the TNTP, the villagers have better and broader choices in their home rather than go to the TNTP for forest products. Although Teluk Pulai Village is the nearest village to the TNTP, the population of the village is not a threat to the TNTP; it used to be, but it is no longer today. Oil palm plantations, in this case, play a vital role in reducing pressure on the conservation areas, which is the TNTP. In addition, the people of Teluk Pulai Village have a strong concern with the sustainability of the TNTP, because they benefit from it; they provide places for swift birds to nest and the birds hunt the food mostly in the TNTP. The nest of the swift birds provides amazing income ranging from IDR 8 million to 167 million a month.

With more stable and higher income, the people of the Teluk Pulai Village have no hesitance to send their kids to big cities for education. The parents realize that the resources in the Teluk Pulai Villages are very limited to accommodate its growing population in the long run. The plausible way to respond to this situation is to equip their kids with better knowledge and technology. For this reason, the people of the Teluk Pulai Village are very enthusiastic in encouraging their younger generation to acquire better education.

Emotionally, self-esteem and self-confidence of the Teluk Pulai villagers have blossomed again. They are no longer seen as people who are at high risk in financial matters. Much credit is offered to them.

4.2 Sungai Melayu Village

4.2.1 The main issues

The population of this village, especially the ones who rely on oil palm for a living has experienced up and down with oil palm plantation. There was a period when the oil palm plantation provided a prosperous livelihood, followed by a catastrophic period, and today the community is enjoying a flourishing period. The rolling livelihood is determined by the price of the fresh fruit bunch (FFB) of oil palm.

The oil palm plantations consist of smallholders' plantations and corporate plantations. The smallholders' plantations include the Transmigration Nucleus Estate (PIR-TRANS) sponsored by the government, and independent plantations developed privately by indigenous people. The participants of the program are indigenous people and the ones coming from Java Island; each household received two ha of oil palm plantation. Moreover, the corporate plantations were owned by PT. Benua Indah. The areas of the nucleus, the plasma, and the independent plantation were 3,810 ha, 17,902 ha, and 3,660 ha respectively.

Besides possessing oil palm plantation, PT. Benua Indah runs a Crude Palm Oil (CPO) mill. The plant processed the FFB supplied by its own plantation and the small holder's plantation. As a matter of fact, the community and PT. Benua Indah were involved in a partnership. In the beginning, PT. Benua Indah was running well and able to purchase the FFBs from the farmers at a profitable price. The livelihood of the community was promising since some of them were self-employed in their own plantation and some others were employed by PT. Benua Indah.

Mismanagement drove the company to the bankruptcy. The plantation and the processing plant were abandoned in 2008. The impact on the livelihood of the people relying on the oil palm plantation was quite devastating because the price of the FFB was very low. Some indigenous people occupied some of the company's plantations for the FFB. This occupation was the seed of the land conflict when the plantation was taken over by another company, which is PT. ISL, in 2015. Some plantations under occupation were well maintained, some plantations were half-maintained, and the rests were merely harvested. Indeed, the quality of the plantations was not uniform when the plantations were taken over.

Although the land conflict was actually pseudo conflict, PT. ISL decided to choose non-litigation processes by showing its willingness to pay compensation. The compensation depended on the quality of the plantation that was categorized into three levels. The compensation was IDR 1,500,000, IDR 1,000,000, and IDR 500,000 for the high, middle, low quality respectively. New plantations replaced the low-quality plantation. When the study was done, some plantations were being assessed for the compensation. Being driven by its own agenda, there was an NGO that calls this kind of situation as a conflict of land between the local community and the company. Instead of helping clarify the real situation, the NGO exaggeratedly capitalized the phenomenon for its own interest.

4.2.2 Role of the corporation

Basically, the role of PT. ISL is to continue what PT. Benua Indah usually did, which to purchase the FFB produced by smallholders' plantations at a reasonable price. PT. ISL is not involved in the management of the smallholders' plantations.

Another role of PT. ISL is to provide employment. The plantations of 6400 Ha in size can accommodate 700 labours, while the CPO mill with a capacity of 45 ton FFB an hour needs 120 to support. However, until 2018 the number of workers absorbed has only reached 250 because most of the plantations are still under construction.

The estate roads built to support the management of the plantation in fact can be used by local communities to access various services, such as health, education, and market. Indirectly, the oil palm plantation contributes to the improvement of the welfare of the community who has no relation to the plantation. The estate roads also facilitate the delivery of electricity service to the villages that used to be isolated. In addition, PT. ISL has also contributed regularly to the maintenance of the main road, which is actually under the responsibility of the local government.

4.2.3 The social-economic impacts

Economic activities increase significantly. The purchase of the FFBs by PT. ISL's CPO mill at reasonable prices has encouraged the local people to grow oil palm independently. There is a land use change toward oil palm in this case. In effect, the demand for land, especially for growing oil palm, increase that in turn jacks the price of land up. This finding is consistent with the one suggested by [7]. Moreover, the activities of growing oil palm have forward and backward linkages, such as provision of fertilizers, oil palm seedling, pesticides, transportation services, and even food. It is well known as the multiplier effect. In Siak of Riau Province, it was found that the multiplier effect of oil palm plantation is 3.01 [8]. In addition, the independent oil palm plantation can be used as collateral for borrowing money from a bank for starting new businesses. Since the oil palm plantations are more reliable collateral compared other types of plantations, the appraisal value of the oil palm plantations is usually also higher. In other words, the oil palm plantations open broader opportunities for investment. In addition, [6] show that oil palm has lower labour requirements than alternative crops (especially rubber), so that adopting farmers are able to manage larger land areas. Labor saved through switching from rubber to oil palm is also used to increase off-farm incomes. Furthermore, [6] also show that oil palm adoption improves household living standards and nutrition.

Higher and more stable income is enjoyed by the member of the community who joins the company. It is quite consistent with [9] who suggest that further expansion of the oil palm area will likely benefit non-farm households through gains in employment income. Employment in oil palm is more rewarding than employment in rubber, so involved in the oil palm sector as a labourer is positively associated with total household income. This positive effects on farmers' livelihoods are also suggested by [10]. Furthermore, the economic gains allow farm households to increase their consumption.

More and better roads make the village have better accessibility leading to lower transportation costs. Some agricultural products that used to be uneconomical to be brought to the market become economical. As a result, the people of the village become more market-oriented. This process also generates additional income for the community.

5. Discussion

The expansion of oil palm plantation seems to be unstoppable until it stops itself due to profitability. So long the oil palm plantation is still the most profitable plant species, then it is difficult to stop its expansion. As suggested by [11], oil palm is one of most profitable land use in the humid tropics attracting local people, corporations, and nations to promote it. In addition, oil palm is widely considered by communities as the best option for meeting their financial needs [5]. Why should we stop it anyway? Like any other plantation, certainly oil palm plantations have negative and positive sides. The fact that oil palm plantation has expanded sharply in recent years, more or less 40% of which is owned by smallholders, it indicates that the growers quite prefer oil palm. The real issue is not how to stop its expansion, but how to make the expansion more beneficial to more people. Oil palm plantation has contributed to the alleviation of rural poverty [11].

While the impact on the welfare improvement of the local communities is so obvious, the development of oil palm plantation is consistently hampered, for example, by introducing a requirement of conducting SIA and FPIC. SIA and FPIC should be applied to forest resource also. So far, the requirement of SIA and FPIC be applied to oil palm plantation but not to forest resource. We might see a community whose income is mostly derived from collecting forest products. Does it mean that the community wants to maintain its lifestyle forever? It does what it could be because it has no other choice or some other reasons. However, certainly, it is misleading to conclude that the community definitely wants to stay in the current lifestyle. As [12] suggest, isolated communities in Indonesia such as Baduy, Kajang, Mentawai, some Dayak groups, Bajau, and others who live in remote parts of the archipelago are all seeking improvements in their material well-being. Therefore, we need to ask such a community to express its aspiration regarding its futures. It is quite possible that the community has been a victim of the tyranny of experts [13].

Since the price of CPO on international market is variable, farm households depending on the oil palm plantation become more vulnerable to external shocks [9]. What happens in the global market will be transmitted rapidly and directly to the households. However, the phenomena have happened with rattan, rubber, cacao, and many other commodities. In other words, fluctuating income and livelihood of farm households are not new phenomena. Nevertheless, any attempt to reduce the vulnerability of the farm households to external shocks is helpful.

Conflict over land is a sensitive issue frequently raised by NGOs. In case of Teluk Pulai Village, there is no issue of land conflict. The community voluntarily hand over the community's land to the company to obtain developed oil palm plantations in return. Meanwhile, different situation was found in Sungai Raya. At glance, it seems that conflicts over land were real. In fact, this issue was raised by NGOs. There were land compensations given by the company to the parties occupying the land. Certainly, those compensations make the conflicts more real. However, when we take a look more closely, we have a different conclusion. Historically, the oil palm plantations were owned legally by a company but then abandoned due to bankruptcy. Some individuals from local community took advantage by occupying the abandoned plantations and harvesting the FFBs. Those occupied plantations could be classified into three categories of management intensity, which were high, medium, and low. A couple of years later, PT. ISL took over the plantations legally and found that some parts of the plantation had been occupied. Instead of choosing hostile approaches, PT. ISL preferred a peaceful approach by offering a compensation according to the management intensity level.

The corporate culture and the local culture do not always match. This is the main reason why employment offered by the company often cannot be capitalized by most of the local people. For example, the company employs a rigid time schedule and a certain level of performance that can rarely be met by most of the local people. Of course, blaming each other will not lead to a settlement. What both sides need to do is to learn together to narrow the gap. The problem is that narrowing such a gap takes a while that puts an additional burden on the company.

An interesting phenomenon was found in Teluk Pulai where oil palm plantations meet a conservation area named TNPT. The TNPT is not being threatened by the plantations, but rather be protected. The people of Teluk Pulai who used to collect forest products from the TNPT choose to stop those kinds of activities because they have the better opportunity by working in the oil palm plantations situated around their settlement. The key success here is the responsible engagement of all relevant stakeholders in addressing the need of the community to have better lives. Perhaps we need to learn more from this little success rather than searching for more complicated strategies [14].

6. Conclusion

In general, the socio-economic impacts of oil palm plantation based on this research, were not as negative as what often were claimed/declared by many NGOs. Conflicts over land claimed by NGOs are not always real conflicts. Many of them are generated by speculative claim over lands that are capitalized exaggeratedly by NGOs. Moreover, an astonishing phenomenon is found in Teluk Pulai Village, where the improvement of livelihood can play a vital role in protecting a conservation area. The oil palm plantations improve the livelihood of the local community as a whole directly as well as indirectly. The direct improvements are materialized mainly through the FFB trade and employment. The indirect improvements are generated by better accessibility and increased economic activities in their community.

In the case of Teluk Pulai Village, the improvement of the livelihood generates positive impacts on the TNTP. The villagers consider the opportunity costs of entering the TNTP for collecting forest products become too high. They prefer to use the TNTP services indirectly by providing a place for swift birds, which hunt food in the national park area, to nest. Meanwhile, their own time is used for farming around their settlements. In short, the improvement of livelihood can play a vital role in protecting a conservation area.

Acknowledgments
We would like to thank Pandhega Paramagama for precious help in preparing the manuscript, Hangga Prihatmaja and Anggiana Ginanjar Adinugraha for collecting field data, and the participants of the conference for useful suggestions. My sincere appreciation goes to reviewers.

References

[1] Bueno, M. A. F. and M. E. M. Helene (1991). Global deforestation and CO2 emission: Past and present. A comprehensive review. Energy & Environment **2** (3), 235–282
[2] Gatto, M., M. Wollni, and M. Qaim (2015). Oil palm boom and land-use dynamics in Indonesia: The role of policies and socioeconomic factors. Land Use Policy **46**, 292 – 303
[3] Susanti, A. and A. Maryudi (2016). Development narratives, notions of forest crisis, and boom of oil palm plantations in Indonesia. Forest Policy and Economics **73**, 130 – 139
[4] Jelsma, I., M. Slingerland, K. E. Giller, and J. Bijman (2017). Collective action in a smallholder oil palm production system in Indonesia: The key to sustainable and inclusive smallholder palm oil? Journal of Rural Studies **54**, 198 – 210
[5] Rist, L., L. Feintrenie, and P. Levang (2010). The livelihood impacts of oil palm: smallholders in Indonesia. Biodiversity and Conservation **19** (4), 1009–1024
[6] Euler, M., V. Krishna, S. Schwarze, H. Siregar, and M. Qaim (2017). Oil palm adoption, household welfare, and nutrition among smallholder farmers in Indonesia. World Development **93**, 219 – 235
[7] Obidzinski, K., R. Andriani, H. Komarudin, and A. Andrianto (2012). Environmental and social impacts of oil palm plantations and their implications for biofuel production in indonesia. *Ecology and Society 17*
[8] Agustira, M. A., R. F. Raola Jr., A. J. U. Sajise, and L. M. Florece (2015). Economic impacts of smallholder oil palm (Elaeis guineensis jacq.) plantations on peatlands in Indonesia. Journal of Economics, Management & Agricultural Development **1** (2), 105–123

[9] Cahyadi, E. R and Laurence, W. (2009). Is contract farming in the Indonesian oil palm industry pro-poor? Journal of Southeast Asian Economics **30**(1), 62-76

[10] Kubitza, C., V. V. Krishna, Z. Alamsyah, and M. Qaim (2018). The economics behind an ecological crisis: Livelihood effects of oil palm expansion in Sumatra, Indonesia. *Human Ecology*

[11] Sayer, J., Ghazoul, J., Nelson, P., and Boedhihartono, A. K. (2012). Oil palm expansion transforms tropical landscapes and livelihoods. Global Food Security **1**, 114-119

[12] Boedhihartono, A. (2017, 03). Can community forests be compatible with biodiversity conservation in Indonesia? Land 6

[13] Easterly, W. (2014). The Tyranny of Experts: Economists, Dictators, and the Forgotten Rights of the Poor. Basic Books

[14] Wilcove, D. S. and L. P. Koh (2010, 04). Addressing the threats to biodiversity from oil-palm agriculture. *Biodiversity and Conservation* **19**, 999–1007

IOP Conf. Series: Earth and Environmental Science 336 (2019) 012005 doi:10.1088/1755-1315/336/1/012005

The performance of oil palm cultivation in Muaro Jambi District

Mirawati Yanita[1]*, Dompak Napitupulu[1], Ernawati HD[1], Gina Fauzia[1], Ira Wahyuni[1]

[1]Department of Agribusiness, Faculty of Agriculture, University of Jambi

*E-mail: mirawatiyanita@unja.ac.id, mherlam@gwdg.de

Abstract. The development of oil palm plantations aimed to eliminate poverty and underdevelopment, especially in rural areas through enhancement of the incomes of the smallholders. However, most of the number of plants have entered the rejuvenation period, which influenced the performance of the cultivation. The objective of this research was to identify the performance of the oil palm cultivations in Muaro Jambi District and analysed the oil palm smallholding incomes. Using survey method, the results showed that age of the smallholder plantation, land size and price, have affected the number of costs incurred and the revenues in old-aged plants. The annual income per smallholder farmer reached IDR 38,211,648 or IDR 18,964,782 per hectare per annum. The research indicated that it would be better to maximize the activities in the use of agricultural inputs and be aware of the most current related innovations and technologies in productivity improvement of oil palms.

1. Introduction

The plantation sub-sector has a wide opportunity to become the mainstay of exports. Development in plantation sector is directed to accelerate the rate of production growth from private, smallholders and state plantations. Among the estate crops, oil palm is perceived not only as a potential crop in the economic development, but also in alleviating poverty [1]. Oil palm has become one of the most important national income sources for the Indonesian economic system. Since 2009, Indonesia has become the largest producer of palm oil in the world [2], [3]. Promoting oil palm plantations have increased farmer's incomes as well as enhancing economic development [4], [5], [6], [7], [8], [9].

Development of smallholder plantations encourages the improvement of farmer welfares and foreign exchange in addition to the absorption of labours in the upstream industry sector, namely the plantations themselves and the downstream industries [10]. The prospect of developing smallholder oil palm plantations is very much determined by the existence of economic policies that favour the people, in order to encourage the realization of the welfare. Oil palm is the main commodity of farmers in Jambi Province.

Muaro Jambi Regency is a regency in Jambi Province with the largest oil palm plantation area that covers 21.22% of the total area of oil palms in Jambi Province and contributes the second largest producer of 181,832 tonnes with the productivity of 2,381 kg/ha/year. The plantations in the Sungai Bahar Sub-district were established in 1983 by PTPN VI of which the latter was established by the Central Government through the Transmigration Programme as a response to the national programme to develop areas outside Java. The transmigration programme was integrated into the oil palm plantation

IOP Conf. Series: Earth and Environmental Science **336** (2019) 012005 doi:10.1088/1755-1315/336/1/012005

development through a partnership programme between the company (large) and the smallholder plantations known as PIR-Plasma Pattern. In such partnership, the term plasma refers to the farmers of the Transmigration Programme. Hence, the partnership of PTPN VI Sungai Bahar Oil Palm Plantation Programme is called the PIR-Trans [11].

The oil palms in Tanjung Sari Village were classified under the category of producing crops (TM) which could be divided into 3 age levels, the 6-10 years, the 11-15 years and e 16-20 years. Each age level will produce different levels of production. Production is one of the factors that influence the level of income in addition to price and production factor. The use of production factors such as capital, land area, infrastructure and labour, are considerations that farmers take into account when conducting cultivation. The income level of the farmers will be determined by the FFB production, the FFB prices and the farming costs to be incurred by farmers in the oil palm cultivation process. The objective of this research is to identify the performance of oil palm cultivations in Muaro Jambi District and analyse the smallholder incomes in cultivating the oil palm.

2. Method

The research was conducted in Sungai Bahar Sub-district of Muaro Jambi Regency. The survey collected data on the: (1) description of smallholder oil palm plantations, (2) oil palm production, (3) plant age, (4) land area, (5) production costs, and (6) price of unity (IDR/kg). Data were collected from primary and secondary sources using direct interviews.

The research location was selected based on purposive sampling. The Tanjung Sari Village was selected since it represented an area with the largest plantation age of 16-20 years. The number of farmers who cultivated oil palms of 16-20 years of age totalled to 270 families. The sampling method used was quota sampling, which is a technique for determining a sample of the population with certain criteria up to the desired sample quota, using the formula from Yamane, T (1967) [6] where the 12.5% precision level of samples obtained is as follows:

$$n = \frac{\mathrm{N}}{\mathrm{N}.(\mathrm{D})^2 + 1}$$

Where:

n = Quantity of the sample
N = Population quantity
D = Precision level (12,5%)

The sample obtained was 52 samples with a plant age of 16-20 years. Data analysis was done descriptively and also quantitatively by calculating the production costs using the following formula [12]:

$$TC = FC + VC$$

Where:

TC = (Total Cost) (IDR/Ha/Year)
FC = (Fixed Cost) (IDR/Ha/Year)
VC = (Variable Cost) (IDR/Ha/Year)

To calculate the income of oil palm cultivation by using the formula as follows:

$$PD = TR - TC$$
$$PD = (Y\ x\ Py) - TC$$

Where:

PD = Income (IDR/Ha/Year)
TR = (Total Revenue) (IDR/Ha/Year)
TC = (Total Cost) (IDR/Ha/Year)
Y = FFB production (Kg/Ha/Year)
Py = FFB price (IDR/kg)

3. Empirical results

Data area, production and productivity of smallholder oil palms in Muaro Jambi Regency, according to the 2016 sub-district as given below in table 1.

Table 1. Total Land Area of Immature Plantations (TBM), Crops (TM), Old Plants (TT), Production and Productivity of Oil Palms According to District in Muaro Jambi Regency in 2016

Sub-district	Total Land Area (ha)				Production (Ton)	Productivity (kg/ha/ year)
	TBM	TM	TTM/TR	Total		
Jambi Luar Kota	563	4,330	-	4,893	16,310	3,767
Sekernan	3,662	13,930	49	17,641	37,394	2,684
Kumpe Ilir	341	12,001	22	12,364	24,679	2,056
Muaro Sebo	4,367	6,680	-	11,047	16,205	2,426
Mestong	318	3,109	-	3,427	6,569	2,113
Kumpe Ulu	1,881	13,027	-	14,908	39,227	3,011
Sungai Bahar	**929**	**19,413**	**11,970**	**32,312**	**46,439**	**2,392**
Sungai Gelam	385	672	43	1,100	1,790	2,664
Total	12,446	73,162	12,084	97,692	188,613	2,578

Source: Plantation Division, Jambi Province, 2016

As seen in table 1, out of the eight sub-districts in Muaro Jambi District, Sungai Bahar has the widest area of oil palm plantation with a total area of 32,312 ha, which was capable of producing 46,414 tons with a productivity of 2,392 kg/ha/year. The number of non-yielding crops/damaged plants, which reached 11,930 Ha, was suspected to be the reason for the existence of large oil palm areas with low productivity. This sub-district has expanded into three sub sub-districts, one of which is the Sungai Bahar Selatan sub-district, whereas the second widest sub-district for oil palm-producing land. The Tanjung Sari Village is the second largest oil palm plantation village after Bukit Makmur Village. The production has reached 1,368 Ha, approximately about 21% of the total 6,567 Ha in the Bahar Selatan District.

Respondents in this study were oil palm smallholder farmers. Some limitations to the research included characteristics that were expected to hinder and influence the willingness and ability of farmers in cultivation, including age, level of education, training experience and total family members. The smallholder farmers comprised of various ages, ranging from 30 to 65 years old. Most of the respondents were around the ages of 43-48 years old under the category of productive age, who still had the physical abilities to work their land. With regards to the length of oil palm farming, it was found that the longest experience in oil palm farming was about 30 years and the lowest was about 7 years with an average activity of oil palm cultivation of 19 years. Data showed that the most of the farmers (43.31%) have been engaged in oil palm farming between 19 to 24 years, while the shortest length of being an oil palm farmer was 7 to 12 years (6.9%). The experience of this effort shows that farmers have had long enough experiences in operating palm oil.

As many as 24.13% of the farmers were high school graduates (SMA), while 52% graduated from elementary school (SD), 16% graduated from junior high school (SLTP), and 9% graduated from universities (PT). The level of education influenced the efforts to implement, process and increase production. The farmers individually owned their oil palm plantations of various land sizes. Of the 58 respondents, 77% had a land area of 1-3 ha, 14% had a land area of 4-6 ha and 9% had a large land area of 7 ha.

3.1. Description of oil palm cultivation

Land as an input that influenced the level of production would have a positive impact. In the study area, the land status was individually owned with a total land area per farmer approximately 3 ha. The age of the plants varied greatly from 6 to 20 years, with the oldest between 16 to 20 years (46%) of whom were Ex-PIR plant or plasma farmers whose plantations were firstly developed by PTPN VI and the production was managed by KUD Dwi Jaya. The cultivation technique carried out by the farmers was of standard, beginning with the preparation of seeds focussing on superiority and quality. The average planting distance was 8x9 m, to allow the growth of the oil palm trees. In carrying out the maintenance, the farmers applied fertilisers to compete with the growth and obtained a maximum harvest outcome, as well as the provision of medicines to fight protect the plants from diseases and weeds. The types of fertilizers used were Urea, KCL, TSP and Kieserite, with the following dosages: 2.0-2.5kg/ph/year for urea, 2.5-3.0kg/ph/year for KCL, 0.7- 1.0 kg/ph/year for TSP and 1.0-1.5 kg/ph/year for Kieserite. Fertiliser was given 2-3 times a year. As for the use of medicines, the average farmers used Gramaxonmore obtained from KUD Dwi Jaya where the payments could be made in instalments. Other maintenance activities included pruning to stimulate plant growth. Pruning was done once every 6 months to one year, which often required 3 to 4 days.

The harvesting process in the study area was carried out twice a month or 24 times in one year. Direct production was marketed to plantation companies, including from Pinang Tinggi Plant and PT. ALS. In 2017, the average production for the 16-20 year old plant was 44,159 kg/year or 16,891 kg/ha/year with the current price of IDR 1,421. FFB marketing was carried out through KUD Dwi Jaya where farmers sell their FFB to KUD and the KUD would channelled them to palm oil mills according to price and distance.

3.1.1. The cultivation revenue

Revenue is the result of the multiplication between production and selling price. The amount of revenue of a business is influenced by the amount of production with the unit price of production. The higher the amount of production and the unit price of production, the higher the revenue and vice versa. Fresh Fruit Bunches are what the farmers produced from their plantations. This research found that for the plants aged 16-20 years, the average production was 34,159/kg/farmer/year and the average per hectare was 48,481 kg/year, which gave the revenue of IDR 48,629,089/farmer/year or an average revenue per hectare of IDR 24,213,563/ha/year. Revenues from the oil palm cultivation in the study area are tabulated below in table 2.

Table 2. Production and Revenue from Oil Palm Cultivation in 2017

No	Description	Revenue farmer/ha/year 16-20 year	
		Per farmer	**Per ha**
1	Production (kg/year)	34,159	18,481
2	Price (IDR)	1,421	1,421
	Total	**48,629,089**	**24,213,563**

3.1.2. Cost of production in oil palm cultivation

Production costs are the values of various inputs in the form of objects and services used during the production process. In general, the costs incurred in the production process consisted of fixed costs and variable costs. Variable costs in farming included the cost of purchasing fertilizer, medicines, and labour costs, while fixed costs included equipment depreciation, road maintenance, etc. The followings were the costs incurred by the farmers in oil palm cultivation.

a. *Fixed Cost*

The fixed cost was the depreciation cost of the equipments through the reduction of the purchase value of the equipments in the remaining value of the tool divided by the economic life of the instrument. The instruments used were hoes, machetes, dodos, egrek, *kep, gancu, angkong, and tojok.* After calculating the depreciation cost of each equipment, the average cost of the depreciation of agricultural equipment used by the farmers totalled to IDR 319,802/year or IDR 160,459/ha/year.

b. *Variable Cost*

Variable costs, are costs incurred due to the amount of products produced, in other words, these costs are influenced by the amount of production. This fee (Table 3) is the cost used in the production process which was calculated for one year. Included in the variable costs, were the cost for purchasing fertiliser, medicine, labour costs outside family members and cost of FFB transportation which included motorbike and gasoline.

Table 3. Variable Cost of Oil Palm Cultivation in 2017

No	Variable Cost	Cost (IDR/farmer/ha/year) 16-20 years	
		IDR/farmer	IDR/ha
1	Fertiliser	4,467,946	2,251,112
2	Medicines	303,077	150,466
3	Labour Cost (outside the household)	4,348,154	2,197,795
4	Other costs	978,462	488,849
	Total	**10,097,639**	**5,088,222**

Data shows that fertilizer costs incurred by farmers are the highest cost among other costs. The optimal use of fertilizer is expected to help plant growth and increase the production of oil palm farming. The Farmers obtained fertilizers through middlemen and KUD. The types of fertilizer used by farmers are NPK phonska, SP-26, UREA, KCL and Dolomite. Fertilizer is given 3 times in one year with an average cost of IDR 4,467,946/year and IDR 2,251,112/ha/year. Research [13] states that the average cost of oil palm fertilizer at the age of 20 years and over in the village of Ujung Tanjung Sungai Bahar Selatan District is IDR 1,287,499/Ha/year. This is not much different from the results of this study with the average cost of fertilizer at the age of plants 16-20 years of IDR 2,251,112/Ha/year.

Medicines cost incurred by the farmers was IDR. 303,077/year or IDR. 150,466/ha/year. While the calculated labour costs were the costs directly incurred by the farmer, the basis of the calculation is the labour costs outside the family where the average labour costs incurred by farmers was IDR 4,348,154. The year and for the cost per hectare is IDR 2,197,795/ha/year. Furthermore, based on the research by [13] states that the average labour cost of oil palm at the age of 20 years and over is IDR 2,775,623/Ha/year, this is not much different from the research area the costs incurred amounted to IDR 2,197,795/Ha/year. For other costs, what is calculated is the cost of motorbike gasoline and road maintenance. The cost of motorbike gasoline is used to refuel motor vehicles that will distribute fresh fruit bunches to the collection site. The average costs incurred was IDR 978,462/year or IDR 488,849/ha/year. The total cost of oil palm farming is the sum of fixed costs and non-permanent costs. Data analysis resulted in the total production cost of IDR 10,417,441/farmer/year or IDR 5,248,781/ha/year for 16-20 years old oil palm trees. The total cost of oil palm farming can be seen in table 4.

Table 4. Production Cost in Oil Palm Cultivation in 2017

Description	Cost (IDR/farmers/Year) 16 – 20 Year	Cost (IDR/Ha/Year) 16 20 Years
Fixed Cost (FC)		
□ Depreciation cost	319,802	160,459
Total (FC)	**319,802**	**160,459**
Variable Cost (VC)		
□ Fertiliser	4,467,946	2,251,112
□ Medicines	303,077	150,466
□ Labour costs (outside the household)	4,348,154	2,197,795

Description	Cost	
	(IDR/farmers/Year)	**(IDR/Ha/Year)**
	16 – 20 Year	**16 20 Years**
□ Other costs	978,462	488,849
Total (VC)	**10,097,639**	**5,088,222**
Production Cost (FC + VC)	**10,417,441**	**5,248,781**

3.1.3. The income generated by oil palm cultivation

The income of cultivation is the difference between revenue and costs incurred during production. In this study, income refers to the income that the farmers received in one year, while the revenue is the product of production with selling prices. The total income earned by each farmer is given in Table 5.

Table 5. The Income Analysis of Oil Palm Cultivation, 2017

Description	Income	
	16 – 20 Year	
	(IDR/farmer/year)	**IDR/ha/year)**
1. Revenue		
a. Production (Kg)	34,159	18,481
b. Price (IDR)	1,421	1,421
Total revenue (IDR)	**48,629,089**	**24,213,563**
Fixed Cost (FC)		
a. Price (IDR)	319.802	160.459
Total (FC) (IDR)	**319.802**	**160.459**
Variable Cost (VC)		
a. Fertilizer	4,467,946	2,251,112
b. Medicines	303,077	150,466
c. Labour cost (outside the household)	4,348,154	2,197,795
d. Other Cost	978,462	488,849
Total (VC) (IDR)	**10,097,639**	**5,088,222**
4. Total Cost (FC + VC)	**10,417,441**	**5,248,781**
5. Income (Revenue – Total Cost)	**38,211,648**	**18,964,782**

The findings were similar to the research by [5] that many smallholders have benefited substantially from the higher returns to land and labour provided by oil palm, although the district authorities and smallholders cooperation played key roles in the realization of benefits.

4. Conclusions

a. Oil palm farming in Muaro Jambi Regency is dominated by plants aged 16-20 years, which were ex-PIR or Plasma plants. At these ages, the production level were very low attributed to several factors.

b. The level of income earned by the farmers was high at a certain level of production. Farmers earned an income of IDR 38,211,648/farmer/year or IDR 18,964,782/ha/year.

5. Recommendation

a. Farmers should maximize using production factors and be able to innovate using current technologies to increase the productivity of the oil palm trees.

b. Farmers must conduct replanting because the production was no longer optimal.

c. The roles of the government in conducting counselling and in replanting activities, were highly recommended to assist the farmers in developing oil palm farming to improve their household welfare.

d. The performance of the smallholder oil palm plantations could be improved by applying Good Agricultural Practices.

IOP Conf. Series: Earth and Environmental Science **336** (2019) 012005 doi:10.1088/1755-1315/336/1/012005

References

[1] Susila WR (2004) Contribution of oil palm industry to economic growth and poverty alleviation in Indonesia. Jurnal Litbang Pertanian **23**:107–114
[2] Krishna, V. (2017). Differential Livelihood Impacts of Oil Palm Expansion in Indonesia. Agricultural Economics **48**(5): 639-653.
[3] Clough, Krishna, V, Corre,MD, Darras, K. (2016). Land Use Choices Follow Profitability at The Expense of Ecological Functions in Indonesia Smallholders Landscapes. Nature Communication 7: 13137-13149.
[4] Feintrenie, L.,W.K. Chong,P. Levang. (2010). Why do farmers prefer oil palm? Lessons learnt from Bungo District, Indonesia. Small-Scale Forestry **3**, 379-396.
[5] Rist, L; Feintrenie, L; Levang, P (2010). The Livelihood Impacts of Oil Palm: Smallholders in Indonesia. April 2010. Biodiversity and Conservation **19**(4):1009-1024. DOI: 10.1007/s10531-010-9815-z
[6] Yamane, Taro. 1967. *Elementary Sampling Theory*. Eagle Wood Cliffs, Prentice Hall.
[7] Castiblanco,C, A. Etter, A. Ramirez. (2015) Impact of Oil Palm Expansion in Colombia: What to Socioeconomics Indicator Show? Land Use Policy. Vol **44**, March. 31-43
[8] Gatto, Wollni, M, Asnawi, R, Qaim, M. 2017. Oil Palm Boom, Contract Farming, and Rural Economic Development: Village level Evidence From Indonesia. World development **95**: 127-140. Elsevier.
[9] Jelsma, I, G.C. Schnoveld, A. Zoomers, A.C.M. van Western. 2017. Unpacking Indonesia's Independent Oil Palm Smallholders: An Actor Dissaggregated Approach to Identifying Environmental and Social Performance Challenges. Land Use Policy. **69**. 281-297
[10] Directorat General of Plantation, (2015). Plantation Statistic. Jakarta
[11] Dinas Perkebunan. Statistik Perkebunan Provinsi Jambi Tahun 2017 Jambi
[12] Pratama, Raharja dan Manurung, Mandala. 2008. *Teori Ekonom Mikro.* Jakarta: FEUI
[13] Setiawan, S. 2016. Analysis of Oil Palm Plantation Income in Ujung Tanjung Village Subdistrict of Bahar Selatan Muaro Jambi Regency, Faculty of Agriculture Jambi University

IOP Conf. Series: Earth and Environmental Science **336** (2019) 012006 doi:10.1088/1755-1315/336/1/012006

Linkages of smallholders' attitudes with the implementation of principles and criteria RSPO certification independent smallholder forum MRM Jambi Province, Sumatera Indonesia

Rosyani[1], Asmadi Saad[2] and Edison[3]

[1]Master Study Program in Environmental Science University of Jambi
[2]Department of Agroecotechnology Agriculture Faculty University of Jambi
[3]Department of Agribusiness Agriculture Faculty University of Jambi

*E-mail: rosyani-pertanian@unja.ac.id,rosy1762@yahoo.com

Abstract. Oil palm plantation industry in Indonesia is growing rapidly. Extensive development took place during 1999-2016. The total plantation area reached 11,118,795 hectares by 2016. Some 5,754,718 hectares was managed by private companies, 4,656,648 hectares by smallholders and 707,429 hectares by state-owned enterprises. Issues of oil palm plantation that are not environmentally friendly are continuously exposed. Therefore, Forum Petani Swadaya-MRM independent smallholder RSPO has obtained certification, meaning that this research is important to be carried out. Respondents in the study were 82 people. Data were analyzed using non-Parametric Chi-Square statistical tests. Smallholders' attitudes and knowledge are cognitively 47.6%, affective 57.3% and conatively 63.4%. Achievement of RSPO principles and criteria; 1). Commitment to transparency is 69%, 2). Compliance with applicable laws and regulations 84%, 3). Commitment to long-term economic and financial viability 64%, 4). Use of appropriate best practices by plantations and factories 47%, 5). Environmental responsibility and conservation of natural resources and biodiversity 41%, 6). Responsible consideration of employees, and of individuals and communities affected by growers and mills 64%, 7). Responsible development of new plantings 71%, and 8). Commitment to continuous improvement in key areas of activities 82%. The relationship between attitude and implementation of principles and criteria for RSPO certification. From the results of parametric statistical tests using Chi-Square, 3.42 $t_{hit} > t_{table}$ 1.68, the meaning of refusal H_0 thank H_1 means that there is a significant relationship between attitudes and implementation of RSPO certification principles and criteria, there is a tendency for a positive attitude, the implementation of the RSPO principles and certification is high. The assessment between the attitude and application of the principle and criteria shows the sustainability status for the forum certifications.

1. Introduction

Data on the development of oil palm area from 1999 to 2016, the growing land area of oil palm plantations in Indonesia. The area of oil palm plantations from 1999 increased rapidly until 2016. In 1999, the area of oil palm plantation was 3,901,802 hectares. In the sixteen-year period it increased dramatically to 11,118,795 hectares. The total area managed by private companies covered 5,754,718

ha, 4,656,648 ha was managed by smallholders, and the remaining 707,429 ha was managed by state-owned enterprises, it means the number of smallholders who cultivate oil palm plantations in Indonesia is quite large (36.5%) [1], but not many of these smallholders have attempted to practice environmentally sound and long-term management of oil palm plantations [2]. However, the MRM Forum Petani Swadaya is one group of farmers applying oil palm plantation practices that are environmentally sound and have obtained RSPO certificates in 2017. Therefore, this research is important to be carried out. The problems in this study are:
1). What is the attitude and knowledge of farmers towards the RSPO certification program
2). What is the level of application or achievement of the principles and criteria for RSPO certification
3). What is a relationship between attitudes and the implementation of RSPO certification principles and criteria
4). What is the sustainability status of RSPO certification

The purpose of this research is to:
1). Know farmers' attitudes and knowledge of the RSPO certification program
2). Know the level of application or achievement of the RSPO certification principles and criteria
3). Analyze attitudes and implementation of principles and criteria RSPO certification
4). Assess the sustainability status of RSPO certification

2. Materials and methods

The data taken in the field are RSPO principles and criteria data that have been implemented by smallholders receiving RSPO certificates. The method used in the study was a simple random sample. Research locations are in Tanjung Jabung Barat District, Merlung subdistrict and Renah Mendaluh Subdistrict consisting of five villages, namely Merlung Village, Rantau Benar Village, Lubuk Terap Village, Sungai Rotan Village and Pulau Pauh Village. It has been chosen as a research location purposively, the five villages are members of the PS-MRM Forum. with the number of 172 households:

According to reference [3], if the total population is less than 100 then the sample is taken entirely, but if it is greater than 100 then it can be taken 10% - 15% or 20% - 30% or more. To determine the sample will be taken using the formula from Taro Yamane or Slovin in Reference [4] as follows.

$$n = \frac{N}{Nd^2 + 1}$$

Note:
n = Number of sample
N = Number of population
d^2 = Precision (set 10%)

Based on the formula above, the following sample numbers are obtained:

$$n = \frac{N}{Nd^2+1} = \frac{172}{172.0,1^2+1} = 82,20 = 82,20 \quad respondents$$

Methods of data analysis used are:

Through Chi-Square test with 2x2 contingency. According to reference [5], if the cell contains a frequency of ≥5, then the formula used:

$$x^2 = \frac{N[(AD-BC)]^2}{(A+B)(C+D)(A+C)(B+D)}$$

Whereas if there is a cell containing a frequency of <5, the following formula is used:

$$x^2 = \frac{N\left[(AD-BC)-\frac{N}{2}\right]^2}{(A+B)(C+D)(A+C)(B+D)}$$

Description: N = Number of samples

IOP Conf. Series: Earth and Environmental Science **336** (2019) 012006 doi:10.1088/1755-1315/336/1/012006

The table is as follows:
Chi-square test analysis with 2x2 contingencies

Table 1. 2x2 Contingency Analysis

Farmers attitudes	Implementation of the RSPO P & C Certification Program		Total
	Successful	*Not successful*	
Positive	A	B	$A+B$
Negative	C	D	$C+D$
Total	$A+C$	$B+D$	N

Value (X^2) in the free degree table (Db) = 1 at 95% confidence level is 3.84 can be compared between X^2 count with X^2 tables with the following decisions:
Ho: There is no real connection between the attitude of farmers and the implementation of the principles & criteria for RSPO certification
H1: There is a real relationship between the attitude of farmers and the implementation of the RSPO Certification Principles & Criteria Program

With the following conditions:
- Accept Ho reject H_1 if X^2 counts <X^2 table means there is no real relationship between farmer attitudes and the implementation of RSPO certification principles and criteria
- Reject H_1 accept Ho if X^2 count >X^2 table there is a real relationship between the attitude of farmers and the implementation of principles and criteria

Furthermore, to measure the relationship between the two variables, contingency coefficients are used with the following formula:

$$C\ hit = \sqrt{\frac{\chi^2}{\chi^2 + N}}$$

Where:
X^2 = Value of Chi-square
N = Number of Samples
C_{hit} = Contingency Coefficient
Weak: 0 – 0,353
Srong: 0,354 – 0,707

Next to calculate the relationship, can use the formula:

$$t_{hit} = r\sqrt{\frac{N-2}{1-(r)^2}}$$

Where ;
$H_0\ ; r = 0$
$H_0\ ; r \neq 0$
If t counts ($\leq$ t table = ($\propto = 5\ \%\ db = N - 2$) Accept Ho
If t counts ($>$ t table = ($\propto = 5\ \%\ db = N - 2$) Reject Ho

3. Results and discussion

In attitude there are three components, namely cognitive, affective and conative [6]. Cognitive is a component that contains a person's understanding of what is true or true for the object of attitude, the Affective component is a person's subjective emotional toward an object of attitude. Affective in question leads to a level of agreement - disagree, and the conative component is a tendency to behave in a person that is related to the object of the attitude he faces. The theory is if the attitude is positive then the application of the principles & criteria will be high. While the RSPO principles & criteria have eight points, namely: which will be used as a standard for member farmers to be implemented on the plantation level.: 1). Commitment to transparency, 2). Compliance with applicable laws and regulations, 3). Commitment to long-term economic and financial viability, 4). Use of appropriate best practices by plantations and factories, 5). Environmental responsibility and conservation of natural resources and biodiversity, 6). Responsible consideration of employees, and of individuals and communities affected by growers and mills, 7). Responsible development of new plantings, and 8). Commitment to continuous improvement in key areas of activities [7].

The results of the calculation of farmer frequencies from the attitude values were as follows: there were 45 respondents (54.9%) high, farmers who had high cognitive value were 39 respondents (47.6%), farmers who had high affective were 47 respondents (57.3%), and farmers who have a high conative value is 52 (63.4%) (see table 2). This means that not all respondent farmers have a positive attitude towards RSPO principles and criteria. Nevertheless the implementation of RSPO principles & criteria is carried out by farmers (see table 2).

Figure 1. *Linkaged Contingency between attitude and Principle & criteria implementated (T value)*

Table 2. Attitude analysis and separate application for each component

			RSPO Implementation			
			R	T	Total	
Cognitive	R	Count	19	11	30	100%
		% of Total	23.2%	13.4%	36.6%	
	T	Count	18	34	52	
		% of Total	22.0%	41.5%	63.4%	
Affective	R	Count	19	11	30	100%
		% of Total	23.2%	13.4%	36.6%	
	T	Count	18	34	52	

			RSPO Implementation			
			R	T	Total	
		% of Total	22.0%	41.5%	63.4%	
Conative	R	Count	19	11	30	100%
		% of Total	23.2%	13.4%	36.6%	
	T	Count	18	34	52	
		% of Total	22.0%	41.5%	63.4%	

Table 3. Linkaged contingency between attitude and principle & criteria implementated

			RSPO Implementation		Total
			R	T	
Attitude	R	Count	22	15	37
		% of Total	26.8%	18.3%	45.1%
	T	Count	15	30	45
		% of Total	18.3%	36.6%	54.9%
Total		Count	37	45	82
		% of Total	45.1%	54.9%	100.0%

The results of the overall analysis show that the attitude variables are significant (see figure 1 and table 3) with the implementation of RSPO principles & criteria. As mentioned by reference [8] that's the farmers who obtained the RSPO certificate implemented the RSPO principle & criteria and Napitupulu opinion supported by reference [9], that the knowledge of the farmers is related to imlementation of RSPO principle & Criteria. But on the other hand, when separately calculated the cognitive component in this case knowledge has not been related to the implementation of RSPO principles & criteria. This is possible because the new PS-MRM Forum obtained RSPO certification in 2017. Member farmers are still in the process stage towards understanding the RSPO program. However, at the level of the management level, it is seen that understanding and thinking and implementing principles and criteria have shown excellent cognitive. This means that not all members of the RSPO who have RSPO certification in the research location have positive knowledge. At the management level at the PS-MRM Forum can jointly conduct socialization to provide the same understanding to certified RSPO member farmers. As stated in reference [10], land markets did not have significant effects on forestland Appropriation and deforestation by farm households In Jambi Province. Instead, our data suggest that forest land Appropriation and the deforestation were primarily instigated by a booming export sector and facilitated by legal ambiguity and high internal tenure security for appropriated landand then by reference [11] Associated with RSPO certification the biodiversity will enhance, then the farmer group (Gapoktan) has been managed. Furthermore, the profit has been organized and the feedback will return to smallholder. The results are related. In the plantation areas oil palmsare planted byintercropping (plot scale) systems. As stated in reference [12] the evidence of that biodiversity enrichment in oil palm can generate synergies between economic and ecological functions under certain conditions. In the initial phase, EFForTS-BEE significantly increased yields per oil palm, which at the plot scale even compensated for the yield losses from oil-palm removal. With knoledge smollholders are able to impelement how to manage oil palm plantations and the environment in accordance with RSPO principles and criteria.

Furthermore, the status of RSPO certificate sustainability at the PS-MRM Forum is examined, indicating the sustainability status of RSPO certification can still be maintained, because the application of the principles & criteria is already high and supported in totality on the attitude variables of farmers.

IOP Conf. Series: Earth and Environmental Science **336** (2019) 012006 doi:10.1088/1755-1315/336/1/012006

4. Conclusion

From the calculation of the frequency of farmers have a positive attitude with a high attitude value of 45 (54.9%), farmers who have high cognitive value is 39 (47.6%), farmers who have an affective value of 47 (57.3 %), and farmers who have a conative value of 52 (63.4%). Overall the value of attitude is positive. Then followed by the implementation of the high RSPO principles & criteria. In total, there is a significant relationship between attitudes and implementation of RSPO principles & criteria. Then the results of the sustainability of certification at the PS-MRM Forum indicate the status that can continue. This means that the positive attitude of farmers has encouraged the implementation of RSPO principles and criteria, which is followed by the sustainability of RSPO certificate status in the PS-MRM Forum.

References

[1] Ministry of Agriculture, 2017. Indonesia Oil Palm farmers Data. Published Indonesia Republica Agriculture Ministry

[2] Keraf, 2010. Krisis & Bencana Lingkungan Hidup Global. Kanisius. Yogyakarta

[3] Sugiyono, 2014. Business Research Methodes, Alfabeta. Bandung

[4] Riduan. 2007. Formulas and Data in Statistical Applications for Research. Alfabeta. Bandung

[5] Siegel, Sidney. 2011. Nonparametric Statistics for Social Science, Gramedia Indonesia

[6] Azwar, S. 2015. Human Attitude, Theoryand Measurement. Jakarta, Library Student

[7] RSPO, 2014. Principles and Criteria for the Production Sustainable of Palm Oil. Endorsed by the RSPO Executive Board and accepted at the Extraordinary General Assembly by RSPO Members on April 25th 2013 (Including Major Indicators endorsed by RSPO Board of Governors on 15 November 2013)

[8] Napitupulu, D. Rosyani, Ernawati, Rukayah Rofiq, 2017. RSPO Certification Impacts on Oil Palm Smallholders welfare in Jambi Province. Jurnal Analisis Kebijakan Pertanian Vol. **15**. No. 2, Desember 2017. 127-149

[9] Rosyani, Dompak Napitupulu, Ermadani, dan Zuhdi, 2015. Impact of RSPO Certification to Economy, Environment and social of oil palm plantation with Tanjung Sehati Smallholder Jambi, Sumatera, Indonesia

[10] Krishna, Vijesh V, Christoph Kubitza, Unai Pascual, and Matin Qaim, 2017. Land Markets, Property Rights, and Deforestation; Insights From Indonesia. World Development Vol xx-pp xxx-xxx, Published by Elsevier Ltd

[11] Rosyani, Bambang and Ermadani, 2018. Independent Smallholders Sustain Biodiversity and Economy) in Jambi Province, Sumatera, Indonesia, Submitted Biodiversity Journal (Not yet Published)

[12] Gerard, A., Wollni, M., Hölscher, D., Irawan, B.,, Sundawati. L., Teuscher, M., Kreft, H., (2017). Oil-palm yields in diversified plantations: Initial results from a biodiversity enrichment experiment in Sumatra, Indonesia.Agriculture, Ecosystems and Environment 240 (2017) 259

IOP Conf. Series: Earth and Environmental Science **336** (2019) 012007 doi:10.1088/1755-1315/336/1/012007

The impact of oil palm revitalization partnership at PT. Brahma Bina Bakti to farming participants revenues in Sekernan District, Muaro Jambi Regency

Ernawati HD[1]*, Zakky Fathoni[1], Ira Wahyuni[1], Gina Fauzia[2], Rio Swantomy Tamba[2]

[1]Lecturer at Faculty of Agriculture, Jambi University, Jambi Province
[2]Agribusiness Study Program Faculty of Agriculture, Jambi University

*E-mail: ernahamid@unja.ac.id

Abstract. This study aims to determine the application of partnership Plantation Revitalization Program (PRP) and reviewing farm income of palm oil farmers participating in a partnership with PT. Brahma Bina Bakti in District Sekernan. This research was conducted in the village of Suak Putat, Tanjung Lanjut with the study of side income. In its application, the pattern of development of the garden (70%: 30%) is 70% of the land to farmers and 30% of the land to the company. While the pattern of results (30%: 70%) is of the total sales value of farmers' land TBS then 70% cut by the company through cooperatives and 30% delivered to farmers through farmer group administrator. This pattern of results is done through credit/interest paid. The result of the calculation is known that the average income of farmers at the farm level is Rp 6,291,397.04 / ha/year while the rate of KUD is Rp 6,831,633.66 / ha/year. Income farming palm at the farmer level is smaller than the palm farm income at the level of KUD. This is caused by differences in the structure of production costs where other costs are not covered sacrificed farmers and cooperatives that the cost of medicines, family workers and agricultural tools.

1. Introduction

The role of the oil palm plantation company is felt to positively affect communities around the plantation both on village development, especially on demographic, economic and social aspects. The development of oil palm plantations to improve the welfare of oil palm farmers has been carried out, specifically through partnership activities. The partnership is a business strategy carried out by two or more parties within a specific timeframe to achieve mutual benefits with the rules of mutual need and mutual growth [1].

In general, there are three partnership patterns in Indonesia, namely the PIR, KKPA and PRP patterns [2]. So far, the weakness of the PIR pattern partnership and the KKPA pattern are on the techniques of oil palm cultivation, repayment of credit and when the land is handed over to farmers. Farmers are unable to manage their land after the partnership period has over independently. Whereas in repayment of credit, the non-smooth return of credit is a classic problem found. The presence of this PRP partnership is projected to improve problems both technically and non-technically in the previous partnership.

The Partnership of Plantation Revitalization Program (PRP) is a plantation partnership pattern built up by the government in 2007 as a determination to quicken the development of smallholder plantations through expansion, rejuvenation and rehabilitation of plantation crops supported by investment credit and interest subsidies by the government by involving companies in the plantation sector as partners [3].

Manaek (2013) [4] conducted the research with the title Income Analysis of Oil Palm Farmers Plantation and the Influenced Factors in Sekernan District, the result of the research showed that there is no significant differences between independent and plasma farmers. By the partnership program, the cooperation members can manage their oil palm plantation sufficiently and independently from the beginning until post-harvest processing.

Muaro Jambi Regency is one of the regencies in Jambi Province. The number of oil palm plantation companies in Muaro Jambi according to the data of Muaro Jambi District Plantation Office in 2014 was 17 companies. However, only 1 (one) company has implemented this PRP pattern partnership, namely PT. Brahma Bina Bakti in Sekernan Subdistrict. Sekernan Subdistrict is one of the sub-districts located in Muaro Jambi Regency where most of the population must live as oil palm farmers.

The implementation of partnerships in Sekernan District by PT. Brahma Bina Bakti was held in 2007 precisely in Tanjung Lanjut Village, Suak Putat Village and Bukit Baling Village. The company in the case of the determination of the agreement must still be considered by the local government and related agencies which must continue to prioritize the principle of partnership. This partnership is required to have the capacity to improve technical and non-technical constraints on the previous partnership pattern, particularly in terms of improving the welfare of farmers participating in the partnership.

Based on the description related to the partnership matters and oil palm farming income, the farmers participating in the partnership, then the authors are interested in conducting research with the title "The impact of revitalization partnership on PT. Brahma Bina Bakti of oil palm farming participants revenues in Sekernan District, Muaro Jambi Regency". The purpose of this study is to determine the application of partnership at the level of the farmers in the partnership and to examine the farm income of palm oil farmers participants of partnership PT. Brahma Bina Bakti in Sekernan District.

2. Material and Methods

2.1. Scope of Research

The object of this research is the farmer participants of the partnership who actively involved in the implementation of oil palm, as well as located in in the study area with 5 years of age or already has Plantation but not produced yet (TBM) or Crop Yield (TM), and the plantation has been turned over to farmers or group of farmers. The focus of this research is on partnership activities, especially on oil palm farming activities carried out by farmers including partnership pattern, cost structure, production, cost of production, income, credit (instalments) and other data related to this research.

2.2. Data Collection

The data consists of primary data and secondary data. The primary data were taken from the farmers partnered with the age of the plant in the top 5 of the year. The data were taken from production data last year. While the secondary data were from the Central Bureau of Statistics (BPS) Provinsi Jambi, Dinas Perkebunan Jambi Province, Dinas Perkebunan Kabupaten Muaro Jambi, the office of the chief of the village, cooperatives and farmer groups. Data collection methods used in this research consists of interviews, namely the collection of data derived from interviews based on a list of questions directly with the farmers of the palm oil partnership, the observation, namely by direct observation systematically to the activity oil palm growers, and the literature study.

2.3. Sampling Methods

Survey method is used in this research, which research used a sample from a population as a source of data by using the questionnaire as the main data collection [5]. The number of farmers participating in the partnership is 328 people. So, the number of respondents taken is 10% of the total population or 33 respondents.

2.4. Data Analysis
To analyze the income of oil palm farm, partnered mathematically used the following formula:

π = TR - TC.......... TR = Production x Price of FFB (Fresh Fruit Bunches) (1)
TC = Fixed Costs (FC) + Cost (VC) (2)
π : Farm Income
TR : Total Revenue
TC : Total Cost

The size of cutting the land credit in a partnership that applied by PT. Brahma Bina Bakti is 70%:30%, meaning that 70% of the sales value is deducted by the company through KUD including; production costs and land loan instalments, and/or other agreed costs, while 30% are handed over to farmers as net income.

3. Results and Discussion
3.1. The Implementation of Partnership of PT. Brahma Bina Bakti
The implementation of partnerships implemented by PT. Brahma Bina Bakti is divided into two patterns/schemes; plantations development patterns and profit sharing patterns. The pattern of plantation development is a land sharing system agreed upon at the beginning of the partnership between the company and the farmers participating in the partnership. Whereas the profit sharing pattern is a profit-sharing system agreed/implemented by the company through KUD after the land has been handed over to the farmers participating in the partnership and is still in partnership or repayment of land credit.

Within its implementation, the management of the plantation to the area of land submitted by farmers at the beginning of the partnership (2,661.90 ha), the company in its management refers to the Regional Regulation No. 21 of 1999 Batanghari District, the net land received by farmers participating in the project was 70% or 1,863.33 ha of the land area provided that each farmer be entitled to 2 ha or 4 ha. While 30% or 798.57 ha of the land was handed over to the partner company for the purposes of facilities, infrastructure, public facilities, substitutes for gardens that were not worthy of submission and for the nucleus estate.

For profit sharing from FFB sales, the company sets a profit-sharing pattern (70%: 30%), which is from the farmer's net land area of 1,863.33 ha (70%) of which KUD cuts 70% of the FFB sales value and the remaining of 30% is delivered to farmers who are members of partnership participants through farmer group administrators. This revenue sharing policy refers to Regulation No. 21 of 1999 Batanghari Government. This is done up to the cost of developing a plantation or called credit /interest is complete.

3.2. Revenue Level of Palm Palm Partnership Participant
The value of sales of FFB in partnership is not a net revenue of farmers participating in the partnership. Farmer revenue is the value of selling FFB after deducting agreed costs (profit sharing). The profit sharing pattern in this study is 70%: 30%, that farmers' income is the sales value of FFB after deducting 70% of the total value (table 1).

Table 1. Production Data, Prices and Revenue of Respondent Farmers of Partnership Participants at the KUD Level

	Production (Kg/Ha/years)	**Price (Rp/Kg)**	**FFB Value (Rp/Ha/Years)**	**Revenue 70% KUD (Rp/Ha/Years)**	**Revenue 30% Farmers (Rp/Ha/Years)**
Highest	15.000	2.029,71	26.027.250	18.219.075	7.808.175
Lowest	10.740	1.362,15	18.635.511	13.044.858	5.590.653
Average	**13.124**	**1.735,15**	**22.772.108**	**15.940.475**	**6.831.633**

Based on table 1, the average price of FFB in the study area is Rp 1,735.15/kg, the average production of FFB is 13,124 kg/ha/year. The average value of total FFB sales is Rp.

22,772,108/ha/year. The value of sales of FFB then is divided by the pattern that has been agreed between farmers and KUD, which is 70% for KUD and 30% for participating farmers. The revenue of farmers participating in the partnership at the KUD level is Rp. 6,831,633/ha/year and the average KUD revenue is Rp. 15,940,475/ha/year.

3.3. Cost of Palm Oil Partnership Participants

According to Soekartawi (2010), there are two types of costs in farming, namely; 1) Fixed costs is the cost of which is not used up in one production period, for example, land tax, water tax, depreciation of agricultural equipment and buildings, heavy equipment (tractors), etc.; 2) Variable costs (variable costs) are costs incurred during the production process, for example, the cost of fertilizers, seeds, pesticides, labour or wage labour, harvest costs (procurement of work tools and experienced labour) and land rent.

In this study what is meant by the cost at the KUD level is the deduction made by KUD on the value of selling FFB from the net land received by farmers after passing through the plantation development period. Costs deducted by KUD are included in the Partnership Agreement (SPK). In principle, KUDs in regulating and managing farmers' finances in partnership participants must remain to prioritize the principles of partnership, which are mutual and open to each other (table 2).

Table 2. Cutting Costs of FFB Sales Value Applied to Farmer Groups Partnership Participants Per Harvest Period in the Research Area

No	Cutting Points	Information	
1	Interest Credits	30 % (from selling values)	
2	Balanced Cost (Fertilizer deposit)	10 % (from selling values)	
3	Certification Installments	-	±30
4	Group Loans	-	
5	ADM & SPW fees	-	
6	Others (Roads Maintenance Fees)	-	
	Total	**70%**	

The costs incurred by farmers at the farmer level are the cost of pesticides and herbicides, the cost of depreciation of production equipment and family labour costs. Based on the calculation of the cost of pesticides and herbicides is IDR 201,426/ha/year, the cost of depreciation of production equipment is IDR 31,927/ha/year and family labour costs are IDR 306,883/ha/year (table 3).

Table 3. Comparison of Cost Structure at the Farmer Level and the KUD Level in the Research Area

No	Costs of Farmers Level		Costs of Partnership		
	Uraian	**(Rp/Ha/Tahun)**	**No**	**Uraian**	**(Rp/Ha/Tahun)**
1	**Production Cost**		1	KUD costs (70%)	
	• Equipment depreciation	31,927.27		• Interest credit (30%)	6,831,633
	• Fertilizers	- ---→		• Fertilizer deposit (10%)	2,277,211
				• Management fees (30%)	6,831,633
	• Family Labors	306,883.35			
	• Non-family Labors	**(3,693,670)** ------→			**3,693,670**
	• Pesticides and herb.	201,426.00			
2	**Total Production Cost**	**540,236.62**	2	Total	12.802.514
3	**Revenue/ Profit sharing**	**6,831,633.66**	3	**Revenue/ Profit sharing**	**15,940,475.94**

In table 3 shows the difference in costs at the level of KUD and the level of farmers participating in the partnership. At the farmer level, the costs borne by the farmers themselves are; depreciation of agricultural equipment, labour in the family and the cost of pesticides and herbicides with an average of Rp. 540,235.62/ha/year. Fertilizer costs and non-family labour costs are borne by KUD. The amount of costs incurred by KUD is the number of credit costs, fertilizer costs and labour costs, which is Rp. 12,802,514/ha/year.

3.4. Income of Palm Oil Partnership Participants

In this study, the farmer income of the participating participants was calculated at the level of the farmers participating in the partnership. At the time of the research, each farmer who joined in the farmer group was still in the period of repayment of credit/land interest. So that the income of the farmers participating in the partnership is still relatively small due to the KUD cut. However, after the credit period is paid off, farmers' income will increase, formulated as follows:

- Income of Farmers at the payback periods

Π_p = TR – TC(30% land credit + 30% other costs + 5% management fee)

Other costs: Certification instalment fees, group loan costs, road maintenance costs.

- Income of Farmers after the payback periods

Π_p = TR – TC (5% management fee)

Therefore, the income of farmers after the credit is paid off will increase by 65%. KUD's deduction is only 5% for ADM fees and management. However, the increase in the income of farmers participating in the partnership is not absolute, because, after the repayment period, the costs of production and maintenance will be fully borne by the farmers.

Table 4. Farm income of oil Palm Farmers in the Research Area

No	Income at Farmers Level		No	Income at KUD Level	
	Description	**(Rp/Ha/Years)**		**Description**	**(Rp/Ha/Years)**
1	Production cost		1	Cost of KUD (70%)	
	• Equipment Depreciation	31,927.27		• Interest credit	6,831,633
	• Fertilizers	-		• Fertilizer deposit	2,277,211
	• Family Labor	306,883.35		• Labor, administration cost, other costs	6,831,633
	• Non-Family Labor	-			
	• Pesticides and Herbicides	201,426.00			
2	Total Production Cost	**540,236.62**	2	Total Cost of KUD	**12.802.514**
3	Production	13,124.00	3	Production	13,124.00
4	Price	1,735.15	4	Price	1,735.15
5	Selling values of FFB	22,772,108.60	5	Selling value of FFB	22,772,108.60
6	Revenue/ Profit sharing	**6,831,633.66**	6	Revenue/ Profit Sharing\	**15,940,475.94**
7	**Income of Farmer**	**6,291,397.04**	7	**Profit**	**3.137.961**

Based on table 4 the income of farmers at the KUD level is Rp. 6,831,633.66/ha/year while net income at the farm level is Rp. 6,291,397.04/ha/year. The large difference in the average income of farmers is due to other costs incurred by farmers such as depreciation of agricultural production equipment and the cost of medicines and labour costs in the family are not included in the dependents/pieces of KUD. The costs borne by KUD are fertilizer costs, labour outside the family and credit/interest costs to the bank amounting to Rp. 12,802,514/ha/year. So KUD's net profit is Rp. 3,137,961/ha/year.

In partnership, if the farmer's land credit has been paid off, the farmer still must sell FFB to the company and the company is also obliged to buy the FFB of the farmer. This is valid for one crop cycle (± 20 years). In general, in terms of credit/interest in land, the farmer's income will increase as well. It can be assumed that the income of farmers after repayment of credit in this study amounted to Rp 6,291,397.04/ha/year increased to Rp 13,123,030/ha/year. However, the problem is with increasing time, then the life of the plant will also cause high production costs. This is in line with research conducted by Soleman Imbiri (2010) [6] with a study entitled "The Impact Analysis of Oil Palm PIR on the Welfare of Surrounding Communities in Manokwari Regency (Case Study on Plasma Participant Farmers from the Arfak Tribe in Prafi District)". The results showed that the oil palm PIR project in the Prafi District after 25 years of operation: still had a direct and positive impact on the additional cash income of Arfak ethnic plasma participating farmers, although the average income derived from oil palm land was relatively smaller at Rp. 395,382,- per month compared to farming land of Rp. 514,693,- per month, and a side business of Rp. 418,909,- per month.

4. Conclusion

Implementation of the partnership pattern of the Plantation Revitalization Program (PRP) conducted by PT. BRAHMA BINABAKTI, in partnership with KUD AKSO DANO, in implementing the pattern of plantation development (70%: 30%), that is from the total land area in partnership with farmers, The land development pattern is applied at the beginning of the partnership while the profit sharing pattern is applied after the land has been handed over to farmers where oil palm plantations have already been produced and are still in partnership. Based on the results of research, oil palm farming income of farmers participating in the partnership calculated at the farm level is Rp. 6,291,397.04/ha/year. While the oil palm farming income of the partnership participants which was seen from the net income from KUD was Rp. 6,831,633.66/ha/year. The large difference in income is due to other costs incurred by farmers but not borne by KUD, namely the cost of pesticides, the cost of depreciation of agricultural equipment and labour costs in the family. Partnership Revitalization in its implementation is not without weakness. In the study area, the low quality of the management of KUD administrators and farmer group officials is a major problem. The weaknesses of KUD in rural areas include management culture that is still feudalistic paternalistic (supervision has not yet functioned). The weak entrepreneurial spirit and the low level of education of administrators. The sense of tolerance still influences the community in determining the management and management of cooperatives, not based on the quality of leadership and entrepreneurship. In addition, the participation of members is still low due to the low quality of supportive services to members and non-members.

Acknowledgement

At this opportunity, the author expressed his gratitude to the Dean of the Faculty of Agriculture, Jambi University and Chair of the Agribusiness Study Program of the Faculty of Agriculture, Jambi University. Also, thanks are also stated for the Head of Business Licensing of the Muaro Jambi District Plantation Service, the Jambi Provincial Plantation Service and the Akso Dano KUD administrator in Sekernan District, Muaro Jambi Regency.

References

[1] Ernawati, Hd. 2011. Implementasi Kemitraan Agribisnis Kelapa Sawit di Provinsi Jambi.http://ebookbrowse.com/ernawati-hd-implementasi-kemitraan-agribisnis-kelapa-sawit-docx-d449860203 accessed on 21 March 2014

[2] Armen, Mara. 2011. Dampak Perkebunan Besar Kelapa Sawit Terhadap Perekonomian Desa di Provinsi Jambi. Laporan Hasi Penelitian Kelompok. Universitas jambi. Jambi

[3] Direktorat Jenderal Perkebunan. 2007. Pedoman Umum Program Revitalisasi Perkebunan: Kelapa sawit, karet, kakao. Departemen Pertanian. Jakarta

[4] Manaek, 2013. Analisis Pendapatan Usahatani Kelapa Sawit Rakyat (Swadaya Murni) dan Faktor yang Mempengaruhinya di Kecamatan Sekernan Kabupaten Muaro Jambi. Fakultas Pertanian Universitas Jambi. Jambi

[5] Singarimbun, Masri. 1995. Metode Penelitian Survei. LP3S. Jakarta

[6] Imbiri, Soleman. 2010. Analisis Dampak PIR Kelapa sawit terhadap Kesejahteraan Masyarakat Sekitar Kebun di Kabupaten Manokwari (Studi Kasus pada Petani Peserta Plasma Asal Suku Arfak di Distrik Prafi).Jurnal Penelitian Progtram Magister PSLP PPSUB. Agritek Vol. **18** No. 2 April 2010

IOP Conf. Series: Earth and Environmental Science **336** (2019) 012008 doi:10.1088/1755-1315/336/1/012008

Socioeconomic and cultural impacts of oil palm plantation development in Indonesia

Tutut Sunarminto[1]*, Joko Mijiarto[1], Eka Dana Prabowo[1]

[1]Forest Resources Conservation and Ecotourism Department, Faculty of Forestry, Bogor Agricultural University, Indonesia

*E-mail: tsunarminto@gmail.com

Abstract. Oil palm industries in Indonesia continue to grow and gave impacts for communities around the plantations. For this cause, it is significant to identify the socioeconomic and cultural impacts on the residential districts around the oil palm plantations and the community's perceptions of the existence of oil palm plantations. The purpose of the study was to assess the social, economic and cultural impacts of oil palm plantations on the local community and their perceptions of oil palm plantations existence. The research was carried out in several oil palm plantations in Central Kalimantan, West Kalimantan, Lampung, South Sumatra, Riau and Merauke during the 2015-2017 period. Data collection was gathered through literature studies, Focus Group Discussions (FGD), Likert Scale questionnaires and observations. Data analysis were carried out using both qualitative and quantitative approaches. The results showed that the existence of oil palm plantations influenced the lives of the surrounding communities. On the socioeconomic aspect, the existence of oil palm plantations had opened up employment opportunities, opened and increased business opportunities, and increased community incomes. On the socio-cultural aspect, the existence of oil palm plantations has helped in building educational, religious, health and other infrastructure facilities. Nevertheless, it could not be denied that positive impacts have also led to various social changes within the society, including changes in the people's lifestyles and reduced community social encounterment.

1. Introduction

Globally, Indonesia is the nation with the largest CPO producer due to the rapid growth of oil palm plantations, which has escalated by 37.7 times during 1980 to 2016. In 1980, the total area of the plantation was only around 300,000 ha, which in 2016 has reached 11.6 million ha. The increase has made Indonesia as the largest CPO producer, as well as placing oil palm as one of the biggest foreign exchange contributors [1] in Indonesia with a value of USD 17.8 billion. In addition, the development of the oil palm industry has also created many jobs for the community and indirectly helped in accelerating the development of village infrastructures around the plantations. This condition ultimately has affected the economic development of the village. As revealed by reference [2], [3], [4], the development of oil palm plantations have contributed to village development.

In the midst of economic benefits, oil palm industry is inseparable from various issues related to environment and culture. Of the environmental aspects, development of oil palm plantations are thought to reduce water quality [5], [6], [7], cause loss of wildlife habitats [8], [9], [10], [11] and induce global climate change [9], [10], [11]. Whereas, the negative social and cultural impacts include the increasingly fading culture especially the role of customary institutions, reducing cohesions of community relations, discrimination against local communities [9], land disputes [12], and loss of access to natural resources [9]. These issues are considered as black campaign against oil palm since oil palm are one of the most oil-producing plants and consumed around the world, hence more profitable compared to other vegetable oils. The advantages of oil palms include high oil productivity, superior potential for food products, and wide application potential [13].

Based on these backgrounds, it is important to identify the community's perspectives related to the development of oil palm plantations. Does the community around the plantations consider oil palm plantations to have negative or positive impacts? Thus, the purpose of the research is to assess the perceptions and impacts of oil palm development in the socioeconomic and cultural aspect.

2. Materials and Methods

The study was conducted in 2015-2017 in the Provinces of South Sumatra, Lampung, Central Kalimantan, West Kalimantan and Papua (table 1). The study focused on the perceptions and impacts of oil palm plantations on the socioeconomic, culture and infrastructure development (table 2). Data were collected using Rapid Rural Assessment (RRA) technique that integrated literature studies, questionnaires, in-depth interviews, FGD and observation.

Table 1. Study location of assessment the perceptions and impacts of oil palm development

No	Province	District	Sub-district	Village
1	South Sumatera	OganKomeringIlir	Sungai Menang	a. Gajah Mati
				b. Sungai Ceper
			Cengal	a. Sungai Jeruju
				b. Sungai Pasir
2	Lampung	Lampung Utara	Abung Selatan	a. Candimas
				b. Klaibening Raya
				c. Bumi Raya
				d. Ratu Abung
				e. Kemalo Abung
				f. Kembang Gading
			Kotabumi Selatan	a. Tanjung Senang
				b. Kelapa Tujuh
3	Central Kalimantan	Kapuas	Dadahup	a. Sumber Alaska
				b. Sri Mulya
				c. Sumber Makmur
				d. Dadahup
				e. TambakBajai
4	West Kalimantan	Sintang	Ketungau Tengah	a. MargaHayu
				b. Swadaya
				c. Senangan Kecil
				d. Wirayuda
				e. Wana Bakti
				f. Mungguk Gelombang
5	Papua	Merauke	Ulilin	a. Belbeland
				b. Mandekman
				c. Rawahayu
				d. Kerely
				e. Kafyamke
				f. Baidub
			Elikobel	a. Sipias
				b. Metaat Makmur
				c. Bupul Indah
				d. Tof-tof
				e. Bouwer
				f. Bumun

Table 2. Study aspects of assessing the perceptions and impacts of oil palm development

No	Aspect	Criteria	Data Collection	Analysis
1	Socioeconomic	a. Jobs opportunities b. Business opportunities c. Income certainty d. Welfare e. Basic material needs	• Literature study • FGD • Questionnaire distribution • Observation	Qualitative descriptive
2	Socioculture	a. Work ethic b. Institutional c. Education d. Togetherness		
3	Infrastructure	a. Electricity b. Accessibility c. Clean water d. Education facilities e. Religious facilities f. Common facilities		

Literature study was carried out by collecting and studying related documents, including reports related to social aspects (EIA Documents, RKL/RPL Documents, statistical figures, various company documents on: Corporate Social Responsibility (CSR), Standard Operational Procedure (SOP) (Environment, Occupational Health and Safety/LK3), and other documents related to the socioeconomic and cultural aspects of the communities.

The questionnaire was constructed using closed-ended type with scoring system. The scoring system is one of the structures that are easy and commonly used in qualitative research, although in its use, there are often errors and weaknesses in the form of consistency in the score structure and weaknesses in the indicator setting of each scoring unit. One method used to overcome this problem is to compose a Likert Scale into a structured score. Although, technically the scale use numerical value from 1 to 5, but following the character of the Indonesian people, the scale should be changed to 1 to 7 [14]. The 7-point scale would also add to the reliability. The value of 1 is given to statements with "strongly disagree" responses, 2 for "disagree", 3 for "somewhat disagree", 4 for "indifferent", 5 for "somewhat agreed", 6 for "agree", and 7 for "strongly agree". The meaningful pattern can be adjusted according to needs, for example, is a range of "very dissatisfied" to "very satisfied", or a range of "very low" to "very high".

In-depth interviews were conducted to explore the people's opinions and views regarding the social impacts that have and were expected to emerge in the future. Interviews were conducted with key persons interviews to dwelve into a problem following the areas of expertise or authority of each key respondent in each village. For this reason, the selection of key respondents was done using purposive sampling. The number of respondents in each village varied according to their needs and conditions. In addition, interviews were also conducted with other related sources who were considered to have understand on the key or social issues, based on previous documents studied.

FGD activities were carried out in the villages in and around the company's area with the aim of:

a. As a forum for problem-solving, if document reviews, observations and interviews found that solutions should be made, and to obtain more in-depth inputs to formulate current social impact studies when the company operates and in the future;
b. Sharpened various issues, problems and needs related to social, economic and environmental impacts on the community around the company; and
c. Determine the perceptions and obtain inputs from stakeholders on the various issues concerning to oil palm plantation management, conflict resolution and social management plans as well as community needs and expectations of the oil palm plantations.

The FGD activities were carried out by involving local leaders and local government officials and other parties who were considered competents with the issue of studies, such as the Chief of Village,

BPD, Village Team, village officials, leaders of local institutions, community leaders, religious leaders and other community members. The FGD were intended to build perceptions of the groups targeted by the activity.

Field observations were very important to have an understanding of the current field conditions and factual situations, related to the object of study or the issue under study; verification of conformity between interview data and actual conditions and ensure visually the extent to which the expected impacts appear can or cannot be overcome; and explore deeper information through direct observation in the field of various matters concerning socioeconomic conditions in and around the company.

Then, the data analysed by descriptive-qualitative to describe the social, economic and cultural conditions of the community. Information on community perceptions of the impact of oil palm plantations were gathered utilizing the following formula [15]:

$$\mathrm{I} = \frac{Distance}{Number\ of\ Classes}$$

Explanation: I=Class Interval, Distance=Highest value reduced by lowest value, Number of Classes= the number of intervals or categories specified so that the interval value is obtained,present in figure 1.

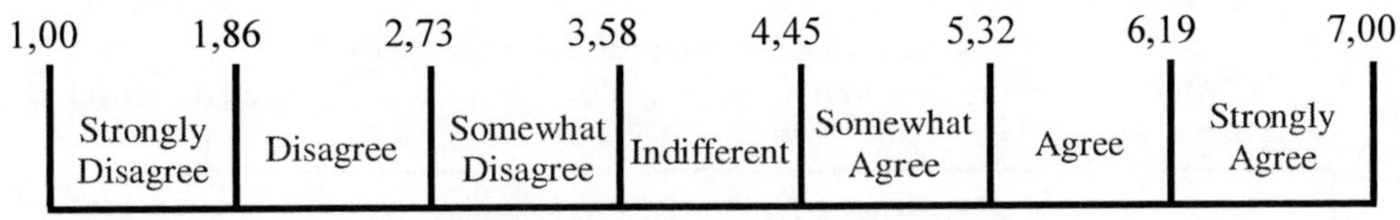

Figure 1. Perception interval scale

3. Results and Discussion

3.1. Development of oil palm plantation in Indonesia

Oil palm was introduced in Indonesia in 1848 by the Dutch and planted in the Bogor Botanical Gardens, which was then commenced to be commercially cultivated in 1910 because these plants flourished and had been tested in several regions. The growth of plantations in Indonesia refered to Act Number 39 of 2014 concerning Plantation (referred to as the Plantation Act). Article 2 of the Plantation Act states that plantations are carried out based on the principle of benefit and are sustainable, integrated, together, openly, and equitable. While in Article 3, it is stated that the objectives of plantation development are to increase community's income and provide job opportunities.

The oil palm industry in Indonesia continues to see rapid development so that in the end it places oil palm as the highest contributor to foreign exchange in Indonesia (figure 2). Oil palm plantations thrived throughout Indonesia (figure 3). In general, according to the ownership, oil palm plantations in Indonesia can be divided into three categories, i.e., private-owned, state-owned and community-owned. Presently, private oil palm is the biggest contributor of CPO in Indonesia (figure 4), which are exported to China, India, Pakistan and Africa.

IOP Conf. Series: Earth and Environmental Science **336** (2019) 012008 doi:10.1088/1755-1315/336/1/012008

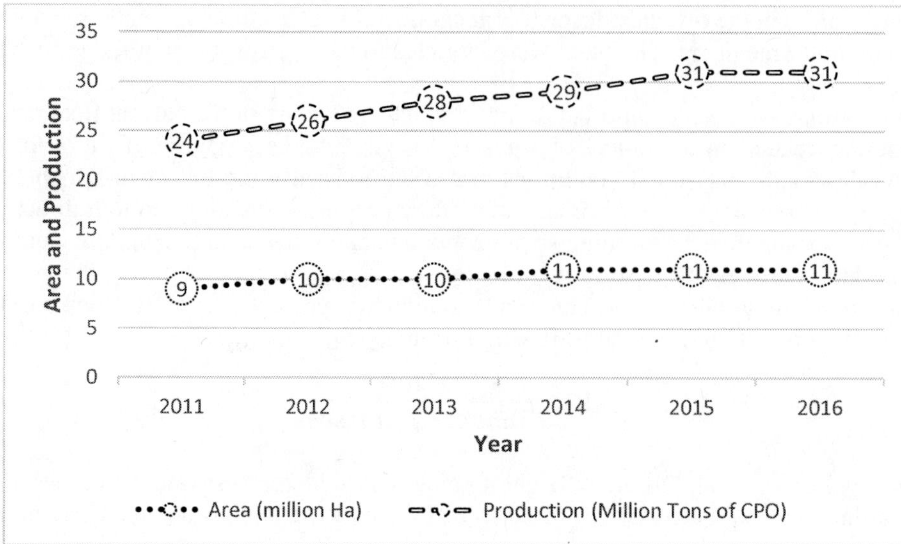

Source: Indonesian oil palm statistics 2016

Figure 2. Oil palm area and production development in 2011-2016 period

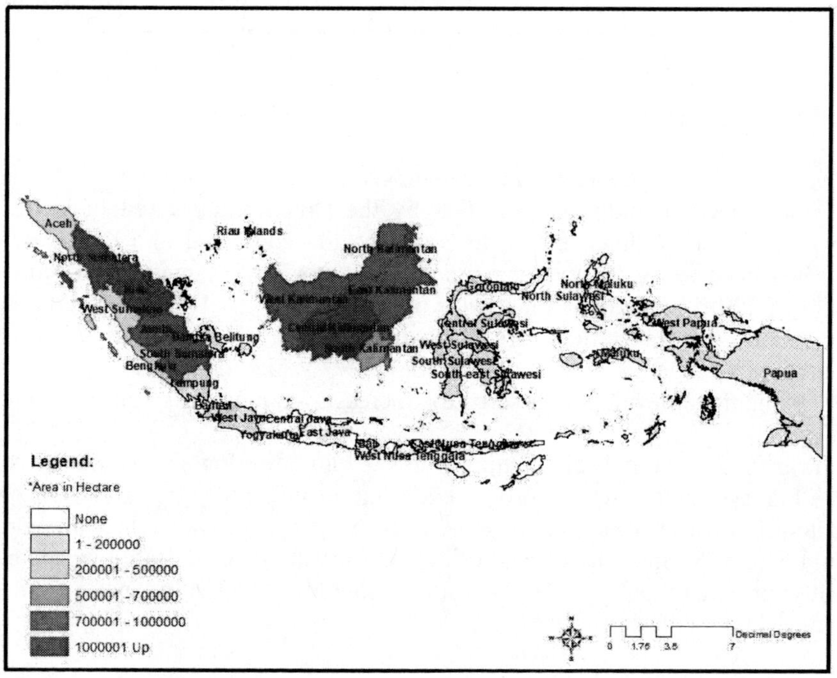

Figure 3. Oil palm plantation area in Indonesia

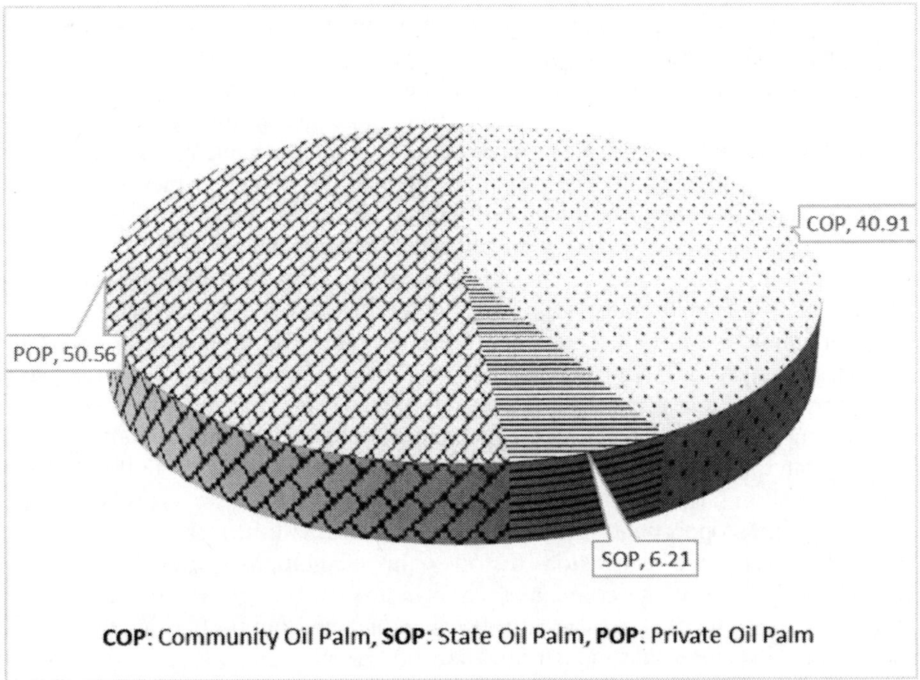

Figure 4. Oil palm plantation area based on ownership

3.2. Socioeconomic impacts of oil palm plantation development

The existence of industries including oil palm plantations provided benefits to the local communities. According to the statements by reference [3] and [16], the industry has provided a multiplier effect [3] and reference [17] states that the oil palm industry brings major changes to rural communities, especially in terms of the economy. Established on the research conducted, it was recognized that the existence of oil palm plantations has positive impacts on the local villages (figure 5). This is in line with research by [18], which also demonstrates the economic benefits of oil palm growth.

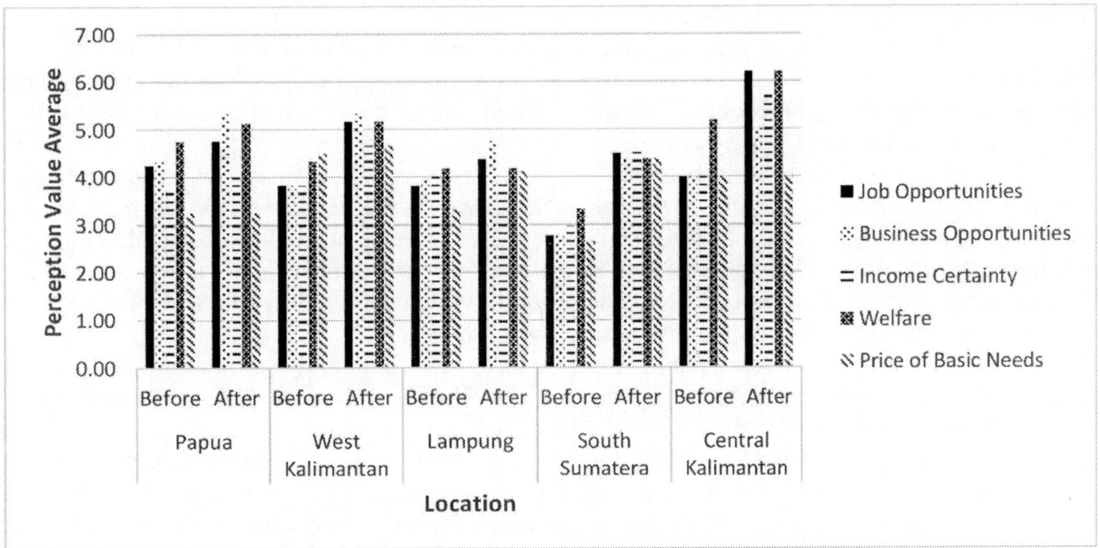

Figure 5. Socioeconomic impacts of palm plantations

The existence of oil palm plantations provided huge amount of work opportunities for the community, starting from daily labour to permanent employees[19], [20]. Nevertheless, most of the people in the study sites, comprised of more daily labourers as harvesters, applicators and other related tasks that did not necessarily require extra skills. This happens because generally, the surrounding communities has low education levels. With more people having permanent jobs, the monthly incomes could be secured, hence enhanced the the economic flow in the village, which increased the people's purchasing power and is turn increases the mobility of goods and services [18]. People who previously worked as farmers did not have certainty in terms of income due to weather factors, amount of production and prices that could change. The existence of the oil palm companies guaranteed the flow of incomes to households from monthly salaries. This was backed up by statements of [21], [22], [23], [24], [25]. The communities would be able to use these revenues for various purposes, especially those that are consumptive, such as buying motorized vehicles, repairing houses, buying clothes, buying household furniture and even sending their children to higher education schools.

Besides providing job opportunities, the existence of oil palm plantations also provided various new commercial enterprise opportunities for the residential areas. The bulk of the community previously only worked in farming, especially the rubber plantation. However, with the presence of oil palm companies, business opportunity grew like stalls for necessities, restaurants, and so on. As told by references [26], [27], the expansion of oil palm plantations provided diverse investment opportunities. The existence of the companies have indirectly brought in workers from outside the village, so the opportunity to work was greater because of market opportunities that have not existed previously. Reference [3] states that industrialization of agriculture can be an attraction in the output development and growth in other economic sectors. Meanwhile, some other business opportunity was to convert rubber into oil palm. A large number of rubber farmers converted to oil palms due to higher oil palm prices, definite market, not affected by weather and safer from theft. Furthermore, oil palms were favoured due to its simplicity of managing [29], [4] and produced higher incomes [26].The existence of the companies, hence have directly affeced the local welfare, due to the strengthening of the economy. As a consequence, there are at least 4 economic sums obtained by the community, namely working as employees, plasma receipts, private oil palm sales and opening of new business. References [29] and [30] state that to strengthen the people's economy, industrialization of agriculture is something that must be practiced.

3.3. Socio-cultural impact of oil palm plantation development

The existence of companies in the study location have directly or indirectly influenced the socio-cultural life of the community, including work ethic, institutional, access to education and community cohesiveness. Contrary to the economic aspects, which almost entirely increased, the socio-cultural aspects faces some declines (figure 6).

Work ethic was one view that changed due to the existence of oil palm plantations. It caused negative impacts, since the certainty of income, has lowered the need to work too hard as before experienced. Meanwhile, it could be positive because there were also people who doubled their work. From morning until noon, the people worked in the company and in the afternoon they worked in the plantation for extra cash. This was in accordance with the statement by [30], whom states that oil palm plantation business allows the community to have other incomes.

The general strategy run by companies to draw close the community is through institutions and influenced at the hamlet layer. Thus, the role of village heads and tribal leaders would be greater, especially in resolving problems that would be faced by companies, such as in terms of land acquisition and other issues related to the interests of the community [5]. The presence of companies also have directly influenced the dynamics of formal institutions that grew and developed in the community, such as the formation of plasma cooperatives. Plasma cooperatives were formed indirectly as the consequence of the existence of plasma plantations managed by plasma recipients.

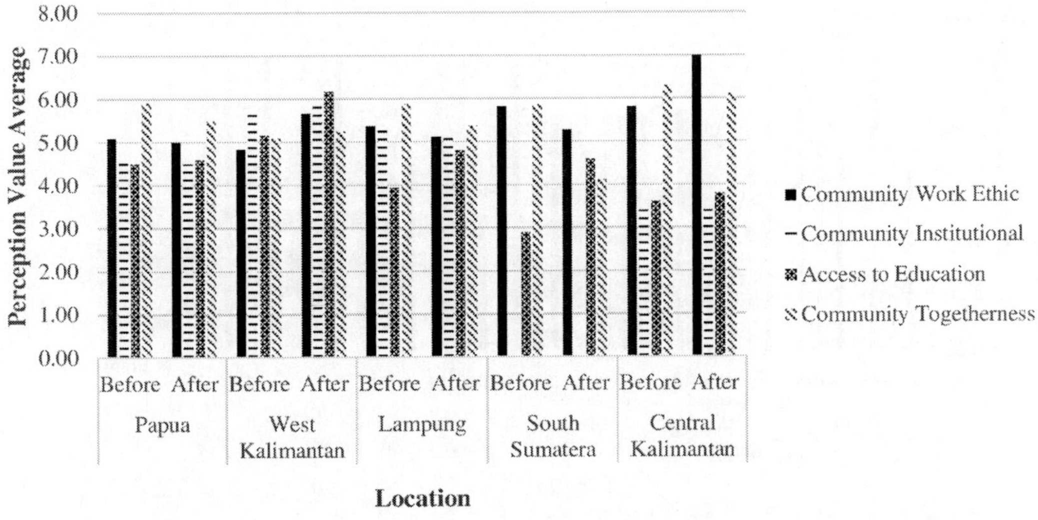

Figure 6. The socio-cultural impacts of oil palm plantation development

As previously explained, the majority of the surrounding community works at a lower level in the company. This has made the parents have higher awareness towards the importance of education for their children. Higher education was expected by the community to open up better opportunities for their children for their future employments, such as staff of oil palm companies (monthly employees). In addition, the existence of scholarships provided by the companies for local dropped out students have encouragement for students. Not only for students, but the company have also employed several honorary teachers in the village. This is in line with research [18] which states that oil palm plantations affect the level of education in the family.

One of the consequences of working in the company was the attachment to working time, which has indirectly reduced community gathering time. Grounded along the outcomes of the FGD, it was recognized that the demands of higher life have increased the awareness of the need to improve welfare by working in companies, thus they consumed more energy and time, and consequently decreased willingness to cooperate and sensitivity to others. This phenomenon often altered the community to become individualistic within the society. Yet, normally, people would take advantage of holidays or Sundays to gather even though it was just for a few days. This is in line with the research result by [24] who finds that the communities close to oil palm plantation loss their opportunities to actively engaged in village activities.

3.4. Impact of oil palm plantation on infrastructure

Infrastructure is everything that form the main supports for the implementation of a process (business, development, projects, etc.). The existence of oil palm plantations through CSR programs has assisted in the construction or upkeep of infrastructure in the surrounding villages. Based on Act 40 of 2007 on Limited Liability Companies, CSR is the company's effort to lift its image in the public eye by creating charity programs, both external and internal, with the objective of increasing the legitimacy of local residential areas and increasing community ownership of the society. Oil palm's impacts of infrastructure development can be ascertained from the growth in infrastructure before and after the existence of oil palm plantations (figure 7). Grounded on this, it could be assured that in the prospect of infrastructure, there was an increase after the existence of oil palm plantations. This was consistent with several surveys, which stated that the existence of oil palm plantations provided CSR assistances [5], [32].

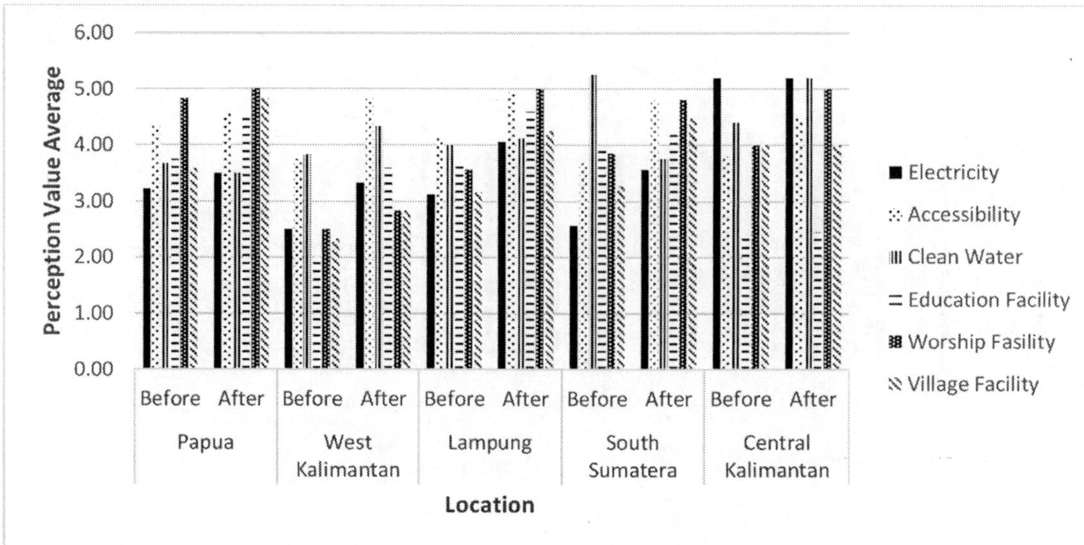

Figure 7. Impact of oil palm plantation development of infrastructure

The impact on electricity infrastructure compared to other aspects of infrastructure was not too significant. This was because not all companies in the study locations assisted in providing electricity infrastructure for their CSR programs. Electricity facilities provided by the company could be in the form of solar panels, construction of electricity poles connected to the factory or in the form of fees for the purchase of gasoline (generator usage).

Accessibility was the most important impact felt by the community. Existence of oil palm company has opened up access to the isolated villages. In addition, the opening of roads has indirectly increased the mobility of the community so that the economy flow was alive. In return, it also provided easier access to school (middle and high school). These conveniences would indirectly increase the motivation of students to go to school diligently and also the parents to send their children for higher education. The access have also increased the mobility of traders from outside the villages that brought household needs, which could not be fulfilled from inside the villages, hence the communities could purchase commodities at lower prices. This is consistent with research by [18] whom states that oil palm plantations facilitate access and provide school buses.

The existence of oil palm plantations was also inseparable from the existence of environmental matters in the form of river water pollution or drought. For this reason, companies generally would build drill wells at certain points that could be used by the community, especially those who were still using river water for washing and latrines (MCK).This is in line with research by [31] that the existence of oil palm plantations makes it easier for communities to obtain clean water.

Related to education, the companies have assisted in both directly building school facilities or offered other educational facilities (such as tables or chairs). In the aspect of worship, companies generally would assist in the form of fundings or goods in the case of construction of religious facilities or activities. Meanwhile, the company also provided assistances to improve village facilities such as development of health centres, sports facilities or village offices.

4. Conclusion

In general, the communities showed positive perceptions towards the development of oil palm plantations, due to the positive outcomes that the society has enjoyed, especially related to socioeconomic aspects, social culture and infrastructure. In the socioeconomic aspects, the benefits of oil palm development for the community include the accessibility of jobs, increased business opportunities, enhanced community income and certainty of income. In the social-cultural aspects, the impingement of the development of oil palm plantation includes enhanced community knowledge and skills, work ethics and community awareness of education and village institutions. In the aspect of infrastructure, the existence of oil palm plantation through CSR programs has provided assistances to the surrounding villages related to infrastructure development, such as in opening access and repairing road conditions, providing facilities for health, education, religion purposes and community needs such as clean water and electricity. Meanwhile, the negative impacts of oil palm development arised due to the problems with land claims and environmental shocks. Fortunately, these problems could be overcome.

Acknowledgment
The authors would like to thank all the oil palm plantation companies that have funded this research, and contributed to data collection and provided the necessary supporting data.

References
[1] Ambiyah A. 2012. the economic and environmental analysis of oil palm expansion in Indonesia: Export Demand Approach and EIRSAM Model. [Disertasi]. Nagoya (JPN): Nagoya University
[2] Mara A,Fitri Y. 2013. Dampak perkebunan kelapa sawit rakyat terhadap pendapatan wilayah desa (PDRB) di Provinsi Jambi. *AGRISEP*. **13** (1):101-110.
[3] Prawoto N. 2010. Pengembangan potensi unggulan sektor pertanian. *Jurnal Ekonomi dan Studi Pembangunan*. **11**(1): 1-19
[4] Susila WR. 2004. Contribution of palm oil industry to economic growth and proverty allevation in Indonesia. *Jurnal Litbang Pertanian*. **23**(3): 107-114
[5] Agustina D. 2014. Analisis lingkungan sosial ekonomi pengelolaan perkebunan kelapa sawit berkelanjutan berdasarkan kriteria ISPO (StudiKasus PT X Kalimantan Selatan)[Tesis]. Bogor (ID): Institut Pertanian Bogor
[6] Norwana AABD, Kunjappan R, Chin M, Schoneveld G, PotterL, Andriani R. 2011. The local Impact of Oil Palm Expansion in Malaysia: An assesement based on a case study in Sabah State. Bogor (ID): CIFOR
[7] Merten J et al. 2007. Expanding oil palm cultivation in Indonesia: changing local water cycles raises risks of drought and floods. DeutschesInstitut fur Entwicklungspolitik and German Development Institut
[8] Henson IE. 2003. Oil palm can it substitute the tropical rainforest. *The planter.***79**(928): 437-450.
[9] Marti S. 2008. Hilangnya Tempat Berpijak :Dampak Ekspansi Kelapa Sawit terhadap Hak Asasi Manusia di Indonesia. Jakarta (ID): Friends of the Earth, Life Mosaic dan Sawit Watch
[10] Rhett A. Butler, Palm Oil doesn't have to be bad for the Environment, 4 April 2007, http://news.mongabay.com/2007/0404-oil_palm.html
[11] Third National Implementation Report (Indonesia) on the Convention on Biological Diversity, www.cbd.int/doc/world/id/id-nr-03-en.pdf
[12] Wakker E. 2005. Greasy Palms: The Social and Ecological Impact of Large Scale Oil Palm Plantation Development in Southes Asia. England (GB): FOE
[13] Hariyadi P. 2014. Mengenai Minyak Sawit dengan Beberapa Karakter Unggulnya. Jakarta (ID): GAPKI
[14] Avenzora R. 2008. Penilaian Potensi Objek Wisata: Aspek dan Indikator Penilaian. Di dalam: Avenzora R, editor. Ekotoursme-Teoridan Praktek. Aceh (ID): BRR NAD Nias
[15] Istijanto. 2009. *Aplikasi Praktis Riset Pemasaran.* Jakarta (ID): Gramedia Pustaka Utama.
[16] Imam Suroso, Arif, 2008, Analisis Daya saing dan Dampak Ekonomi Regional Pengembangan Kelapa Sawit Di Kabupaten Siak, IPB: Universitas Pertanian Bogor
[17] Syahza A. 2007. Percepatan pemberdayaan ekonomi masyarakat pedesaan dengan model agro estate berbasis kelapa sawit. *Jurnal Ekonomi*. **12** (02): 297-310
[18] Sari E. 2017. Persepsi masyarakat sekitar tentang dampak perkebunan Sawit PT. Surya Agrolika

Reksadan PT Adimulia Agrolestari. Bogor (ID): Institut Pertanian Bogor

[19] Drajat B. 2014. Upaya Mengatasi Black Campaign Kelapa Sawit dan Langkah Strategis Kedepan. Bogor (ID): Lembaga Riset Perkeunan Nusantara

[20] Afifudin S, Kusmana SI. 2007. Analisis struktur pasar cpo: pengaruhnya terhadap pengembangan ekonomi wilayah Sumatera Utara. *Jurnal Perencanaandan Pengembangan Wilayah*. **2**(3): 124-146

[21] KomisiSawit Indonesia. 2008. *Indonesia Palm Oil Statistics 2008*. Jakarta ID: Kementerian Pertanian Indonesia

[22] World Growth. 2009. Palm Oil-The Sustainable Oil, Arlington, VA. Tersedia pada: http://www.worldgrowth.org/assets /files/Palm_Oil.pdf

[23] Utami R, Putri EIK, Ekayani M. 2017. Dampak ekonomi dan lingkungan ekspansi perkebunan kelapa sawit (studi kasus: Desa Penyambungan, Kecamatan Melung, Kabupaten Tanjung Jabung Barat, Jambi. *Jurnal Ilmu Pertanian Indonesia*. **22**(2): 115-126

[24] Roanuddin M. 2016. Dampak sosial dan ekonomi aktifitas perusahaan perkebunan sawit PT. Gawi Makmur Kalimantan di Desa Rintik Kecamatan Babulu Kabupaten Penajam Paser Utara. *Jurnal Ilmu Pemerintahan* **1**(1): 12-25

[25] Hendriono W. 2016. Studi dampak perkebunan kelapa sawit terhadap kondisi sosial ekonomi masyarakat di Kecamatan Andowia Kabupaten Konawe Utara.Kendari (ID): Universitas Halu Oleo

[26] Unjan R, Nissapa A, Phitthayaphinant P. 2013. An identification of impacts of area expansion policy of oil palm in Southern Thailand: a case study in Phatthalung and Nakhon Si Thammarat Provinces. *Social and Behavioral Sciences*. **91**: 489-496. http://doi.org/b9xt

[27] Hidayat D. Analisisperananperkebunankelapasawit di Provinsi Riau dalam era otonomidaerah. [tesis] Bogor (ID): InstitutPertanian Bogor

[28] Saputra A. 2013. Faktor-faktoryang memengaruhi konversi tanaman karet menjadi kelapa sawit di Kabupaten Muaro Jambi. *Sosio Ekonomika Bisnis*. **16**(2): 18-25

[29] Saragih B. 2001. *Suaradari Bogor: Membangun Sistem Agribisnis*. Bogor (ID): Yayasan USESE

[30] Muktamarianti AI. 2016. Tipologi kebun sawit swadaya di Provinsi Riau [skripsi]. Bogor (ID): Institut Pertanian Bogor

[31] Masilawati YD. 2017. Dampak keberadaan industri kelapa sawit terhadap tata lingkungan permukiman di Desa Kumasari Kabupaten Mamuju Utara [skripsi]. Makassar (ID): Universitas Islam Negeri Alauddin Makassar.

[32] Wardie J, Taufik EN. 2017. Kajian implementasi program CSR perusahaan perkebunan kelapa sawit kepada masyarakat di Kabupaten Kotawaringin Barat. *Agrisocionomics*. **1**(1):18-25

IOP Conf. Series: Earth and Environmental Science **336** (2019) 012009 doi:10.1088/1755-1315/336/1/012009

Biodiversity loss and gain based on indigenous people's perception. (Case study: Tajok Kayong and Nanga Tayap Villages, West Kalimantan Province)

Farras Imantoko[1]*, Arzyana Sunkar[2], and Yanto Santosa[3]

[1]Bachelor of Forestry from Bogor Agricultural University, Indonesia
[2]Protected Area Management Division, Department of Forest Resources Conservation and Ecotourism, Faculty of Forestry, Bogor Agricultural University, P.O Box 168 Bogor, Indonesia
[3]Ecology and Wildlife Management Division, Department of Forest Resources Conservation and Ecotourism, Faculty of Forestry, Bogor Agricultural University, P.O Box 168 Bogor, Indonesia

*E-mail: farrasimntk@gmail.com

Abstract. Loss and gain on biodiversity of plant species that have been resulted from the establishment of oil palm plantations have often been accused as the reasons for the changing pattern in community means of living. Such a shift has also been highlighted as related to the destruction of the indigenous people livelihoods. Indigenous people were often dependent on the local plants for survival, yet land use change to oil palm plantation has caused changes in plants composition. This research was conducted to identify the indigenous people's perceptions related to gain and loss of plant species diversity due to the establishment of oil palm plantation. The study was carried out in the indigenous Dayak Villages of Tajok Kayong and Nanga Tayap, West Kalimantan Province of Indonesia in March 2018. Data were collected using field observation and interviews. The overall perceptions of Dayak and Malay communities (the two indigenous groups of West Kalimantan) on the plants conditions related to oil palm development, showed similar results, except related to plant diversity, in which 75.00% of the Malays perceived a decreased in plant diversity, while on the contrary, 75.68% of Dayaks perceived it as an increase, in terms of flowering plants and other functional plants. The results indicated that the establishment of oil palm plantation has resulted in the gain of 27% plant species and loss of 73% plant species with functional values (medicine and food). However, the lost species could be substituted with other plants and could also be replaced by chemical medicine from public health service and purchased from the local market. Both indigenous people of Dayak and Malay were not affected by the species loss since before the plantation's establishment, their living patterns had already transformed into semi-modern and modern living styles where the people have lower dependences on their surrounding plants.

1. Introduction

Biodiversity has a very important value for the surrounding community, such as various types of plants that are used by indigenous peoples in West Kalimantan [1]. According to [2] for indigenous peoples, their land or territory is a means for their survival. This is also supported by [3] statement that indigenous peoples in West Kalimantan have a high dependence on natural resources in the forest to live their lives. The existence of various types of plants is influenced by the existing ecosystem conditions. One of the changes in ecosystems that still become world attention for several decades ago is the conversion of forest land into oil palm plantations [4], [5]. Conversion of forest areas into oil palm plantations causes the Dayak community to have difficulty in obtaining various types of plants that they previously used as food and life sources [3].

On the other hand, the results of [6] show that local communities have participated in the process of converting land into oil palm plantations due to economic factors that encourage them to get better results. The results of Siradjuddin's research [7] also show that oil palm plantations make the economic conditions of the surrounding communities become increasingly developed. Various literature states that indigenous people in West Kalimantan are Dayaks, whereas there are also Malay people who are indigenous ethnic groups in West Kalimantan [8]. Dayak and Malay communities have different cultures and habits in living life, especially in terms of the use of natural resources around it [9], so that changes in ecosystems are thought to affect the patterns of life of indigenous peoples. This makes the perception of the two ethnic groups becomes important to know, where reference [10] state that people's perceptions of the types of natural resources are influenced by the benefits that can be felt from that type before. Community perception is also an important factor in determining the success rate of oil palm plantation management in addition to ecological factors [11]. In connection with this matter, it is necessary to study the perception of indigenous peoples on gain and loss of plant species that related to oil palm plantation

2. Instruments and methods

2.1. Time and location research

The research was conducted around NTYE, Tajok Kayong Village and Nanga Tayap Village, Nanga Tayap Subdistrict, West Kalimantan Province. The duration of our research was 1 month, starting from March 12, 2018.

2.2. Instruments

The instruments used in this research consists of an audio recorder, Microsoft Excel, Microsoft Word, camera and GPS (Global Positioning System). We used an interview guide that contains a list of structured questions and five-scale Likert statements.

2.3. Data collection

Methods of data collection used to identify species diversity based on community perception were based on both interviews and field observations, together with literature study. Interviews were conducted through depth-structured interviews. Data were recorded under the form of species composition from the knowledge of the community about the types of plants, both emerging and lost or the type of increase and decrease since the palm plantation was established and also the public perception related to it. The price data of the types of plants were obtained by economic valuation method by doing 3 types of approaches which are market value approach, replacement value approach and regulatory value approach.

In this study, respondents were indigenous Dayak and Malay people who were either related to oil palm cultivation or not with a total of people with a sampling unit of 1 person per family. This is because both indigenous Dayak and Malay people are long-lived and settled in their residential areas around the oil palm plantations so that whatever the type of work they do, they can still feel directly the impacts gain and loss of plant species related oil palm plantation.

The total respondents in Tajok Kayong and Nanga Tayap Village are 110 people. The village of Tajok Kayong has an area of 6.26% of the area of Nanga Tayap Subdistrict, while Nanga Tayap Village has an area of 4.98% of Nanga Tayap Subdistrict [12]. Their types of jobs do vary, with quite high amounts of income as it could be seen from the status of children. There are some who continue studying up to University level, so this proves that the economic prosperity of the people is quite high. Given that one of the most common tribes on the island of Kalimantan is the Dayak community, it is in accordance with the statement [13] that the Dayak people are Kalimantan island residents who occupy a region permanently for a long time This supports the idea that West Kalimantan Province has its uniqueness towards the process of acculturation, because of the Dayak people could face some modern way of living style in several aspects.

3. Results and discussion

3.1. General conditions of study sites

Tajok Kayong Village and Nanga Tayap Village are located in Nanga Tayap District which has an area of 172,812 ha which is 5.47 percent of the total area Ketapang Regency with 3,158,000 ha [14]. Based on their livelihoods, the types of livelihoods of the people in Nanga Tayap Subdistrict can be divided into farmers, fishers, large entrepreneurs, artisans, farm laborers, construction workers, mining workers, traders, transporters, civil servants, ABRI, and farmers. The highest number of population is livelihoods as farmers (11.453 people).

Their types of work are diverse, with a high amount of income, which can be seen from the status of the children they have, there are some who continue to the level of study, so this proves that the economic welfare of the community tends to prosper. Although divided into hundreds of sub-tribes, but they still have some similarities as a distinctive Dayak ethnic, cultural identity, namely the existence of a long betang house, the main pattern of livelihood in the form of rotational cultivation, its view of nature, and a distinctive material culture such as pottery, mandau chopsticks, and Dayak dance. They are inseparable from farming life that represents their customs and beliefs. The Dayak tribe cannot be separated from forest resources which are the source of their livelihood so that Dayaks can be considered as forest nomads [3].

3.2. Indigenous people's perception of oil palm plantation related to gain and loss of plant species

Community perceptions of gain and loss in plants species are important to be studied from Dayak and Malay communities because indigenous peoples are known for their habit of using species of natural resources around them (table 1).

Table 1. Dayak and Malay people's perception of plant species diversity related to oil palm plantation

Statement	Perception (%)							
	Do not know		Not Agree		Doubt		Agree	
	Dayak	Malay	Dayak	Malay	Dayak	Malay	Dayak	Malay
Plant can be used as a source income	0.00	0.00	74.32	100	18.92	0.00	6.76	0.00
There is no harmful plants	0.00	0.00	25.68	18.75	12.16	6.25	62.16	75.00
NTYE oil palm plantation makes community become easier to find medicinal plants	17.57	0.00	75.68	85.41	0.00	0.00	6.76	14.50
NTYE oil palm plantation makes community become easier to find food plants	2.70	0.00	85.41	83.33	8.11	2.08	4.05	14.50
NTYE oil palm plantation makes plants become more diverse	2.70	0.00	20.27	75.00	1.35	4.16	75.68	20.83
NTYE oil palm plantation makes community become easier to plant other plants	0.00	0.00	10.81	0.00	0.00	0.00	89.19	100.00

Most people feel that the existing types of plants do not have an impact on their income, seen with the majority of both Dayak (74.32%) and Malay (100%) stating that they do not agree with the alternative aspects of community income. Malay people give a full response (100%) because they are already first and thorough in buying some of their needs in the market, slightly different from Dayak people where there are some (6.76%) who agree that the availability of plants can be used as an alternative to their income. This happens because the community does not do economic activities involving the plants that are around them. Then, since the existence of NTYE's oil palm plantations caused some types of understorey such as weeds which were found quite a lot by the community, with (62.16%) Dayak people and (75%) Malay people feeling that they did not feel any undersea plants which were felt to be detrimental since the existence of oil palm plantations NTYE, because they both consider that the type of weeds does not cause harm to their daily lives.

In the aspect of medicinal use, both tribes showed a similar response, can be seen with most Dayak people (77.78%) and the Malay community (85.41%) felt that medicinal plants were becoming

IOP Conf. Series: Earth and Environmental Science **336** (2019) 012009 doi:10.1088/1755-1315/336/1/012009

difficult to find. However, even though the community felt the medicinal plants were becoming more difficult to find, they did not feel lost because they were used to buying drugs at the nearest health center and also the public hospitals around their homes. This is in accordance with the results of the study [15] that traditional knowledge of the community is reduced due to socio-cultural changes that generally affect social values, where the younger generation is looking for more practical alternatives. It also shows that there is a pattern of changes in habits of Dayak and Malay communities that have changed before the impact of increasingly difficult medicinal plants are found so that this makes indigenous people feel that there is no significant influence they feel.

Likewise, the aspects of food plants that received similar responses from the two tribes, namely the Dayak community (85.19%) and Malay (83.88%) who gave a disagreeable response and showed that food plants became increasingly difficult to find. This statement shows that both Dayaks and Malays are equally aware of changes in the types of plants that surround them, even though there are no effects that are directly felt and hinder the fulfillment of their needs because they are used to buying food to the market or taking plants directly yard or field crops from their crops. According to [16], [17] yards are usually planted with a variety of seasonal and perennial crops for daily use. Home yards are often referred to as living barns, living stalls or living pharmacies [18]. The community's perception of this matter was also supported by a large number of Dayak (89.19%) and Malay (100.00) people giving a response agreeing on the easy aspect of planting other plants around their homes.

This can happen because in the Dayak community, they previously had a habit of living closely with plant species so that they would be more easily aware of knowing that new types of plants were gained in species and individual. It is different from the Malay people who basically have a close pattern of life with trade in the market, so they do not pay too much attention to the condition of the types of plants that are around their homes. Thus, as many as 5 of the total 6 statements from Dayak and Malay respondents were found the same results except in the aspect of a variety of plants, where Dayak people felt that plants were becoming more diverse while not for Malay people. Thus, all people in general feel that there is a change in the type of plants around them, but they can take advantage of new types and substitute the loss species with the gained species or by buying in the market.

3.3. Gain and loss of plant species based on indigenous people's perception

Table 2 indicates that there has been acquisition and loss of plant species, as explained by the perspective regarding the increase in the variety of plants by Dayak people (75.68%) and the decreasing variety of plants by the Malay community (75.00%).

Data related to prices are also obtained from the knowledge of the people who did use these types of plants so that ultimately economic valuations can be done based on the prices obtained from the knowledge of the community. The valuation carried out aims to find out how much value there is in biodiversity because the existence of natural resources itself is the initial factor of nature in the process of economic activities that involve these natural resources in the future [19].

Table 2 shows plants species that can no longer be found by Dayak people include *Shorea sp. (Meranti), Neolamarckia cadamba (Jabon), Shorea sp. (Tengkawang), Durio kutejensis (Pekawai), Durio sp. (Kusik), Momordica sp. (Pare Hutan), and Artocarpus anisophyllus (Mentawa).* In meranti species, the community uses the woody plants for the needs of buildings and others. This species of meranti can no longer be found by the community, but the community does not experience a significant loss because they feel they need a lot of time and energy to enter the area and take the wood. This is consistent with the results of the study [20] that distance is a factor that influences the rate of encroachment and positive interaction. Thus, this is in harmony with the people who feel heavy to travel to the location of taking woody plants due to consideration of time and energy spent.

Then, people who have experienced a more advanced life change have made their habit of buying building materials even bigger, so they prefer to spend money to build their houses and buildings like public facilities around them. Unlike the Malay community, they did not use the species of woody plants mentioned by the Dayak community. This is because they have already been able to fulfill their living needs by buying goods produced by trading on the market.

Table 2. Loss and gain of plant species based on indigenous people's perception

No.	Loss Species	Average Prices	Gain Species	Average Prices
1.	Shorea sp.[1,4]	IDR 2,400,000/m^3	*Diplazium sp.* [2,4]	IDR 2,000/bundle
2.	Neolamarckia cadamba [1,4]	IDR 950,000/m^3	*Limnocharis flava* [2,4]	IDR 1,500/bundle
3.	*Tengkawang* (wood) (Shorea sp.)[1,4]	IDR 450,000/m^3	*Ipomoea aquatica* [2,4]	IDR 2,000/bundle
4.	*Tengkawang* (fruit) *(Shorea sp.)* [2,4]	IDR 3,000/Kg		
5.	Durio kutejensis [2,4]	IDR 55,000/Kg		
6.	*Durio sp.* [2]	IDR 100,000/Kg		
7.	*Momordica* sp. [2,3]	IDR 3,000/Kg		
8.	*Artocarpus anisophyllus* [2]	IDR 5,000/Kg		
	TOTAL	IDR 3,966,000		IDR 5,500

Note: ([1]building material; [2]food needs; [3]medicinal needs; [4]source income)

Although, overall they still realize that the existence of medicinal plants and food around them is reduced. Malay people also do not feel that the surrounding plants are becoming more diverse, in contrast to the Dayak people who actually realize that there are several species of plants that are increasingly found and even new species can be found. The species of plants that gained consist of *Diplazium sp.(Pakis), Limnocharis flava (Genjer)* and *Ipomoea aquatica (Kangkung).* This plant is used by the Dayak community because it has only been found since the existence of oil palm plantations, where previously the community had never found such species. Kaplan et.al [15] explains that in the ethnobotany discipline there is a set of assumptions about the relationship between behavioural patterns and socio-cultural arrangements that are integrated into language, cognitive systems, rules and local cultural ethics. This shows that the pattern of human behaviour in using several types of plants becomes something that will affect the emergence of values for these species. With the existence of new species of plants or new natural resources, humans will automatically find ways to use them. This happens because humans try to use natural resources and the environment based on their experience and knowledge [21].

The economic value derived from the new species found by the community is important because this indicates the presence of oil palm plantation can also have a positive impact that felt directly by the people living around it. Even though valuation has lost more economic value, it does not affect because people can replace these food needs from other sources such as yards, field crops, to markets while for building materials they can buy mild steel which is considered more efficient than taking wood directly from the forest.

3.4. Gain and loss of plant individual based on indigenous people's perception

In addition to the gain and loss of biodiversity, there are also several species of plants that have gained and also loss individually (Table 3). Several species of medicinal plants such as *Curcuma zanthorrhiza (temulawak), Eurycoma longifolia (pasak bumi), Tinospora cordifolia (brotowali)* which considered loss individually by the community, do not give effect to their livelihood because the level of dependence on the used of medicinal plants is very low, and also they realize that the reason is because the use of generic medicines is felt obviously more efficient and saves more both energy and time.

Table 3. Loss and gain of plants individual based on indigenous people's perception

No	Loss Individu	Average Prices	Gain Individu	Average Prices
1.	*Calamus rotang* [1]	IDR 1,500/Kg	*Imperata cylindrica* [0]	IDR 0
2.	*Cucurbita sp.* [2]	IDR 8,000/Kg	*'Jankos' (Volvarariella volvacea)* [2]	IDR 12,500/Kg

3.	*Curcuma zanthorrhiza* [3]	IDR 8,000/Kg	*Turnera ulmifolia* [0]	IDR 0
4.	*Eurycoma longifolia* [3]	IDR 40,000/Kg		
5.	*Tinospora cordifolia* [3]	IDR 35,000/Kg		
6.	*Eusideroxylon zwageri* [1,4]	IDR 16,500,000/m^3		
7.	*Shorea laevis* [1,4]	IDR 9,750,000/m^3		
	TOTAL	IDR 26,342,500		IDR 12,500

Note : ([1]building material; [2]food needs; [3]medicinal needs; [4]source income)

In addition, the existence of public health services and also public hospitals around their homes makes them becomes prefer to go directly to treatment rather than looking for medicinal plants into the forest that requires much energy. This is consistent with the results of the study [already cited on 18] that the ineffective taking of plants in the forest makes people begin to think about planting or cultivating medicinal plants around settlements or yards. Thus, changes in the pattern of life from indigenous people already happened before the existence of the NTYE oil palm plantation. Public perception about the use of medicinal plants is influenced by one of their patterns of life that has become more modern day by day, so this is in accordance with the statement [22] that the existence of modern life will change the indigenous knowledge of Indonesian tribes outside Java.

So with the knowledge regarding the species of plants above, Dayak and Malay communities have experienced some changes in their pattern of life systems before the existence of NTYE oil palm plantations, such as they have switched their habit to buy their needs on the local market, and also use local financial institutions such as existing local cooperatives.

4. Conclusion

In general, Dayak and Malay communities agree that there are changes in the species of plants that exist, although (75.68%) the Dayak community stated that there were several species that consider gained, while the Malay community (75.00%) felt that there were no any gained plant species. There are 8 species of plants that considered loss with the function as food and building material, while there were 3 species of plants that consider gained with the function of food based on the perception of indigenous peoples. The loss species doesn't affect their livelihood because they can substitute it with other species on their surrounding fields or local markets for food needs, and also through public health services for medicinal needs.

Acknowledgment

This research was supported by PT. SMART Tbk. Thanks are also forwarded to all the Directors and Staffs in NTYE (Nanga Tayap) oil palm estates for the permissions and facilities provided during the study period.

References

[1] Budiasih, LA. 2005.Inventerisasi Jenis Tumbuhan yang Berkhasiat sebagai Obat di kawasan Hutan Baning Sintang Kalimantan Barat, Skripsi, Fakultas Pertanian Universitas Tanjung Pura Pontianak

[2] Soni CP. 2012. Kearifan Lokal Masyarakat Adat Dayak Kanayaitn dalam Pengelolaan Hutan Adat (Marang) di Kampung Sidas Data Kecamatan Sengah Temila Kabupaten Landak Kalimantan Barat [skripsi]. DIY (ID) : Universitas Negeri Yogyakarta

[3] Sudrajat J. 2016. Sejarah Pemanfaatan Sumberdaya Hutan dan Pergeseran Nafkah Suku Dayak di Kalimantan Barat: Paramita : Historical Studies Journal

[4] Azhar B, Chong LP, Zakaria M, Hassan N, Arif M. 2014. Effects of monoculture and polyculture practices in oil palm smallholdings on tropical farmland birds. Basic and Applied Ecology **15**:336-346

[5] Obidzinski K, Andriani R, Komarudin H, Andrianto A. 2012. Environmental and social impacts of oil palm plantations and their implications for biofuel production in Indonesia. Ecology

and Society **17**(1): 25
[6] Feintrenie L, Schwarze S and Levang P. 2010. Are local people conservationists? Analysis of transition dynamics from agroforests to monoculture plantations in Indonesia. *Ecology and Society*15(4): 37. [online] URL: http://www.ecologyandsociety.org/vol15/iss4/art37/
[7] Siradjuddin I. 2015. Dampak Perkebunan Kelapa Sawit terhadap Perekonomian Wilayah di Kabupaten Rokan Hulu. Jurnal Agroteknologi **5** (2) : 7-14
[8] Yusriadi. 2015. Identitas Orang Melayu di Hulu Sungai Sambas. Jurnal Khatulistiwa (Volume 5). Fakultas Ushuluddin, Adab dan Dakwah IAIN Pontianak
[9] Batubara SM. 2017. Kearifan Lokal Dalam Budaya Daerah Kalimantan Barat (etnis Melayu dan Dayak), Jurnal Penelitian Ipteks, Universitas Muhammadiyah Pontianak (IDN).
[10] Homer, P. M. and Kahle, L. R. (1988). A structural equation test of the value-attitude-behavior hierarchy. Journal of Personality and Social Psychology, **54**(4), 638–646.
[11] Aikanathan S. 2013. Determinants for sustainable management of the palm oil plantation industry: a Malaysian a case-study [tesis]. Kuala Lumpur (MY). University Malaya
[12] [BPS] Badan Pusat Statistik. 2017. *Kabupaten Ketapang dalam Angka 2017*. Ketapang (ID): BPS Kabupaten Ketapang
[13] Darmadi. 2016. Dayak asal-usul dan penyebarannya di Bumi Borneo (1). Vol.3, No. 2. Sosial Horizon : Pontianak
[14] [BPS] Badan Pusat Statistik. (2017). *Kecamatan Nanga Tayap dalam Angka 2017*. Ketapang (ID): BPS Kabupaten Ketapang
[15] Kaplan D, & RA. Manners (2003). Teori Budaya (terjemahan) Landung Simatupang. Pustaka Pelajar. Edisi III. 294 hal
[16] Danoesatro, (1980), Tumbuhan Obat Keluarga, Penebar Swadaya, Jakarta
[17] Giono, W, (2004), Budidaya Tumbuhan Obat di Perkarangan, Agromedia Pustaka, Jakarta
[18] Lovadi I, R. Linda, Meliki. (2013). Etnobotani Tumbuhan Obat oleh Suku Dayak Iban di Desa Tanjung Sari Kecamatan Ketungau Tengah Kabupaten Sintang. Pontianak (ID) : Protobiont Vol **2** (3): 129 – 135
[19] Fauzi A, (2004). Ekonomi Sumberdaya Alam dan Lingkungan : Teori dan Aplikasi. Gramedia. Jakarta
[20] Sunkar A, Santosa Y, Hasibuan SH. (2017). Identifikasi Peubah Sosial Ekonomi Penentu Keberhasilan Konservasi Keanekaragaman Hayati. Media Konservasi Journal
[21] Walujo EB, (2011). Sumbangan Ilmu Etnobotani dalam Memfasilitasi Hubungan Manusia dengan Tumbuhan dan Lingkungannya. Jurnal Biologi Indonesia **7** (2): 375-391
[22] Waluyo EB. (1991). *Perkembangan pemanfaatan tumbuhan obat di luar Pulau Jawa*. Prosiding Pemanfaatan Tentang Tumbuhan Obat dari Hutan Tropika Indonesia. Institut Pertanian Bogor, Bogor

IOP Conf. Series: Earth and Environmental Science **336** (2019) 012010 doi:10.1088/1755-1315/336/1/012010

Smallholder farmers agency in the face of palm oil "negative campaign"

Delima Hasri Darmawan[1]

[1]Center for Socio-Economic and Agricultural Policy, Ministry of Agriculture Jl. Tentara Pelajar Road No. 3B Bogor 16111 Indonesia

*E-mail: delimahasridarmawan@yahoo.co.id

Abstract. It is well known that the palm oil industry in Indonesia faces many challenges. Despite having the Indonesia Sustainable Palm Oil (ISPO) as the national effort to set standards on operating palm oil industry within the environmental ethics, Indonesia recently face the possibility of having its palm oil exports to the European Union (EU) banned in the following years. This is based on their argument that palm oil in Indonesia has created deforestation of rainforests, which they pledge to prevent based on their Resolution on Palm Oil and Deforestation of Rainforests. As a result, protests from different stakeholders in the sector have been flooding, including from the government, exporters, plantations, as well as small-holder farmers. This paper focuses on the political agency of smallholder farmers in advocating their interest against EU's plan – which to many has been called as a "negative campaign" against palm oil. We define the term "political agency" as the capacity of small-holder farmers to participate in matters that concerns their well-being and have access to governance and policymaking. A key feature of smallholder's political agency is their participation in the discussion of policy making through Rural Producers Organizations (RPOs) or farmers associations. We test whether previous claims on the unequal power between small-holders and large corporations hold when it comes to advocating their respective interests. Specific questions that are addressed in this paper are: (1) Do small holders advocate their concerns towards the negative campaign? Or are they "free riding"?; (2) If they independently advocate their interest, how?; (3) Does their advocacy lead to a clear response or solution from the government?; and lastly (4) How does their political agency affect their socio-economic welfare?

1. Introduction

Palm oil production holds an important role to Indonesia's economy as it is Indonesia's source of foreign reserves and employment opportunities. Within Indonesia's agriculture products, palm oil is one of the most important commodities, accounting for 1.5-2.5 percent of Indonesia total GDP. The tree's versatile oil is used for many uses such as cooking, cosmetics, and biofuel. It is highly lucrative where the plant produces more oil per hectare than any other oil plants. With production over 40 million tons in 2017, Indonesia has become the world's largest producer and consumer of palm oil, accounting for half of the world's supply [1]. The majority of its export is shipped to China and India, where they themselves account for a third of the world's palm oil imports. According to the Indonesian Palm Oil Association (*Gabungan Pengusaha Kelapa Sawit Indonesia* or GAPKI), the production of palm oil is expected to rise in the coming decades due to the increasing world population and overall living standards per capita. Indonesia also plans to expand its market outside the current market by developing biodiesel consumption. Therefore, palm oil will continue to have significant role in Indonesia's economy in the future.

The palm oil sector, however, is under scrutiny when it comes to its environmental and social impact. For many years, palm oil sector in Indonesia have received many criticism as it plays a role in deforestation and hurting the natural habitats of many endangered species. Deforestation have also resulted in making local communities vulnerable as their livelihood is affected. Such criticism has been widespread, coming from both national and international actors. Internationally, the European Union has been a prominent advocate against the palm oil sector and has been promoting bans towards palm oil.

Out of all Indonesia's production, smallholder farmers make up a large share of the palm oil production in Indonesia and Malaysia, the world's main palm oil producing countries [1]. Because palm oil trees can produce its fruit for more than 30 years, the sector can provide a source of employment needed for poor rural communities. They owned or managed around 3.1 million ha of palm oil, accounting more than 40% of the total palm oil area in the nation. They are also spread across different provinces in Indonesia, making it important not only for the development of national palm oil, but also regional development. A negative campaign against palm oil sector will therefore significantly affect smallholder farmers in the palm oil sector (figure 1).

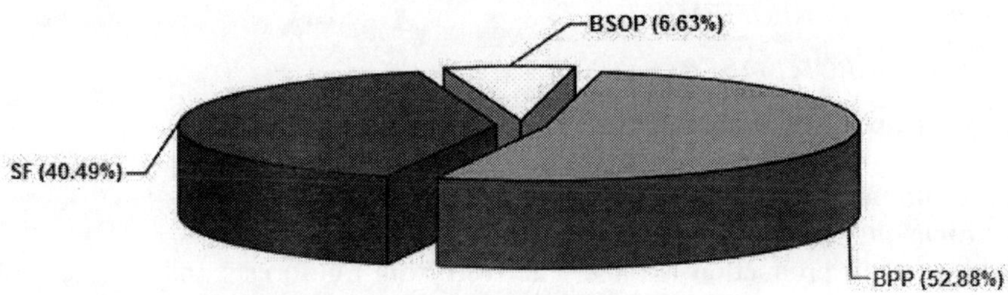

BSOP = Big State-Owned Plantations
BPP = Big Private Plantations
SF = Smallholder Farmers
Source: Ministry of Agriculture Republic of Indonesia

Figure 1. Palm Oil Production Share in Indonesia

The objective of this paper is to evaluate how smallholder farmers in Indonesia participate in discussions against palm oil negative campaign. Moreover, this paper focused on the role of *Rural Producers Organizations* (RPOs) or farmers associations to promote smallholder farmers' agency on advocating their interest. Specific questions that are addressed in this paper are: (1) Do smallholders advocate their concerns toward the negative campaign? Or are they "free riding"?; (2) If they independently advocate their interest, how?; (3) Does their advocacy lead to a clear response or solution from the government?; and lastly (4) How does their political agency affect their socio-economic welfare?. Answers to these questions are based upon a series of Focus Group Discussions and interviews with palm oil smallholder farmers conducted during the year 2017 in the Bengkulu, South Sumatera as one of the main palm oil growing district in Indonesia.

2. Methods

To answer the research questions, a survey of 100 smallholder palm oil farmers randomly selected in Bengkulu, South Sumatera is conducted. The sample is selected through a stratified sampling, where from the list of smallholder producers, 50 smallholder farmers who are not member of an RPO and another 50 smallholder farmers are derived from 10 RPOs which 5 farmers are randomly selected from each RPOs. This way, the sample is not bias towards only those smallholder farmers who are members of an RPO, which is a perquisite to generate inference on the role of RPO membership on smallholder farmers' agency.

The survey is conducted to obtain information on their perception and how they advocate in the face of a negative campaign in palm oil, on certification, how their membership in an RPO increase their success in obtaining palm oil sustainable certification, and how their agency affects their economic welfare. Many of the survey questions uses binary data, for example, on RPO membership, the survey asked whether farmers belong to RPOs or not, where it is coded dichotomously. If they are currently a member of an RPO, they are coded as "1", and if they are not a member of an RPO, they are coded as "0". Focus group discussion was also conducted to gain more insights on smallholder palm oil farmers experiences.

To generate causal explanation, quantitative analysis were also made, mainly to see whether membership in RPOs lead to more agency (more smallholder farmers advocate their concerns against the negative campaign on palm oil), certification, and consequently their welfare.Quantitative studies is chosen for causal explanantion purpose and to see the magnitude of the causal relationship between main variables. The following model is constructed and is estimated using a Probit regression model using the survey data collected.

$$Agency_i = \beta_o + \beta_1 RPOMEMBER_i + \beta_2 X_i + \varepsilon_i \quad (1)$$

$$Certification_i = \beta_o + \beta_1 RPOMEMBER_i + \beta_2 X_i + \varepsilon_i \quad (2)$$

$$Welfare_i = \beta_o + \beta_1 RPOMEMBER_i + \beta_2 X_i + \varepsilon_i \quad (3)$$

Where X_i is a set of control variables that can explain the relevant outcomes.

3. Negative campaigns against oil palm

3.1 Environmental and social impact of palm oil production

Indonesia's palm oil production is mostly derived from rainforests and therefore plays a key role in deforestation and other environmental impacts [2]. Since agricultural land is limited, land use for palm oil plantation needs to clear out other cultivations or tropical forest (Figure 2). Illegal logging has become widespread providing threat to the environment. According to the World Wildlife Fund (WWF), an area the equivalent size of 300 football fields of rainforest needs to be cleared out each hour to make way for palm oil production. Forest destruction was estimated at 0.84 Mha of primary forest between the year 2000 to 2012. As a consequence of this, biodiversity and ecosystem stability have also been threatened. Deforestation has been argued to killing endangered species, uprooting local communities, and contributing to the release of climate-warming gases, and therefore many environmental advocates have emphasized the significant costs palm oil has brought. The Center for Orangutan Protection, for example, has indicated that the great ape in Central Kalimantan has become extinct as the palm oil plantation grows within the region [3]. Other species such as orangutan could become extinct in the wild within the next 5-10 years, and Sumatran tigers less than 3 years. A "conservation emergency" has been established by the UN as these endangered species have been or nearly come to extinction. Deforestation for palm oil production also contributes significantly to climate change. The loss of timber and other parts of the forest has emitted immense quantities of smoke into the atmosphere, which makes Indonesia as a country with higher greenhouse emission than any other countries besides China and the United States.

3.2 Impact on local communities

The expansion of the palm oil industry also has a huge social impact to local communities. On one side, driven by its profitability, the sector has the potential to develop new jobs and improve the standards of living of people and smallholders farmers. According to UNDP, around 16 million jobs that depend on the palm oil sector. On the contrary, palm oil plantation has also been argued to bring negative effect for endangered indigenous tribes and local communities because it destroys their living spaces or land appropriation. The environmental effect of palm oil also has brought health risks to its surroundings, which eventually impacted the livelihood of local people. In regions like Kalimantan, for example, local livelihood of Dayak clan are undermined by the development of palm oil production [4]. Human rights violations and confrontation in particular between large-scale producers and local communities often occur, where land disputes commonly occurred in the palm oil production in Indonesia.

With plantations systematically destroying the rainforest land that the local people depend on, communities are continuously finding themselves with no choice but to become plantation workers. Poor work conditions have made them vulnerable towards large companies and vulnerable to world market prices. In many cases, child labour has also become common in the palm oil production with children made to carry large loads of heavy fruit, weed fields and spend hours every day bent over collecting fruit from the plantation floor. All this with little or no pay for their efforts.

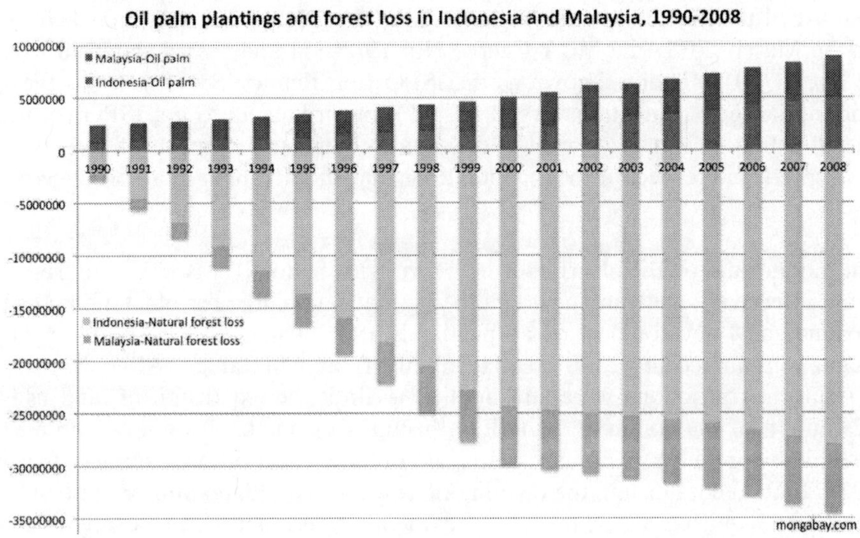

Source: FAOSTAT

Figure 2. Oil palm plantings and forests loss in Indonesia and Malaysia 1990-2008

3.3 EU' Plan to ban palm oil

The EU is the second-largest palm oil importer on earth after India and therefore is a key palm oil trading partner for Indonesia. In 2009, it implemented a renewable energy directive and set a target that 20% of its energy should come from renewables as well as 10% of its transportation should also run on renewables. In early 2018, the EU draft another directed which bans the import of biofuel from crops on land converted from forest or food crops. This is based on the fact that roughly one-third of the palm oil imported into the EU goes into biodiesel production, while the rest predominantly to the food and chemical industries. The EU Parliament approved the ban in January 2018 and planned to implement the ban in 2021. The base of this directive is the negative social and environmental effect of palm oil, which motivate producers to switch to sustainable agricultural practices.

The plan, however, has been rejected in mid-2018 by The European Commission as they knew a ban was unworkable from a trade policy perspective because it will halt EU's plan in pursuing ongoing trade relations with Indonesia and Malaysia. Following that, however, the EU will finalize the "EU Action Plan" on deforestation which will introduce new measures to limit palm oil imports. Ideas are more focused on trade agreement and licensing system for palm oil that uses sustainable certification.

4. Results and discussion

4.1 Certification

Efforts have been made by the government and different stakeholders in facing the negative impact of palm oil. At the same time, Indonesia continues to promote and expand its palm oil production as many livelihoods depend on the sector as well as its importance to the Indonesian economy. Sustainability certification has been one of the main efforts by putting social and ecologic aspects on the agenda of companies, and thus inducing technical progress in processes for lowering the greenhouse gas footprint and increasing awareness about social conditions [5]. Due largely to greater international awareness, global business leaders have joined environmentalists to meet a more-sustainable oil palm production by aligning with the *Roundtable on Sustainable Palm Oil* (RSPO) criteria, which aim to increase the food supply while keeping in mind the goals to safeguard social interests, communities and workers as well as to protect the environment and wildlife [6]. In 2011, Indonesia established the *Indonesian Sustainable Palm Oil* (ISPO) that was launched by the Ministry of Agriculture in March 2011 (the MoA Decree No. 19/2011) and was updated in March 2015 (the MoA Decree No. 11/2015) is now known as the ISPO Certification System. It has the same spirit as RSPO, the Indonesian government has urged all palm oil producers to get ISPO certification. It is a mandatory certification scheme to ensure the quality and the respect of norms regarding the environment, workers and respect of local populations that should apply to all producers [7].

4.2 Moratorium

A government moratorium on the clearing of new forest for palm oil was effective from 2011 to 2015. The moratorium effectively halts any new land being made available for plantations for palm oil. The government banned new development on all peatlands after swathes of carbon-rich peat were drained for use as plantations in recent years, creating highly flammable areas. After the first stage of the moratorium, stringent restrictions were still applied to limit the expansion of land use for palm oil, where only 2 million of the country's 25 million hectares of peatland are now eligible. In 2018, the Indonesian president continues to sign a moratorium on new palm oil development that will last three years until 2021. In this moratorium, the opening of new palm oil plantations will be delayed to reduce conflicts, as well as requiring all central and provincial governments to re-evaluate current permits. The moratorium aims to improve the governance of sustainable palm oil plantations, provide legal certainty, increase the productivity of smallholder palm oil plantations, maintain environmental sustainability, and contribute to the reduction in greenhouse gases. The moratorium also allows the government to resolve issues of land conflict and overlapping permits.

4.3 Against EU's palm oil ban

At the same time, Indonesia needs to defend its Palm Oil industry. EU's plan to ban palm oil import has been argued to violate the World Trade Organization (WTO) rule in which discrimination has been done against palm oil when compared to similar vegetable oils. This is due to the fact that other similar industries contribute far more to deforestation, but are not affected by the EU ban the same way it was applied to palm oil. However, the EU ban does not affect those industries in the same way. As part of this effort, there could be two scenarios, which is trade retaliation or trade cooperation. As part of the trade cooperation, the use of Indonesia Sustainable Palm Oil (ISPO) certification as the basis of importation procedure in EU has been proposed.

The ban against palm oil is not economically feasible if applied. According to Prof. Gernot Klepper [8], there will be a number of reactions if palm oil ban is applied to palm oil in EU. First, lower prices will raise palm oil demand from outside the EU, hence a large part of the palm oil will not be exported to the EU, but will be exported somewhere else. This will unlikely slow the deforestation, which is against the spirit of the ban in the first place. Second, lower palm oil prices will widen the wedge between other vegetable oils. Palm oil users outside the European biofuel sector will have an increased incentive to substitute other vegetable oils with palm oil, thus increasing palm oil demand. Deforestation therefore will not be reduced as production moves towards other uses. Lastly, in the

European biofuel market, demand for biodiesel will not change. Instead, soy, canola, rapeseed, and other vegetable oils will serve as feedstock for biodiesel increasing their market share. Therefore, banning palm oil will not reduce deforestation, as it increases the use of other types of vegetable oil, which also contributes to deforestation (figure 3). The best way to move forward, according to Prof. Klepper, is therefore to move towards a responsible and sustainable palm oil production. At the same time, the sustainability requirements should also be extended to all vegetable oils.

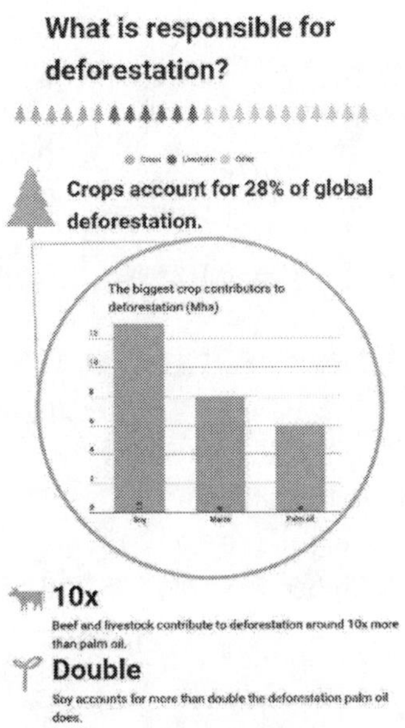

Source: ASEAN Today

Figure 3. Sectors Contributing to Deforestation

4.4 Discussion

4.4.1 Smallholder farmers' agency in facing palm oil's negative campaign

Indonesian smallholder oil palm farmers own and/or manage at least 3.1 million ha of oil palm, accounting for more than 40% of the total oil palm area and generating an estimated 35% of total crude palm oil production nationwide. On average, each farming household manages about 2 ha of land, compared with private companies that manage about 4,000 ha. Additionally, in comparison to private companies and state-owned enterprises, smallholder farmers have lower productivity per hectare. It is therefore critical that they form part of the ongoing transition to a more sustainable palm oil value chain. Although higher standards are promoted globally as well as nationally, small players are difficult to meet the required sustainability standards and will take a more organized effort as they are more spread across different places. Most small players cannot afford to overhaul their working practices and focus on sustainability. Smallholders farmer are also more vulnerable towards oil prices and financial access as well as having limited inclusive land use planning and knowledge and experience in best management practices and financial capital to accomplish sustainable farming. Due to these challenges, smallholder farmers have limited agency in advocating their need in the palm oil sector. The survey data indicates that the majority of smallholder farmers have difficulties in advocating their concerns, where only 54% of the survey participants advocate their concerns. Meanwhile, 76% of the participants indicate that they are negatively affected by the negative campaign exist in the palm oil sector. This indicates that while they are significantly affected, in general, many of them have little power to raise their concerns (table 1).

Table 1. Summary Statistics of All Variables in from Survey Conducted

Variables	Obs	Mean	StdDev	Min	Max
Affected by negative campaign (1=participate)	100	0.76	0.44	0	1
Holds sustainable certification (1=participate)	100	0.32	0.68	0	1
Advocate their concerns to the government (1=participate)	100	0.54	0.24	0	1
RPO Membership (1=member)	100	0.74	0.37	0	1
Receive support from government or external actors in obtaining certification (1=Yes)	100	0.54	0.3	0	1
Education (1=no education, 2=elementary school, 3=middle school, 4=high school, 5=higher education)	100	2.55	1.22	1	5
Age	100	42.4	13.2	19	72
Land Ownership (1=own land)	100	0.71	0.39	0	1
Member of other civic organization (1=member)	100	0.34	0.49	0	1

As discussed above, an important part of being successful in the current palm oil sector is the ability to comply with the sustainability standards as in the RSPO or ISPO. As of 2017, however, less than 1 percent of independent smallholders' farms were certified as sustainable by the RSPO and ISPO. There are many reasons for this. Since independent smallholders are not linked to any particular company or mill, they do not receive training, supervision or support from companies, and only receive limited support from the government. Because they receive limited information about good agricultural practices, this led to lower productivity and a lower concern for sustainability by rural smallholder farmers. Due to the limited information, small farmers often purchase cheap, low-yield seedlings and burn land to make way for crops, which eventually hard to develop their palm oil marketing. In addition, smallholders' independence leaves them vulnerable to exploitation. Middlemen acting as fruit brokers between farmers and mills often charge exorbitant fees for their services or take a large chunk of the profits from farmers' oil palm sales. Meanwhile, independent smallholders who do want to pursue sustainable palm oil production may find it prohibitively expensive. At the same time, the direct cash income benefits from premium fees for certified and higher quality fruits are limited. Certification is profitable only if independent smallholders consistently receive high premium prices and fees (Larsen et al, 2018). In practice, the premium price is set by intermediaries which oftentimes smallholders find themselves in a lower bargaining position, on top of already high price fluctuation.

There are several programs that the government have helped to promote sustainable certification for smallholder farmer. The Ministry of Agriculture Republic of Indonesia, for example, has joint cooperation with the UNDP is assisting independent smallholders to be certified by ISPO. The pilot project was held in Riau, South Sumatera and West Kalimantan and they plan to certify 2,200 smallholders. The Secretariat ISPO Commission- MoA also have a joint project with International Finance Corporation (IFC) in assisting independent smallholders to be certified by ISPO. The pilot project was held in Rantau Prapat at South Sumatera where the secretariate of ISPO Commission will also establish the socialization and give ISPO training for the local government, extension workers, and palm oil associations and farmers. Another joint project between the Secretariat ISPO Commission and the MoA was with the Conservation International (CI) is assisting independent smallholders to be certified by ISPO. On March 2017, they held training for different stakeholders in Sorong and Manokwari, in the province ofWest Papua.

The survey data confirms the common scheme that the majority of smallholder farmers does not hold sustainable certification, where only 32% of the respondents receive certification (table 1). Even if they received help from the government or NGOs, the effect is not likely last as it is difficult for them to maintain their sustainable certification, once support from an NGO or company ends. RSPO's 2017 Impact Report, for example, shows a 38 percent decrease of certified independent compared to the previous year largely due to licenses expiring. While this indicates that independent smallholders can operate sustainably, they simply lack the resources to afford and maintain certification.

4.4.2 Effectiveness of Rural Procedures Organizations (RPOs) in smallholder farmers 'agency

As smallholders typically cannot directly access certification individually [9], and need support from external actors to comply with the standards [10], participation in organizations is essential for their success. Regarding the RSPO, smallholders are also obliged to join a group certification and establish a group manager who is responsible for an internal control system (ICS) to monitor smallholders' performance. These organizational requirements create a new challenge for rural farmers and will determine whether they can enter the market. The role of Rural Producer Organizations (RPOs) such as farmers cooperative, Village Unit Cooperative (*Koperasi Unit Desa* or KUD), and farmer groups, are therefore pivotal. After joining the certification, all plantation activities ranging from input supply and credit support to FFB selling, are centralized in RPOs. RPOs also provides a forum for sharing and communicating problems as well as the possible solutions related to palm oil plantation. RPOs become a forum for sharing knowledge and information on a smaller scale, as well as supervise all oil palm plantation activities, including fertilizer application, harvesting, sorting, and distributing the product. Certification will eventually lead to an increase of smallholders' financial capital and hence to contribute positively to the livelihood and for a higher income. When compared to uncertified smallholders, most certified scheme and independent smallholders believe that they get a higher price, just based on the fact that the quality of a certified products is better than those without no certification.

Table 2 shows how membership in RPOs is more likely to be successful in advocating their concerns and receiving the certification. Thus, the farmer's organizational strength is a requirement for successful smallholder farmers. The estimation indicates that membership of RPOs have higher participation in advocating their concerns and receive certification.

Table 2. Estimation Results in Explaining Participation in Policy Discussion

Political Agency	Member	Non-member
Advocate their concerns Chi2 (1, N=100) = 13.22***	76.3%	51.1%
Does not advocate	23.7%	48.9%
Receive certification Chi2 (1, N=100) = 2.338**	75%	50.5%
Does not receive certification	25%	49.5%

*Notes: *p<.05, **p<.01, ***p<.001*

Most cooperatives, however, could not compete in open-market economies and are today weak, dormant or dead [11]. With the decline of cooperatives and other farmers' organizations, many farmers lack a collective voice. They cannot to access affordable production inputs such as finance, technology, land and water, and are locked out of markets. As a result, a large number of small-scale farmers live in poverty and cannot influence policies that affect their livelihoods. Strong and vibrant farmers' organizations can provide opportunities for farmers to effectively play a role in the market economy and benefit from it. However, identifying and promoting authentic farmers' organizations that empower smallholders, is a big challenge for governments and their development partners. Most groups are hastily formed, often with no regard for the social-cultural and economic structures of the farming communities. Such groups are not viable and incapable of serving as channels through which farmers can take part in decision making. Thus, only a small amount of strong and vibrant farmers' organizations who can provide opportunities to farmers to effectively play a role in the market economy and benefit from it. Table 3 shows the estimation result between being a member of an RPOs and advocating to their preferences, obtaining certification, and increasing welfare.

The result indicates that being a membership have a positive and significant effect to smallholder farmers' participation in advocating their interest and concerns against the negative campaign in palm oil sector, as well as obtaining sustainable certification, and consequently their income. Indeed, insights from the FGD that was conducted, membership of an RPOs is seen a very important role in increasing farmers' agency, although the strength of RPOs varies across places. Participants indicate that several factors might contribute to the success of RPOs, namely: (1) having a strong organizational structure; (2) education level of its members; (3) number of members, in which the smaller the membership, the more easier to achieve collective action; (4) ethnic homogeneity; and (5) age.

Table 3. Estimation Results in Explaining Participation in Policy Discussion

Independent Variables	**Dependent Variables**		
	Advocating Interest	Obtain Certification	Increase welfare
RPO Members	0.3536	0.498	0.797
	(0.236)*	(0.275)*	(0.469)*
Education	0.32	0.232	0.3876
	(0.0827)**	(0.0934)**	(0.1570)**
Age	0.0542	0.0003304	-0.0006
	(0.005)	(0.0082)	(0.0138)
Land Ownership	0.25	0.166	0.297

Independent Variables	Dependent Variables		
	Advocating Interest	Obtain Certification	Increase welfare
	(0.213)	(0.255)	(0.428)
Market Distance	-0.004	-0.002	-0.003
	(0.002)	(0.001)	(0.002)
Member of Political Party	0.902	0.805	1.462
	(0.468)***	(0.3030)***	(0.563)***
Member of Other Civic Organization	-0.078	-0.009	0.029
	(0.102)	(0.207)	(0.354)
Contstant	2.353	0.962	-1.624
	(0.876)	(0.636)	(1.056)
Number of observation	N=100	N=100	N=100

* Significant at 10%; ** Significant at 5%; *** Significant at 1%

5. Conclusion

Smallholder farmers play an important role in the Indonesian palm oil sector, and therefore are highly affected by the negative campaign against palm oil. This paper shows that being a member of a Rural Producer Organizations (RPOs), such as cooperatives and farmers association, smallholder farmers are able to increase their political agency. That is, their ability to participate in discussions with the government and advocating their concerns as well as receiving government support. Since sustainable certification is a huge part of palm oil sector, this paper test whether membership in RPOs will deliver more certification to smallholder farmers. The findings show that farmers who are members of cooperatives or farmers groups, they are more successful in advocating their interests, receiving certification and support from the government, and consequently increase their welfare. Despite its importance, RPOs, however, are still difficult to develop, and thus we see low political agency among smallholder farmers in general. This shows that smallholder farmers in general are struggling to face compliance with sustainable requirements in the palm oil sector, while RPOs being the most reliable vehicle for them to cope with these challenges are relatively weak. Thus, an important takeaway from this research is the importance of institutional support to strengthen RPOs in supporting smallholder farmers. In other words, smallholder farmers are able to cope with challenges within the palm oil sector, including negative campaigns, when RPOs are strong.

References

[1] [FAOSTAT] Food and Agriculture Organization of the United Nations Data. 2018. Rome: FAO. Accessed 12 September 2018. http:// faostat3.fao.org/home/E
[2] Gerber JF. 2011. Conflicts over industrial tree plantations in the South: Who, how and why? Global Environmental Change **21**:165–176. doi:10.1016/j.gloenvcha.2010.09.005
[3] WWF. 2016. Palm Oil Buyers Scorecard Measuring the Progress of Palm Oil Buyers. http://wwf.panda.org/what_we_do/footprint/agriculture/palm_oil/solutions/responsible_purchasing/palm_oil_buyers_scorecard_2013/
[4] Budidarsono S, Susanti A and Zoomers A. 2013. Oil Palm Plantations in Indonesia: The Implications for Migration, Settlement/Resettlement and Local Economic Development. Chapter 6 in Biofuels - Economy, Environment and Sustainability, Prof. Zhen Fang (Ed.), InTech, DOI: 10.5772/53586
[5] KPMG Sustainability. 2013. Improving smallholder livelihoods: Effectiveness of certification in coffee, cocoa and cotton. http://www.sustaineo.org/
[6] RSPO. 2013b. RSPO Standard for Group Certification Final – approved July 2010 (Amendment

April 2013). http://www.rspo.org/

[7] Rist L, Feintrenie L and Levang P. 2010. The livelihood impacts of oil palm: Smallholders in Indonesia. Biodiversity and Conservation **19**:1009–1024. doi:10.1007/s10531-010-9815-z.

[8] International Sustainability and Carbon Certification (ISCC), 2018, *https://www.eco-business.com/opinion/winners-and-losers-from-the-eus-proposed-ban-on-palm-oil/*

[9] Bitzer, Verena, Pieter Glasbergen, and Bas Arts. 2013. Exploring the potential of intersectoral partnerships to improve the position of farmers in global agrifood chain. Agriculture and Human Values **30** (1):5-20. doi: DOI 10.1007/s10460-012-9372-

[10] Lemeilleur, Sylvaine. 2013. Smallholder Compliance with Private Standard Certification: The Case of GlobalGAP Adoption by Mango Producers in Peru. International Food and Agribusiness Management Review **16** (4):159-180

[11] Enabling The Business of Agriculture. 2015. World Bank Progress Report

Palm oil spills on the performance of non-palm plantations in Indonesia: an empirical analysis

Harianto[1]

[1]Department of Agribusiness, Faculty of Economics and Management, Bogor Agricultural University, Bogor 16680, Indonesia

*E-mail: harianto.ipb@gmail.com

Abstract. Rapid development of oil palm plantations is predicted to cause various problems, one of which is the conversion of land and the shift of non-palm oil businesses to palm oil businesses. This research is aimed at analyzing the impacts of palm oil industry, through the price of palm oil in the international market, to the area and production of non-oil palm plantations (rubber, cacao, and coffee). The study employed time series data from 1991 to 2015. An econometric model was built to connect variables that affect the area and the production level of non-palm plantation and use multiple regression methods to determine its effect. The results showed that the price of palm oil had a positive effect on the expansion of rubber, cocoa, and coffee plantations. The effect of the price of palm oil on the production of non-palm plantations also shows a positive direction. Thus, the development of oil palm plantations does not have a negative impact on the development of non-palm plantations. The policy implications of the results of this study are the need to utilize the positive impacts of development of the palm oil industry for the development of plantations in the broadest sense.

1. Introduction

Palm oil is increasingly important in the international oil and fat market. The characteristics of palm oil that are suitable for various uses in the food and non-food industries make the demand for palm oil tend to increase from year to year. The dominance of palm oil in the international market is mainly due to the higher competitiveness of palm oil than other sources of vegetable oil. The role of palm oil in the oil and fat trade continue to increase with the growing use of palm oil for non-food industry needs, especially for increasing biofuel production, and surging demand for palm oil in China and India.

Crude oil prices in international markets are fluctuating and tend to increase, and coupled with the awareness to reduce greenhouse gas emissions has made the petroleum substitute industry sourced from renewable raw materials increasingly important. Palm oil is one of the raw materials for the biofuel industry whose use continues to increase. Increased use of palm oil for energy has led to debate about its impact on food security and land use change. The use of palm oil products for energy production leads to an increase in food prices. Increase in palm plantation due to the increase in palm oil price will ultimately reduce the area dedicated for non-oil plantations. If indeed the development of oil palm plantations reduces the area for other plantation crops, it can have a negative impact on export diversification that also means increasing the risk of long-term foreign exchange earnings.

The price increase of palm oil in the international market is one of the important factors in increasing the area of oil palm plantations in Indonesia. Based on data from the Central Statistics

Agency (BPS), in 1991 the area of oil palm was 0.77 million hectares and continued to increase to 11.91 million hectares in 2016. At the same time, the production of palm oil also increased from 0.55 million tons to 6.64 million tons. The rapid growth of the palm oil has made palm oil a very important role in the Indonesian economy. Palm oil exports tend to continue to increase and occupy the largest share of total non-oil and gas exports. Oil palm plantations and its derivative industries also make a significant contribution to poverty alleviation and provide employment opportunities as in [1].

With the Indonesian government's decision in 2018 requiring the use of 20 percent palm oil-based biofuel (B20), oil palm plantations expected to continue to increase in the future. Increasingly oil palm plantations have an impact on the emergence of various social, economic and environmental problems. The growth of oil palm areas considered to have neglected the interests of local communities, reduced income diversification in the regions, and reduced biodiversity and wildlife. Reference [2] indicated that increasingly limited land resources could cause oil palm expansion to replace land use for other uses. The wider area of oil palm can have an impact on forest encroachment, the use of peatland, or land conversion. Increasing palm oil prices and the area of oil palm plantations that continue to expand expected to have negative impacts on the area of non-palm plantations that also become a mainstay for Indonesian exports.

This research is aimed to find out how the influence of palm oil development, represented by the price of palm oil on the international market, the area and production of rubber plantations, cocoa, and coffee. Rubber, cocoa, and coffee selected as cases because they are considered that these 3 products do not have a direct relationship in their demand or consumption with oil palm. However, these 3 crops, namely rubber, cocoa, and coffee have a production relationship that competes for land use. This is different from coconut plantations that have links with oil palm, both in terms of demand or consumption and aspects of supply or production.

2. Literature review

Various studies conducted to determine the effect of the price of a commodity on the prices of other commodities. Reference [3] shows a close link between oil prices and commodity prices. Using pairwise Granger causation test, he found a direct relationship from the price of wheat and soybean to the price of ethanol. Reference [4] shows that there is a close relationship between the price of petroleum and the price of maize, and the price of maize has a spill impact on related agricultural products. Reference [5] using weekly data from 2003 to 2007 found a significant relationship between the price of petroleum and soybean prices. The relationship between these 2 prices changed from negative throughout 2003-2006 to be positive throughout 2007. However, with the same set of data in the same year [5] found no relationship between the price of petroleum and the price of corn. Previously, reference [6] using the Granger test and impulse response also failed to find a significant relationship between the price of petroleum and the price of edible oils. From these various studies concluded that the empirical closeness and direction of the relationship between energy prices and commodity prices vary between commodities and between periods.

Increased crude oil prices and awareness of the need to prevent an increase in greenhouse gases have caused biofuel production to increase. Increased biofuels production can certainly affect commodity prices to be used as raw materials for biofuels that will indirectly cause land use changes. Various studies show that there is a significant relationship between biofuels production and changes in land use for commodities. Reference [7] using the CGE model shows how biofuels production becomes a driving force for land use change and deforestation. Reference [8] founds the need for corn plant expansion to meet increasing industrial demand for corn-based ethanol in the US. The expansion of the corn area will replace the soybean area and cause an increase in the price of wheat. The results of their study found the elasticity of the addition of area corn with an increase in wheat price of 1. The results of this study are in line with the research of [9] who examines the relationship between corn prices and the area of land reserved for production.

Changes in land use for the needs of various commodities expected to occur if there is a relative price change between them. Reference [10] documented a shift in land use between rangeland and cropland in response to change in commodity prices in the US. Reference [11] examined the effect of biofuels on land use cash rent in Iowa. They examined the relationship between cash rental rates for cropped land (hay) and non-cropland (pastureland) use. The study found that the relationship between crop prices and the amount of cropland explained by the influence of product market prices on the share of cropland in the total farmland. Reference [11] analyzed using regression panel model with explanatory variables as expected corn price, soybean price, and feed for cattle prices. The results of their analysis show that increasing corn prices causes an increase in land conversion from non-cropland to corn production. Reference [12] also showed the impact of changes in agricultural commodity prices on deforestation. The results of various studies mentioned above implicate that changes in relative prices between commodities can affect land use and the level of production of various commodities.

This research is also based on the finding of [13] in his testimony for the US Senate Committee on Homeland Security, and Governmental Affairs stated that food prices and fuel prices are highly correlated. Thus, the corollary idea is that an increase in the area of oil palm estimated to affect the area of other plantations. Changes in land use from non-palm oil crops to oil palm plantations naturally occur because of the relative changes between the prices of palm oil and non-palm oil prices. Changes in relative prices not only affect the area of non-palm oil plantation but also expected to affect the level of production of non-oil palm.

3. Research methods

The non-palm plantations selected in this study are rubber, cocoa and coffee plantations. The selection of these 3 commodities is based on the consideration that the 3 commodities are also the mainstay of Indonesia's non-oil and gas exports. Indonesia is the largest palm oil producer in the world, and for rubber, cocoa and coffee commodities Indonesia rank second, third, and third in the world respectively. These 3 commodities are also important employment providers. The share of smallholders in the area of rubber, cocoa and coffee is the largest compared to the area owned by government own enterprises or large scale privately own enterprises. In 2015 the share of smallholder plantation areas from the total area for rubber, cacao and coffee was 84, 97 and 96 percent, respectively.

The performance of rubber, cocoa, and coffee plantations analyzed in this study is the area and production level of each commodity. The effect of palm oil on the performance of non-palm oil plantations is represented through changes in palm oil prices on the international market. The international palm oil price employed is the world price of crude palm oil (CPO) and the world price of palm kernel oil (PKO). While the price for each non-palm commodity will be represented with the price received at the exporter level. The price at the exporter level calculated from the value of exported commodity divided by the quantity exported. As a control for the influence of commodity price variables, the research model also included the variable wage rates for labour work in plantations and credit interest rates in Indonesia.

The models used in this study are as follows:

$$AREA_{i,t}=\beta_0+\beta_1 WAGE_t+\beta_2 INTRATE_t+\beta_3 PWCPO_t+\beta_4 PWPKO_t+\beta_5 PEXP_{i,t}+U_t \quad (1)$$

$$PROD_{i,t}= \beta_0+\beta_1 WAGE_t+\beta_2 INTRATE_t+\beta_3 PWCPO_t+\beta_4 PWPKO_t+\beta_5 PEXP_{i,t}+U_t \quad (2)$$

Where:

β_0 = *Intercept (constant)*
$\beta_1 - \beta_5$ = *Coefficient regression of each variable*
U = *Error term*
i = *Rubber, Cocoa, Coffee*
t =*Time (year)*

The description of the variables used in equation model and data sources can be looked at table 1. The data used in this study are time series data from 1991 to 2015. All data employed in this study are secondary data.

Table 1. Description of the variables in the model, its unit value, and data sources

No.	Symbol	Variable name	Unit value	Sources of data
1.	$AREA_i$	Area (rubber, cacao, coffee)	Hectare (Ha)	Directorate General of Estate Crops (DGEC)
2.	$PROD_i$	Quantity of production (rubber, cacao, coffee)	Ton	DGEC
3.	*WAGE*	Monthly wage of plantation worker	USD/month	BPS
4.	*INTRATE*	Credit interest rate	%	BPS
5.	*PWCPO*	World price of crude palm oil	USD/ton	Oil World
6.	*PWPKO*	World price of palm kernel oil	USD/ton	Oil World
7.	$PEXP_i$	Export price (rubber, cacao, coffee)	Rubber - USD/ton Cacao – USD/ton Coffee – USD/ton	DGEC

Multiple regression method used to estimate the influence of predictor variables on variable area and variable production of each commodity. Multiple regression method employed to analyze the influence of several independent variables on a dependent variable. The use of multiple regression method can determine the relative influence of each independent variable on the dependent variable, based on a complex data set, the results of which can be wrong if the analysis conducted incorrectly. In this research the model for each commodity treated as a single equation and not analyzed simultaneously. That is why in the section of results and discussion also carried out separately for each commodity.

To ensure the goodness of fit of the model used in explaining the variations that occur in the area and production variables, the determination coefficient indicator (R^2) and F test are used. The classic assumption test carried out to determine whether the coefficient obtained is truly unbiased and efficient. The Durbin Watson test used to see the existence of autocorrelations. In this study the Glejser test employed to detect the presence of heteroskedasticity, and *VIF* indicators to ensure the absence of multicollinearity. To be able to compare the ranking of the influence of each explanatory variable on the dependent variable (area and production area), then the standardization of the estimation result coefficient is done.

4. Results and discussion

4.1. Rubber

After estimating the rubber area model, there is a high autocorrelation phenomenon with a Durbin Watson (*DW*) statistic of 0.617. To eliminate autocorrelation in the model, variable dummy crisis is added. Dummy crisis is rated 1 for the year of the 1997 financial crisis in Asia and the 2008 and 2009 world economic crisis. The new model can explain variations in the area of rubber with the coefficient of determination (R^2) of 0.517 and significant F test values at 0.002. All variables have a *VIF* value below 10 that indicates that there is no serious multicollinearity. After the Glejser test, heteroskedasticity phenomenon was not found in the model. The Glejser test did not find explanatory or independent variables that significantly affected the error term U with a significance level of less than 0.05.

The regression model for the quantity of rubber production also has an autocorrelation problem that is the same as the rubber area model with the *DW* value of 0.543. In the model of production level also was added variable dummy crisis. The result of the model has an R^2 value of 0.791 with a statistical F test value of a significance level of 0.000. None of the *VIF* values for each explanatory variable is greater than 10, so there is no serious multicollinearity problem in the model. Likewise, after the Glejser test can be concluded the model does not have heteroskedasticity problems, because there are no explanatory variables that significantly affect the error term *U* with a significance level smaller than 0.05. In summary, the estimation results in the area model and the number of rubber production model provided in table 2.

CPO world price variables give positive and statistically significant effects on the level of error of less than 1 percent to the area of rubber plantation. Even the world price of oil palm has the greatest influence on the area of rubber compared to other variables. Wages have a significant negative effect on the area. Increasingly expensive labour can be an obstacle to efforts to increase the area of rubber plants. Likewise, the economic crisis can cause a decrease in the area of rubber. A declining world economy can cause the reduce need for rubber raw materials, and indirectly have a negative impact on rubber plantations.

Table 2. The results of estimation of factors affecting the area and production of rubber

Explanatory variables	Var. dependent: $AREA_{rubber}$		Var. dependent: $PROD_{rubber}$	
	Unstandardized coefficients	Standardized coefficients	Unstandardized coefficients	Standardized coefficients
1. *WAGE*	-207.201**	-0.510	-1046.068***	-0.464
2. *INTRATE*	-2715.382	0.172	-17436.868	-0.199
3. *PWCPO*	662.697***	1.105	704.017	0.211
4. *PWPKO*	20.903	0.051	-1.143	0.000
5. $PEXP_{rubber}$	-88.044*	-0.734	207.278*	0.311
6. D_{CRISIS}	-67370.311*	-0.208	-64811.016	0.036
7. *INTERCEPT*	3134189.832***	-	1906701.510***	-

Description: *** significant at 1% level; ** significant at 5% level; * significant at 10% level

Despite being positive, the CPO world price had no statistically significant effect on Indonesian rubber production. Rubber production is mainly influenced by the level of labour wages. The higher the wage level, the more rubber production will decrease. Rubber plantations are very intensive in using labour. The number of workers working on rubber plantations in 2016 reached 2.47 million people. Plantation smallholders contributed the largest workforce of 2.23 million people and followed by private companies amounting to 0.16 million people.

4.2. Cocoa

The coefficient of determination (R^2) is high, both for the area and production of cocoa models of 0.859 and 0.867, respectively, indicating the ability of the model to explain the variation in the area's of cocoa and the level of cocoa production. *DW* values are 1,457 and 1,591 indicate that there is no serious autocorrelation problem. The *VIF* value for each explanatory variable in the area model or level of production model is much smaller than 10, and even the largest *VIF* is only 3.889. In both models, no serious multicollinearity symptoms were found. Likewise, based on the results of the Glejser test, there was no problem with heteroskedasticity in the model, where no explanatory variables had a significant effect on the error term *U* of the model. The results of the estimation of area and production models in summary are presented in table 3.

Based on the estimation of the cocoa area model and cocoa production model, it can be seen that the CPO world price has a positive and statistically significant effect at the 10 percent level. However, for the PKO world price variable there is no significant effect on the area or the production level of cocoa. Price of cocoa at exporter level had a significant effect on the area but not on the level of production. Based on the size of the standardized coefficient, it can be seen that the influence of the CPO world price on the area and production level is lower than the effect of the wage and interest rate. However, the influence of world price of CPO to cocoa production level is greater than the exporter price of cocoa.

Table 3. The results of estimation of factors affecting the area and production of cocoa

Explanatory variables	Var. dependent: $AREA_{cocoa}$		Var. dependent: $PROD_{cocoa}$	
	Unstandardized coefficients	Standardized coefficients	Unstandardized coefficients	Standardized coefficients
1. *WAGE*	-553.376***	-0.339	-507.411***	-0.697
2. *INTRATE*	-16684.655***	-0.263	-6966.369**	-0.246
3. *PWCPO*	324.136*	0.135	252.463*	0.235
4. *PWPKO*	-137.336	-0.083	-32.088	-0.043
5. $PEXP_{cocoa}$	279.654***	0.522	1.343	0.046
6. *INTERCEPT*	846866.031***	-	646349.727***	-

Description: *** significant at 1% level; ** significant at 5% level; * significant at 10% level

Variable amount of production is the amount of cacao production in the form of cacao bean. The $PEXP_{cacao}$ variable does not affect the level of production, possibly due to the use of cacao bean prices received by exporters as a substitute for prices received by cacao producers. The smallholders sell cocoa they produced in the form of unfermented or low quality of cocoa bean to the traders or collectors, and by the collector, cocoa bean would be upgraded before being sold to exporters. Thus, the price of cocoa bean received by producers is likely not correlated or not influenced by the price at the level of exporters.

4.3. Coffee

Based on the test of the classic assumption of the regression model it can be concluded that the model employed does not have serious problems with autocorrelation, multicollinearity, and heteroskedasticity. *DW* values for the area model and production level model are 1.689 and 1.407, respectively. Likewise, no serious multicollinearity was found because the maximum *VIF* value of the explanatory variables for the model area and production model were 3.849 and 3.562. The model was also clean from heteroskedasticity problems because based on the Glejser test of both models, no explanatory variables were found that significantly level 5 percent of the error term *U* from the model. Based on R^2 value, area model and production model also have high goodness of fit because it can explain the variation of area variable and production variable of 0.722 and 0.833 respectively. The *F* test of the model also shows a high level of significance with a probability of error less than 1 percent.

The world price of CPO has no significant effect on the area of the coffee plantation, but has a positive effect at a 1 percent level of significance on the level of coffee production. The PKO world price has no significant statistical effect on the area and the level of coffee production. Price of coffee at exporter level significantly affects the area and also the level of coffee production. Area variables and production level variables are also negatively affected by plantation labour wages and interest rates.

Table 4. The results of estimation of factors that affect the area and production of coffee

Explanatory variables	Var. dependent: $AREA_{coffee}$		Var. dependent: $PROD_{coffee}$	
	Unstandardized coefficients	Standardized coefficients	Unstandardized coefficients	Standardized coefficients
1. *WAGE*	-140.629***	-0.594	-276.503***	-0.800
2. *INTRATE*	-3312.855***	-0.360	-3523.744***	-0.262
3. *PWCPO*	74.163	0.212	187.955***	0.368
4. *PWPKO*	-38.637	-0.160	-41.868	-0.119
5. $PEXP_{coffee}$	-42972.539*	0.444	-3301.138**	0.236
6. *INTERCEPT*	1367490.055***	-	679672.637***	-

Description: *** significant at 1% level; ** significant at 5% level; * significant at 10% level

Coffee plantations in Indonesia are dominated by small-scale smallholders and using labour intensively. A large workforce is needed for the maintenance of plantation and harvesting. The green beans of coffee usually are harvested by farmers manually. Labour wages are an important factor in determining the amount of labour used by coffee farmers. In 2016 there were 1.79 million households operating coffee plantation. These small-scale farmers generally experience working capital limitations so that loan interest rates can negatively affect farmers' access to capital and ultimately have a negative effect on the level of production. On average the coffee farmers in Indonesia are only able to produce green coffee beans of 0.704 tons per hectare, far lower than in Brazil which amounted to 1.51 tons per hectare in 2016. Capital constraints are a factor that prevents Indonesian coffee farmers from being able to apply better coffee cultivation technology.

The rapidly growing palm oil industry expected to bring changes to physical, social and economic infrastructure. The expansion of oil palm plantations and its derivative industries is certainly accompanied by the construction of physical infrastructure for roads, warehouses and ports. Reference [14] indicated that palm oil development has a major influence on the infrastructure of the main growing regions of Indonesia that led to significant distortions in the transportation network of the islands on which it is grown. The development of the palm oil industry is also accompanied by institutional developments in the form of cooperation between the nucleus plantation and the plasma plantation. The palm market that penetrates to remote areas in Indonesia predicted to be able to improve the business capabilities of the community.

On the public policy side, the increasingly dominant role of palm oil in the economy has encouraged the government to provide support and greater attention to the palm oil industry. Various public policies such as the transmigration program, 20 percent oil palm biofuel obligations, and other marketing policies are directly and indirectly aimed at the growth of the palm oil industry as in [15]. Rubber plantations, cocoa, and coffee industries seem to get positive spill over effects from the growth of the palm oil industry. The growth of the palm oil industry also encourages the growth of land area and production of rubber, cocoa, and coffee. The expansion of oil palm land is allegedly not in the form of conversion of land for rubber, cocoa, or coffee plantations into palm plantation but instead there is a tendency to expand towards the reducing of forest areas, peatland, or land for other allotments. Reference [16] has found that the expansion of palm oil plantations in Papua and Kalimantan has caused deforestation and negative impacts such as water pollution, air pollution and soil erosion.

5. Conclusion

Based on the estimation results of the planting area model and production model of rubber, cocoa, and coffee concluded that in general, the world price of CPO has a positive influence on the harvested area and on the level of production. The results indicate that an increase in the area of oil palm due to the increasing competitiveness of oil palm will not cause a reducing area of rubber, cocoa and coffee.

The positive influence of world palm oil prices on area and production even tends to be greater than the influence of the own price of rubber, cocoa, or coffee. The effect of palm oil prices is only inferior to the influence of plantation labour wages in cocoa and coffee commodities models. Cocoa and coffee are industries that use labour intensively so that the increase in wages of labour will bring negative effects to the performance of the industry. In other words, the development and competitiveness of plantation in Indonesia still rely heavily on cheap labour.

The policy implications of the results of this study are the importance of making good use of the momentum created by the development of the palm oil industry to build other highly competitive plantation industries. Improvements in physical, social and economic infrastructure that are likely to accompany the growth of the palm oil industry can certainly also be used for the development of the rubber, cocoa and coffee industry. However, the results of this study indirectly warn of the need to take into account the negative impacts of the growth of the palm oil industry on land use change in forestland, petland, and other land purposes.

Acknowledgments
The author is grateful to the participants of ICNREC 2018 for their valuable comments and suggestions. The author also expresses thanks and appreciation to anonymous reviewers for their correction and thoughtful suggestions. However, none of them are responsible for the views expressed here or for any errors and omissions in the final product.

References
[1] Susila WR. 2004. Contribution of Oil Palm Industry to Economic Growth and Poverty Alleviation in Indonesia. *Jurnal Litbang Pertanian* **23**(3): 107–14
[2] Petrenko C, Paltseva J, Searle S. 2016. *Ecological Impacts of Palm Expansion in Indonesia.* White Paper. The International Council on Clean Transportation. July 2016. Washington, DC 20005 USA
[3] Saghaian S. 2010. The Impact of the Oil Sector on Commodity Prices: Correlation or Causation? *Journal of Agricultural and Applied Economics* **42**(3): 477-485
[4] Muhammad A, Kebede E. 2009. The Emergence of an Agro-Energy Sector: Is Agriculture Importing Instability from the Oil Sector? *Choices* **24** (1): 12–15
[5] Campiche J, Bryant H, Richardson J, Outlaw J. 2007. Examining the Evolving Correspondence between Petroleum Prices and Agricultural Commodity Prices, Paper presented at the American Agricultural Economics Association Annual Meeting, Portland, July 29 – August 1 2007
[6] Yu TH, Bessler D, Fuller S. 2006. Cointegration and Causality Analysis of World Vegetable Oil and Crude Oil Prices. Paper presented at the American Agricultural Economics Association Annual Meeting, Long Beach, July 23–26, 2006
[7] Hertel T, Rose S, Tol R. 2009. Land Use In Computable General Equilibrium Models: An Overview. In Economic Analysis of Land Use in Global Climate Change Policy. Routledge Explorations in Environmental Economics. United Kingdom: Routledge
[8] Elobeid A, Tokgoz S, Hayes D, Babcock B, Hart C. 2006. The Long-Run Impact of Corn-Based Ethanol on the Grain, Oilseed, and Livestock Sectors with Implications for Biotech Crops. *AgBioForum* **10**(1): 11-18
[9] Fabiosa JF, Beghin JC, Dong F, Elobeid A, Tokgoz S, Yu TH. 2009. Land Allocation Effects of the Global Ethanol Surge: Predictions from the International FAPRI Model. Working Paper, Center for Agricultural and Rural Development, Iowa State University
[10] Mills K, Dicks MR, Lewis D, Moulton R. 1992. Methods for Assessing Agricultural-Forestry Land Use Changes. OAES Research Report P-928, Oklahoma State University
[11] Du X, Hennessy D, Edward WA. 2008. Does a Rising Biofuels Tide Raise All Boats? A Study of Cash Rent Determinants for Iowa Farmland under Hay and Pasture. *Journal of Agricultural & Food Industrial Organization*6:1-23
[12] Morton DC, DeFries R, Shimabukuro Y, Anderson O, Arai E, Espirito-Santo F. 2006. Cropland Expansion Changes Deforestation Dynamics in the Southern Brazilian Amazon. Proceedings of the National Academy of Sciences of the United States of America, **103**: 14637–14641
[13] Rosegrant M. 2008. Biofuels and Grain Prices: Impacts and Policy Responses. Testimony for the U.S. Senate Committee on Homeland Security and Governmental Affairs, May 7 2008. Washington, DC
[14] Gellert P. 2010. Extractive Regimes: Toward a Better Understanding of Indonesian Development. *Rural Sociology* **75**(1): 28-57
[15] Pramudya EP, Hospes O, Termeer JAM. 2017. Governing the Palm Oil Sector Through Finance: The Changing Roles of the Indonesian State. *The Bulletin of Indonesian Economic Studies* **53**(1):57-82
[16] Obidzinski K, Andriani R, Komarudin H, Andrianto A. 2012. Environmental and Social Impacts of Oil Palm Plantations and Their Implications for Biofuel Production in Indonesia. *Ecology and Society* **17**(1): 25. http://dx.doi.org/10.5751/ES-04775-170125

Replanting policy of Indonesian palm oil plantation in strengthening the implementation of sustainable development goals

Jan Horas Veryady Purba[1]

[1]Sekolah Tinggi Ilmu Ekonomi Kesatuan, Jl. Ranggagading No 1, Bogor, Indonesia, 16143

*E-mail: janhorasvpurba@gmail.com

Abstract. Indonesia's palm oil industry is one of the National strategic commodities. This industry is inclusive and has a very broad impact on Indonesia's economic development, regional development, poverty alleviation, the country's foreign exchange resources and attempted to help overcome Indonesia's balance of payments deficit. Along with the end of the first 25 years of age in most of oil palm plantations in Indonesia, the replanting policy has become an important momentum, because it concerns the future of the Indonesian oil industry, then continued with the second milestone of the 2040s. With replanting, Indonesia also improves productivity and is followed by improvements in technical culture. Indonesian oil palm productivity is directed to reach the 35-26 scenario, which is the production of 35 tons of FFB per ha and 26 percent yield so that Indonesia's palm productivity will shift from productivity levels of 4.4 tons of CPO per hectare to 9 tons of CPO per hectare. With this level of productivity, Indonesia's CPO production will increase from 77.17 million tons in 2050 to 94.64 million tons. The purpose of this study is not merely to look at oil palm plantations from an economic standpoint because in a broader scope, oil palm plantations have four multi-function agriculture, namely green functions, blue functions, yellow functions, and white functions. The replanting policy of Indonesian palm oil also plays a role in strengthening Indonesia's palm oil industry in the achievement of the Sustainable Development Goals, which includes the economic fields (8 SDGs), social fields (6 SDGs) and the environment (3 SDGs).

I. Introduction

The oil palm plantation has played an important role in encouraging Indonesia as the world's largest CPO producer and exporter. This also shows one of the strategic roles of the People's Palm Oil Plantation in the Indonesian economy, both in the acquisition of foreign exchange, regional development. Poverty alleviation, increasing regional GRDP and its important role in "feeding the world" [1].

In 1980, Indonesian palm plantations began to grow increasingly rapidly. In the 1990s, which was driven by smallholder oil palm plantations, both plasma, and independent farmers. This success has driven the acceleration of Indonesia's oil palm plantations to reach 14.3 million ha in 2018, and more than half (53%), are smallholder oil palm plantations [2]. This needs to be underlined, that smallholder oil palm plantations have an important role in the Indonesian palm industry, bringing Indonesia as a major producing country with a 55.7 percent share of world CPO production (Malaysia is in second

IOP Conf. Series: Earth and Environmental Science **336** (2019) 012012 doi:10.1088/1755-1315/336/1/012012

place with a 28.9 percent share) in the same year, [3] and a major CPO exporter on the global market where in 2017, Indonesia's palm oil exports reached USD 31.05 billion [4].

The current issue is the moratorium on Indonesian oil palm plantations as stipulated in Presidential Instruction (Inpres) No. 8 of 2018, which covers two main things, namely Evaluation of Licensing for Oil Palm Plantations and Increased Productivity of Oil Palm Plantations. This policy leads to a replanting strategy, which is related to the Indonesian palm oil industry in the future. This policy is also not just a natural development to replace oil palm plants that have reached a 25-year cycle, as happened in a number of provinces in Indonesia but are carried out by design, and no longer practice expansion. In the perspective of vegetable oils in the global market, Indonesia also faces a variety of very strong world pressures in various dimensions. While the world market is in dire need of CPO and even becomes the largest source of vegetable oil in the world.

This shows the point of the problem faced in saving the future of the Indonesian palm oil industry. When viewed in the most extreme conditions replanting failure can potentially reduce 4 million hectares of Indonesian palm oil, and the area of Indonesian palm oil plantations will decline drastically, and will be under the position of Malaysia and no longer as a major CPO producer on the world market.

At present, the productivity of national oil palm plantations has only reached 3.69 tons CPO/ha. State oil productivity (PTPN) of 3.97 tons CPO/ha; Private plantation productivity is 3.37 tons CPO/ha and the productivity of new oil palm plantations reaches an average of 2.82 tons CPO/ha.

Indonesian palm oil commodities not only play an economic role, but these commodities are also seen as Indonesia's strategic industry, and cannot be separated from various aspects of national development. Therefore, in addition to the internal problems facing the Indonesian palm industry, the policy of developing the Indonesian oil industry must also be in line with world development issues, namely sustain development goals (SDGs). With these two aspects, the research question is: how is the replanting policy on oil palm plantations in Indonesia and how is the implementation of SDGs in the development of the palm oil industry in Indonesia?

The goal of this study was (1) to analyze replanting policy of oil palm plantations in Indonesia and (2) to analyze the support for the development of the Indonesian palm oil industry in implementing SDGs.

2. Materials and methods

2.1. Conceptual analysis

The analytical method in this study is a descriptive method, and attempts to answer the research problems above from various aspects comprehensively. There are two important things in this study, namely the replanting and support of the palm oil industry in the implementation of SDGs. Conceptual analysis of replanting can be summarized as follows.

Oil palm plantations have a productive cycle of up to 25 years, or one production cycle. At the age of 3 to 4 years, this plant can already be harvested. High productivity will be achieved at the age of 8 to 20 years, and after that it begins to decline. Thus it is easy to see that the production of oil palm is inverted U shape. Until the age of 25 years, then enter the replanting stage.

From an economic standpoint there are two ways to increase Indonesia's palm productivity, the first way is through increasing the partial productivity productivity of oil palm plantations in harvested area that shift the productivity curve of oil palm plantations currently (P1) to the new productivity curve (P2). In practice, this first method is carried out through the best management practices of existing TM gardens.

The second way is through the use of seeds that are superior when replanting. Until now, some of the seed varieties produced by palm oil companies have a productivity potential of around 35 tons of FFB and 26 percent yield or can produce around 9 tons of oil per hectare. (figure 1)

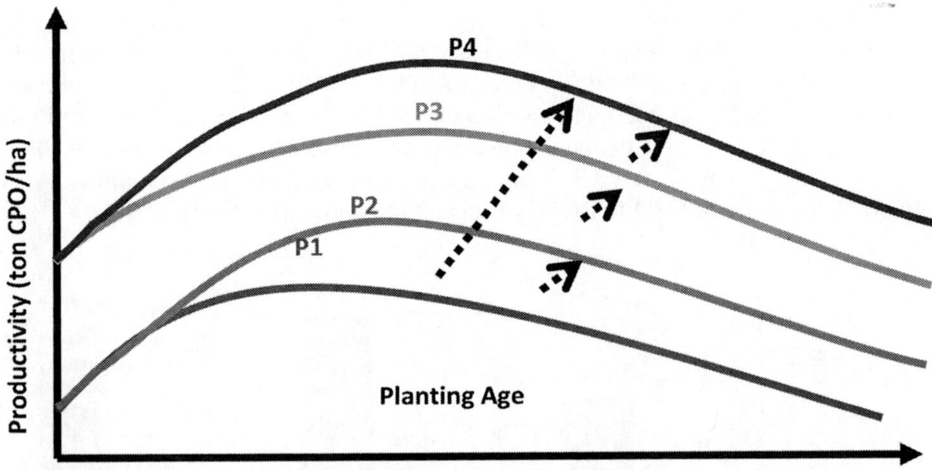

Figure 1. Increased productivity due to improvement of technical culture (P2 and P4) and total productivity factors (P3) Source:[5]

In order for the ideal plant composition, the standard norm of replanting each year is an average of 4 percent of the area of the garden. So that nationally with an area of 10 million hectares, it is expected that there will be around 400 thousand hectares replanting every year. The second way will shift the productivity curve from P1 to P3.

The combination of these two methods, namely the improvement of technical culture in harvested area and replanting using superior seeds, will result in continuous increase in total productivity. This combination will shift the productivity curve from P1 to P4 and so on. Thus, the replanting program is a new milestone in advancing the Indonesian palm industry, and this is very relevant at this time in line with the current moratorium policy so that the replanting program is an implementation of policies that will determine the future of the Indonesian oil industry.

2.2. Data

Type of data used in this study are secondary data on Indonesian palm oil statistics from the Ministry of Agriculture and BPS until 2017/2018, and the method of data analysis used is descriptive analysis by exploring relevant data on government policies and reviewing the results of the results of empirical studies from UNDP , GAPKI, PASPI, and other sources of reference.

3. Result and discussion

3.1. Result

The purpose of this study was to analyze replanting policy of oil palm plantations in Indonesia and to analyze the support for the development of the Indonesian palm oil industry in implementing SDGs.

a. ReplantingPolicy

The PASPI projection [6] shows that the combination method will increase national oil productivity from 4.4 tons of CPO/ha in 2020 to 7.42 tons of CPO/ha in 2050. Productivity of smallholder oil palm plantations will increase from 3.39 tons CPO/ha (2020) to 6.38 tons CPO/ha (2050). While the productivity of the private oil palm plantation will increase from 4.93 tons CPO/ha (2020) to 8.32 tons CPO/ha (2050). While the productivity of stated owned oil palm plantation will increase from 4.16 tons CPO/ha (2020) to 7.3 tons CPO/ha (2050). (Figure 2). If it is assumed there is no expansion of oil palm plantations, and is only affected by an increase in the level of productivity as the data above Indonesia's CPO production will reach 45.76 million tons in 2020 and 77.17 million tons in 2050. Malaysia's CPO production projection is expected to reach 21.36 million tons in 2020 and 40.82 million tons in 2050.

The consumption of vegetable oil per capita in the world will be remembered, along with the increase in income (GDP per capita), will affect the demand for world vegetable oil. The world population in 2020 will reach 7.49 billion people and 9.078 billion in 2050. It is estimated that global CPO consumption in 2020 and 2050 will each reach 73.47 million tons and 154.84 million tons.

This shows that world CPO demand continues to grow. The lack of global CPO supply will encourage the development of new CPO producing countries, both in Asia and in Africa and other ROWs. In 2020, the share of CPO in ROW will reach 8% and will grow to 23%.

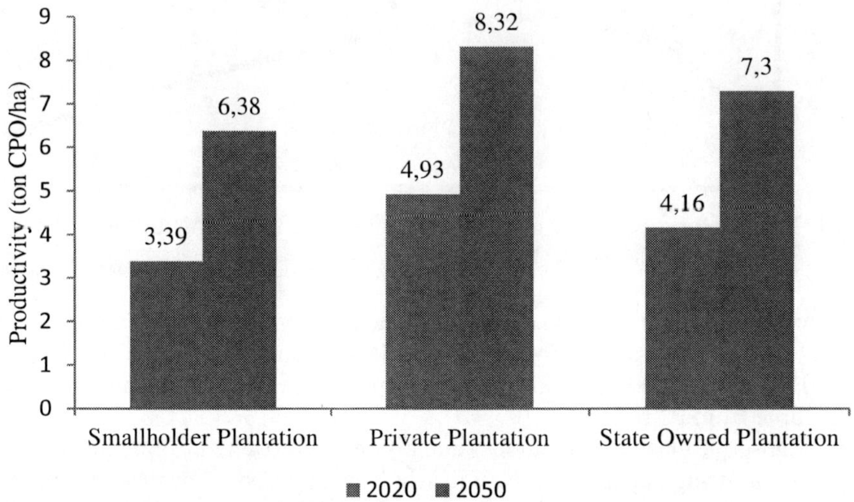

Figure 2. Projected productivity of smallholder, state and private palm oil plantations in 2020 and 2050

With better preparation, then in the 2040s, it is the second milestone in improving the productivity of Indonesian CPO (along with the second replanting, after the 25-year cycle ends). Indonesian palm productivity is directed to reach the 35-26 scenario, which is the production of 35 tons of FFB per ha and 26 percent yield so that Indonesia's palm productivity will reach 9 tons of oil per hectare, and Indonesia's CPO production can be increased from 77.17 million tons in 2050 to 94.64 million tons.

b. The support of the Indonesian palm oil industry in implementing SDGs

The World Bank (World Bank, 2012) proposes a simple concept of Growth, Green, Inclusive. Economic growth must be green but also inclusive (broad impact). Conversely, environmental conservation (green) must guarantee economic growth and be inclusive. Therefore, development management needs to guarantee all three things. The big objectives of SDGs can be classified into the following three fields.

Table 1. SDGs in the perspective of sustainability, economic, social and ecological dimensions

No	Purpose	SDGs
1	Economy	(1) removing poverty in various forms and all places, (2) removing hunger, malnutrition and building inclusive food security, (3) building sustainable energy, (4) inclusive and sustainable economic growth, (5) sustainable industrialization, (6) reduction of inequality, (7) consumption and sustainable production, and (8) global cooperation in sustainable development.
2	Social	(1) health, (2) inclusive quality education, (3) gender equality, (4) availability of clean water that is inclusive, (5) city development that is inclusive and selling and (6) inclusive social peace.
3	Ecology	(1) overcoming global climate change, (2) sustainability of the sea and waters, and (3) forest sustainability and biodiversity (biodiversity)

Source: [13]

Given the important role of the government in realizing sustainable development [14], [15] proposes that the concept of sustainable development is not only 3-P (Profit, Planet, People) but 4-P (Profit, Planet, People, Policy) Government policy variables are considered very important in addition to determining development policies also specific to each country. In the implementation of SDGs, it will certainly vary in each country which in this case is determined by government policies in each country.

3.2. Discussion

a. Replanting is a Milestone to Support Implementation of SDGs

Replanting of Indonesian palm oil plants provides a high opportunity for Indonesia to maintain Indonesia's role as the largest producer and exporter of oil palm to meet the demand for world vegetable oil. This is in line with the policy of intensification through increasing productivity two to three times the current level of productivity.

However, replanting is not limited to replacing new oil palm plants. Replanting has a more strategic role in Indonesia's new economic development. This is supported by the following important things: (1) The national palm oil industry is inclusive and has a very broad impact on Indonesia's economic development, as well as regional development and one of the country's foreign exchange sources; (2) Indonesian palm oil in a wider scope has four functions, namely green function, blue services/function, yellow services/function and white function [7], [8].

"Indonesian Palm Oil: More Sustainable" is a commitment that Indonesia is currently developing, both in the long-term goal of developing the Indonesian Palm Industry, as well as an effort to succeed in Palm Oil Diplomacy to the European Union and at the same time converting negative accusations from anti palm oil NGOs to the development of the oil industry Indonesian palm oil.

In a broader perspective, the development of a sustainable Indonesian palm oil industry is in line with and supports the success of Indonesia in achieving the Sustainable Development Goals (SDGs), whose scope is wider than the Millennium Development Goals that have been implemented in the previous time period.

Indonesia's palm oil industry is one of Indonesia's strategic commodities, both in economic development, regional development and increased GDP as well as poverty reduction in the center of palm oil producers, and is able to overcome Indonesia's trade balance deficit through export activities.

Sustainable development is a more inclusive and quality development process. This development paradigm has been made as a United Nations as a 2015-2030 global development platform known as the 2030 Sustainable Development Goals. As a global development platform for every country, sector, region, industry is expected to adopt and contribute to sustainable development.

Sustainable development [1], [10]) has three pillars namely economic, social and ecological aspects which are often called 3-P (profit, people, planet) (figure 3).

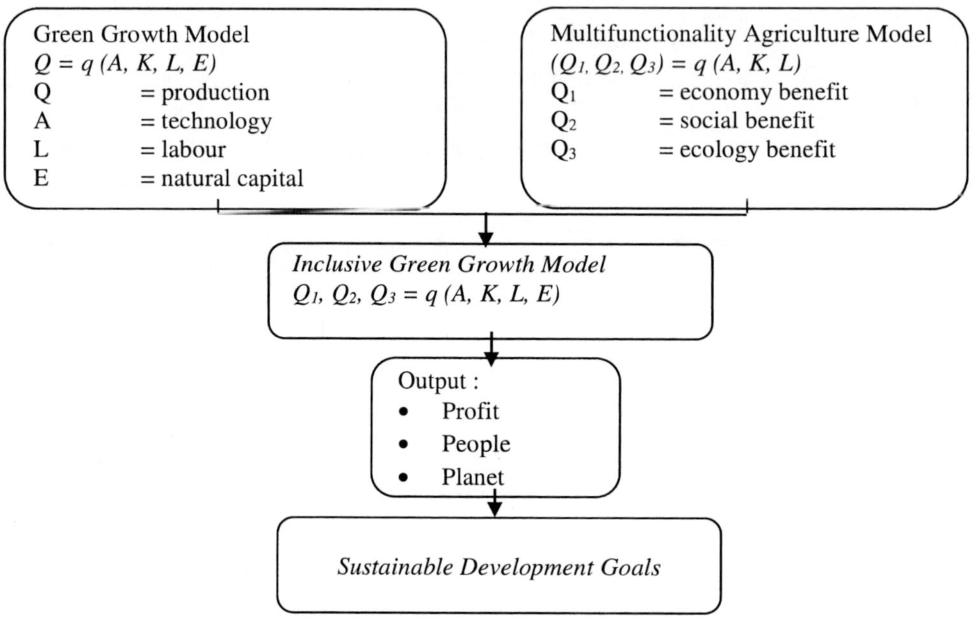

Figure 3. Research Teorotical Concept of Sustainability Dvelopment Goals in Oil Palm Industry [10a]

Sustainable development not only enough or exclusive produces economic benefits but also provides social benefits and ecological benefits across generations. Sustainable development is a relative matter and is country, sector and industry specific [11], [12]). Related to the dimension of sustainable development, the development of Indonesian palm oil plantations includes 3 important pillars, namely sustainability in the economic, social and ecological dimensions.

The paradigm of sustainable development is the root of SDGs. The goals of the SDGs are inseparable from each other. Even to achieve these SDGs requires synergy between fields (ICSU/ISSC, 2015). Elimination of poverty for example requires synergy between inclusive food security (SDGs 2), inclusive economic growth (SDGs 8), reduction of inequality (SDGs 10) even with climate change mitigation (SDGs 13).

b. Contribution of palm oil industry

The contribution of the palm oil industry to the achievement of the SDGs is related to the multifunction of the plantation itself. Multifunction in the agricultural sector, including plantations, has long been taking place across generations since the beginning of human civilization on earth. But the issue of multifunctional agriculture began to be discussed at the international level at the Rio Earth Summit in 1992. The Declaration of the Committees of the Ministers of Agriculture The Organization of Economic Coordination Development [16] defines multifunctional agriculture as follows "Beyond its primary function of producing food and fiber, providing environmental benefits such as land conservation, sustainable management of renewable natural resources and preservation of biodiversity, and contributing to the socio-economic viability of many rural areas. Agriculture is multifunctional when it has one or several functions in addition to its primary role of producing food and fiber. "This means that in addition to the main function of the economic function (producing food and fiber), agriculture also has social functions and ecological functions.

The **green functions** consist, among others, of landscape management and the landscape of landscape amenities, wildlife management, wildlife creation and animal welfare, the maintenance of biodiversity, recycling and limitation of carbon sinks. Other public beneficiaries that can be created by agriculture are the **blue services** and contain water management, improvement of water quality, flood control, water harvesting and creation of (wind) energy. A third kind are **yellow services** and peripheral to the role of farming for rural cohesion and vitality, ambience and development, exploiting cultural and historical heritages, creating a regional identity and offering hunting, agro-tourism and agro-entertainment. Finally, many authors acknowledge the **white function** produced by agriculture, such as food security and safety. The four functions of agriculture/plantations internationally are often called 3-P namely profit (white function), people (yellow service), and planet (green function and blue service). (Figure 3)

Multifunctional plantations in Indonesia are also recognized in Law No. Article 4 of the 18 of 2004 (has been amended to Act No. 39/2014), that plantations have three functions, namely: 1) economic function (increasing prosperity and prosperity of the people and strengthening regional and national economic structures), 2) ecological functions (increasing conservation soil and water, carbon sequestration, oxygen providers and buffer protected areas and 3) socio-cultural functions (as an adhesive and unifying nation).

Empirically the economic contribution of the palm oil industry has been widely proven by various experts, including encouraging economic development, foreign exchange sources and state revenues [17], [11], [18]. Judging from the forward and backward input-output linkages and various multiplier indicators, it shows that oil palm and downstream industries are one of the locomotives of the national economy [19]. As a foreign exchange source, Indonesia's export of palm oil industry products is the largest source of foreign exchange (net exports) in non-oil and gas exports [20].

The increase in palm oil production in the center of oil palm plantations has led to a significant increase in the district's GDP [21], which then has an impact on the economic development of the region concerned. Economic growth driven by increased palm oil production is not only enjoyed by those involved in oil palm plantations but also enjoyed by people who are not directly involved in the production process of oil palm plantations [19]. Increased income of oil palm farmers is higher than other commodity farmers [22]. Even the economic benefits of palm oil are also enjoyed by the European Union community. CPO imports carried out gave great benefits both to GDP, government revenues and the work opportunity of the European Union [23].

The contribution of the palm oil industry in the social aspects has also been empirically proven, among others, its role in rural development (improving quality of life) and poverty reduction [24]. As a pioneer sector in remote areas, oil palm plantations also increase the availability of rural infrastructure and increase the availability of education and health facilities. In aggregate, the increase in national palm oil production decreases rural poverty [25], [26] in the centers of oil palm plantations such as North Sumatra, Riau, South Sumatra and Central Kalimantan, which significantly reduces poverty [27].

Various studies also prove that the ecological role of oil palm plantations includes preserving the cycle of carbon dioxide and oxygen (photosynthesis, which absorbs carbon dioxide from the Earth's atmosphere and produces oxygen to the earth's atmosphere), restoration degraded land soil and water conservation, increased biomass and carbon stock land, reducing greenhouse gas emissions/peatland restoration. Each hectare of oil palm plantations absorbs carbon dioxide from the earth's atmosphere by 161 tons/ha and produces 18.7 tons of oxygen/ha [28], [29], [30]. Oil palm plantations also increase biomass (organic material) land which is increasing with older plants [31]. Oil palm plantations on peatlands also reduce greenhouse gas emissions [32], [33], [34]. In terms of the role of the various water hydrological indicators such as evapotranspiration, groundwater reserves, forwarding of rainfall to the soil surface, the rate of infiltration of the solum layer and humidity [29] between oil palm plantations and forests are relatively the same.

The environmental services produced by oil palm plantations such as oxygen cycle sustainability, the sustainability of the hydrological cycle and the sustainability of the carbon dioxide cycle are important parts of global ecosystem functions. Carbon dioxide released by the global community and its activities that consume fossil fuels such as industry, transportation, housing and others, by oil palm plantations are absorbed (through photosynthesis) and then stored in the form of biomass and produced oxygen for human life in the earth. The wider and more widespread palm oil plantations are increasingly large and the more widespread the absorption of carbon dioxide, the production of biomass and the production of oxygen from oil palm plantations.

Thus, the contribution of the palm oil industry in achieving Indonesia's SDGs 2030 covers economic, social, and environmental aspects. Indonesian palm oil plantations that are currently developing in 190 districts in the country will contribute significantly to the achievement of the SDGs 2030, especially the districts/provinces of oil centres in Indonesia.

c. *Replanting roadmap in Indonesia*

Malaysia's success story in replanting was carried out in the period 2000-2005. This success is supported by national policies (MPOB) and requires the replanting problem as an imperative replanting. By consciously, replanting aims to shift the productivity curve upwards or to improve the level of productivity. This commitment is also reflected in replanting scenario with accelerated replanting.[9]

The replanting strategy of oil palm plantations in Indonesia includes two important things, namely (1) strategies to increase productivity, to increase crop productivity by improving technical culture, and (2), replanting strategies with the latest improved varieties (productivity of about 8 tons per hectare) for non-crop productive old age. Increasing palm oil production through improved oil productivity through the improvement of technical culture and sustainable replanting, will bring Indonesia's oil palm plantations to a more quality and sustainable phase, namely in the capital-driven phase as well as the innovation-driven phase.

In 2017, the President of the Republic of Indonesia was directly involved in initiating this replanting movement in several palm oil centers in Indonesia, accompanied by financial support especially for smallholders. However, the realization of replanting has not run as expected, and is not supported by the availability of superior seeds.

Ideally, assuming a 25-year palm oil crop production cycle requires replanting around 4 percent per year, so the ideal composition is immature area 12 percent, and harvested area 88 percent (table 2).

Table 2. Age composition of oil palm plants and ideal composition (%)

	Immature area	Harvested Area		
		4-7 years old	8-15 years old	16-25 years old
Nasional	20	21	38	21
Ideal	12	16	32	40

Sources: PASPI, 2016 (processed)

The current condition of unharvested area reaches 20% and is still far from ideal conditions (12%). A replanting roadmap for Indonesian palm oil plantations is very necessary accompanied by a replanting acceleration program in the next 5 years.

4. Conclusion

In line with the objectives of this study, the conclusions are as follows:

The replanting policy in Indonesia emphasizes 2 important strategies, namely to increase productivity through the improvement of technical culture and improved varieties. This strategy is directed to reach the 35-26 scenario, where the level of productivity of Indonesian palm oil plantations will reach an average of 9 tons of CPO per hectare, and the potential for increasing Indonesia's CPO production can reach 94.64 million tons in 2050.

The development of a sustainable Indonesian palm oil industry is in line with and supports the success of Indonesia in achieving the Sustainable Development Goals (SDGs). The contribution of the palm oil industry in achieving the Indonesia's SDGs 2030 covers economic, social and environmental aspects. Ideally, the replanting area is 440,000 ha per year or 4 percent of the total area of Indonesian oil palm plantations. The lagging area that should be slammed now reaches 20% and is very far compared to the ideal condition of 12%. Therefore, it is necessary to prepare The Indonesia oil palm replanting roadmap and be accompanied by replanting acceleration programs in the next 5 years. With this comprehensive paradigm, the Indonesian palm oil industry continues to grow in a sustainable perspective.

Acknowledgment

I would like to thank Prof. Dr. Moermahadi Soerja Djanegara, Dr. Iriyadi, Mr. Nusa Muktiadji for their valuable support in this research. I also would like to extend many thanks to Dr. Tungkot Sipayung as the Executive Director of PASPI, which has facilitated this research and supported me to attend the International Conference.

References

[1] PASPI, 2016. Perkebunan Sawit Rakyat dan Peran Strategisnya di Masa Mendatang. *Jurnal PASPI Monitor* Vol. II No. **15** : p 369-374

[2] Kementerian Pertanian RI, 2018. "Kementerian Pertanian: Lahan Sawit Indonesia Capai 14,03 Juta hektare", Tersedia pada https://ekonomi. kompas.com/read/2018/02/26/203000426/kementerian-pertanian--lahan-sawit-indonesia-capai-14-03-juta-hektare

[3] USDA, 2018. Palm Oil production. Available in https://www.indexmundi.com/agriculture/?commodity=palm-oil&graph=production

[4] GAPKI 2018. Perkembangan mutakhir Industri Minyak Sawit Indonesia. Available in https://gapki.id/news/3971/perkembangan-mutakhir-industri-minyak-sawit-indonesia

[5] PASPI. 2016. Replanting Sawit dan Pembangunan Ekonomi Baru. *Jurnal PASPI Monitor. Analisis isu Strategis Sawit.* Vol. 2 No. **33** : p 487-494

[6] PASPI, 2014. Industri Minyak Sawit Indonesia Menuju 100 Tahun NKRI. Bogor. PASPI. Bogor

[7] Aldington, T.J. (1998), "*Multifunctional Agriculture: a Brief Review from Developed and Developing Country Perspectives*", unknown status. FAO Agriculture Department, Internal Document.2

[8] Huylenbroeck, G. V.; V. Vandermulen, E. Mette Penningen, A. Verspecht. 2007: Multifunctionality of Agriculture: A Review Definition, Evidence and Instruments. Living Review in Landscape Research **1**: (2007) : 3

[9] Idris Omar,Azman Ismail and Chang Lin Chong. Improving Productivity: The Replanting Imperative. *Oil Palm Industry Economic Journal* (Volume **1**/2001) : 21-29

[10] World Bank, 2012. Inclusive Green Growth: The Pathway to Sustainable Development. The World Bank Washington DC

[11] Sipayung, T, & Purba, J.H.V. (2015). *Ekonomi Agribisnis Minyak Sawit.* Palm Oil Agribusiness Strategic Policy Institute. Bogor

[12] Sato Y. 1997. *The Palm Oil Industry in Indonesia: Its Structural Changes and Competitiveness.*

In Waves of Change in Indonesia's Manufacturing Industry (ed: M.E Pangestu and Y. Sato). Institute of Developing Economics. Tokyo
[13] Moon, W. (2012). *Conceptualizing Multifunctional Agriculture from a Global Perpective.* Departement Agribusiness Economics Southern Illinois University
[14] PASPI. 2016. Sustainabilitas Ekspor Minyak Sawit Indonesia Sebagai Industri Strategis Nasional. *Jurnal PASPI Monitor. Analisis isu Strategis Sawit.* Vol. **2** No. 32 : p 479-486
[15] Feher, I *and* J, Beke. 2013. *The Rational Of Sustainable Agriculture.* Iustum Aequum Salutare IX:2013(3) p 73-87
[16] Moon, W. (2012). *Conceptualizing Multifunctional Agriculture from a Global Perpective.* Departement Agribusiness Economics Southern Illinois University. Carbondale IL 62901.
[17] OECD. 2001. Multifunctionality Towards an Analytical Framework. OECD. Paris
[18] Tomich, T.P dan Mawardi, M. S. 1995: *Evolution of Palm Oil Trade Policy in Indonesia* 1978-1991. Elaeis Volume **7** (1): P 87-102
[19] Susila, W. R. 2004. *Impact of CPO Export Tax on Several Aspects on Indonesia CPO Industry.* Oil Palm Industry Economic Journal **4** (2): 1-13
[20] Amzul, R. 2011: The Role Palm Oil Industry In Indonesia Economy and Its Export Competitiveness. PhD Dissertation. University Of Tokyo
[21] PASPI, 2014 : *Industri Minyak Sawit Indonesia Berkelanjutan*. Peranan Industri Minyak Kelapa Sawit Dalam Pertumbuhan Ekonomi, Pembanguunan Pedesaan, Pengurangan Kemiskinan dan Pelestarian Lingkungan. *Palm Oil Agribusiness Strategic Policy Institute*. Bogor
[22] PASPI, 2015. Prognosa Dampak Pungutan Ekspor Minyak Sawit Terhadap Industri Minyak Sawit Indonesia. Jurnal PASPI Monitor Vol. **1** No. 13 : p 91-96
[23] Greig-Gran, M. 2008: The Cost Avoiding Deforestration Update of The Report Prepared For Stern Review Economic of Climate Change. International Institute For Environment And Development
[24] Europe Economics. 2014. *The Economic Impact of Palm Oil Imports In The EU.* Europe economics, chancery house, 53-64 chancery lane. London WC2A IQU
[25] Sumarto, S and A. Suryahadi. 2004: Trade, Growth and Poverty in Indonesia. National Conference Of The University Outreach Network. Bogor
[26] Susila, W. R. 2004. *Impact of CPO Export Tax on Several Aspects on Indonesia CPO Industry.* Oil Palm Industry Economic Journal **4** (2): 1-13
[27] Joni, R. 2012. Dampak Pengembangan Biodiesel dari Kelapa Sawit Terhadap Kemiskinan, Pengangguran dan Pertumbuhan Ekonomi Indonesia. Disertasi. IPB. Bogor
[28] PASPI, 2014. *Industri Minyak Sawit Indonesia Berkelanjutan* : Peranan Industri Minyak Sawit dalam Pertumbuhan Ekonomi, Pembangunan Pedesaan, Pengurangan Kemiskinan dan Pelestarian Lingkungan. PASPI. Bogor
[29] Henson I. 1999. *Comparative Ecophysiology of Palm Oil and Tropical Rainforest.* Oil Palm and Environment A Malaysian Perspective. Malaysian Oil Palm Brower Council. Kuala Lumpur
[30] Harahap, I. Y, Y Pangaribuan, H. H Siregar, E Listia. 2005: Lingkungan Fisik Perkebunan Kelapa Sawit. PPKS. Medan
[31] Fairhurst. T. and R. Hardter, 2004: Oil Palm: Management for Large and Sustainable Yields. Oxford Graphic Printers, Pte Ltd
[32] Chan, K. W. 2002: Oil Palm Carbon Sequestration and Carbon Accounting: Our Global Strength. MPOA
[33] Murayama, S. and Baker, Z. A. 1996. *Decomposition of Tropical Peat Soils. Decomposition Kinetic of Organic Matter of Peat Soils*. Japan Agricultural Research. Quarterly. **30**: 145-151
[34] Melling, L. Hatano, R. and Goh, K. J. 2005. *Soil CO_2 Flux From Ecosystem in Tropical Peat Land of Serawak*. Malaysia. Tell us. **57**: 1-11
[35] Sabiham, S. 2013. Sawit dan Lahan Gambut dalam Pembangunan Kebun Kelapa Sawit di Indonesia. Himpunan Gambut Indonesia

IOP Conf. Series: Earth and Environmental Science **336** (2019) 012013 doi:10.1088/1755-1315/336/1/012013

The impact of ISPO certification on economic, social and environmental aspect in the palm oil plantation

Rodhiah[1], Ifdal[2], Ira Wahyuni Syarfi[2], and Hasnah[2]

[1]Student at Graduate Program of Agricultural Economics, Faculty of Agriculture, Andalas University, Campus Limau Manih, Padang 25163, Indonesia
[2]Lecturer at Graduate Program of Agricultural Economics Faculty of Agriculture, Andalas University, Campus Limau Manih, Padang 25163, Indonesia

*E-mail: rodhiah2ty@gmail.com

Abstract. The Indonesian Government requires all Palm Oil Plantation Companies to have an *Indonesian Sustainable Palm Oil (ISPO)* certificate. Indonesia issued a policy known as ISPO so that Indonesia management of palm oil plantations pay attention to aspects of economy, social, and environment. ISPO certification obligations are financially burdensome for businessmen because large certification costs certainly affect the company's production costs, sales and revenues of palm oil companies. The research objective was to describe the process to get ISPO certification for Palm Oil Plantation Companies, to determine the impact of the application of ISPO certification on the economic, social, and environmental aspects of palm oil companies. Data analysis was carried out by comparing the impact ISPO implementation to PT. Perkebunan Nusantara VI Ophir at the time before (2013, 2014, 2015) and after (2016, 2017) ISPO certification. The results showed that stages of procedures for recognizing PT. Nusantara Plantation Ophir VI's ISPO certification is in accordance with the mechanism as stipulated in Minister of Agriculture Regulation No. 11/Minister of Agriculture Regulation/OT.140/3/2015. The impact of the application of ISPO certification on PT. Perkebunan Nusantara VI Ophir has an effect on increasing production costs, production levels, selling prices, sales, and income. The results of the calculation of the revenue to cost ratio, obtained by the R/C value is greater than one, which means that the application of ISPO is financially efficient over the costs incurred. The impact of ISPO implementation aside from the financial aspects received by the company after obtaining ISPO certification affects the aspects of operational, social community relations, the company's concern for the environment, as well as trading CPO and Kernel. ISPO certification received by PT. Perkebunan Nusantara VI Ophir affect operational aspects, relationships with the social community, the company's concern for the environment, as well as trading CPO and Kernel.

1. Introduction

1.1 Background of the study

The agricultural sector has an important role in Indonesia's economic activities. One sub-sector that has a large potential is the plantation sub-sector. One of the plantation commodities that has an important role in economic activities in Indonesia is palm oil. Palm oil is one of Indonesia's export commodities which is quite important as a foreign exchange earner in addition to oil and gas. Indonesia is the world's largest producer and exporter of palm oil [1].

The development of the palm oil industry has caused controversy in the international community. On the one hand, the development of palm oil and the palm oil industry provides welfare for the community and the state, on the other hand, it has a social and environmental impact. The growing criticism of the development of palm oil plantations in the world has triggered the emergence of negative campaigns launched by a number of western countries. Several European countries and the United States have boycotted CPO products not being able to enter the EU market under the pretext of palm oil as a cause of an increase in Greenhouse Gases. These problems need to be addressed so that sustainable palm oil development can be realized. The indicator for management that is used in Indonesia at this time is the ownership of ISPO certificate (*Indonesian Sustainable Palm Oil*) [2].

Through ISPO, the Government of Indonesia wants to protect and promote sustainable palm oil plantation business in accordance with market demands, meet global demands and increase the competitiveness of Indonesian palm oil in the global market, as well as support Indonesia's commitment to reduce greenhouse gas emissions, and pay attention to the preservation of resources natural and environmental power. With the ISPO arrangement, it is expected that all palm oil plantation businessmen in Indonesia can increase awareness of the importance of producing sustainable palm oil which can affect the increase in competitiveness of Indonesian palm oil in the world market [3].

1.2 Problem of the study

Sustainable development should pay attention to 3 aspects (3P), namely *profit* (economy), *people* (social), and the *planet* (environment). However, businessmen tend to only take into account the economic aspect. Palm Oil plantations provide substantial revenues for the central and regional governments as a result of which much forest area is allocated for palm oil development [4]. The consequences of palm oil development include negative impacts on the environment and social conflict [5]. Non-environmentally friendly practices such as land clearing techniques with forest burning and uncontrolled waste disposal have created a bad image for the Indonesian palm oil industry [2].

Indonesia issued a policy known as ISPO so that Indonesia management of palm oil plantations pay attention to aspects of economy, social, and environment. However, businessmen tend to only take into account the economic aspect. The ISPO certification obligation turned out to be a polemic that financially burdened palm oil businessmen due to certification fees and bureaucratic obligations for obtaining ISPO approval. One of the main considerations for plantation companies to obtain an ISPO certificate or not to take care of an ISPO certificate is the high cost both for the fulfilment of the requirements and for the processing of the certificate. The cost of making a large certificate certainly affects the production costs incurred by plantation companies in producing palm oil.

Fulfilment of ISPO certification, in addition to generating costs, can also provide revenues received to plantation companies. And a wider market reach, as well as the level of income received by plantation companies. The impact of implementing ISPO aside from the financial aspects can be seen in terms of operations, social community relations, environmental concerns, and trading of CPO & Kernel products.

Based on the background and problem of the study, it is deemed necessary to conduct a more in-depth study, especially related to the impact of implementing ISPO certification on economic, social and environmental aspects for palm oil plantation companies.

1.3 Purpose of the study

The purpose of this study is:

1. Knowing the process to get ISPO certification for Palm Oil Plantation Companies?
2. Analyzing the impact of implementing ISPO certification on economic aspects for Palm Oil Plantation Companies?
3. Analyzing the impact of implementing ISPO certification on social and environmental aspects for Palm Oil Plantation Companies?

2. Research Method

2.1. Research location

The study area is selected by purposively based on the consideration that out of a total of 346 palm oil businesses in Indonesia that have received the ISPO certificate, 9 (nine) of them are oil palm businesses in West Sumatra, and the majority of them are palm oil plantation businesses located in West Pasaman Regency. Table 1 shows that in West Pasaman District, there are 5 (five) palm oil plantation companies that have received ISPO certificates and have applied ISPO. The research was conducted at PT. Perkebunan Nusantara VI Ophir Business Unit. This research was conducted for 1 (one) month starting from June 1 until June 30, 2018.

Table 1. List of Oil Palm Companies recieved ISPO certificate in Pasaman Barat

No	Palm Oil Plantation Companies	The Certified Unit	Certification Period
1	PT Agrowiratama	PKS Agrowiratama dan Kebun Sei Aur	16 Jan 2014 to 15 Jan 2019
2	PT Gersindo Minang Plantation	PKS PT GMP, Kebun GMP, Kebun PHP1, Kebun PHP2	16 Jan 2014 to 15 Jan 2019
3	PT Bakrie Pasaman Plantations	PKS Air Balam, Kebun Air Balam dan Kebun Sei Aur	17 Jun 2015 to 16 Jun 2020
4	PT. Perkebunan Nusantara VI (PKS Ophir)	PKS Ophir, Kebun UU Ophir dan Kebun UU Pangkalan Limapuluh Kota	5 Feb 2016 to 5 Feb 2021
5	PT. Bintara Tani Nusantara	1 PKS dan 2 Kebun PT. Bintara Tani Nusantara	6 Dec 2016 to 5 Dec 2021

Source : ISPO, 2017.

2.2. Data collection method

Data collected in this research consist of primary and secondary data. Primary data is obtained from direct observations in the fields and through interviews with key informants of plantation companies that responsible for implementing ISPO certification. Secondary data collection is carried out through literature studies, exploration of available literature and data, and reference books and the internet in accordance with the results of research on palm oil, as well as conducting in-depth interviews with officials of relevant agencies/institutions. This approach is carried out by collecting all documents containing secondary data in the relevant agencies such as; Plantation Office, Food Crop Agriculture Office, Industry and Trade Agency, Central Statistics Agency.

Research data covers several years before and after the application of ISPO certification to palm oil plantation companies. This research data is obtained from the data of PT. Perkebunan Nusantara VI Ophir, at the time before obtaining ISPO certification (2013, 2014, 2015) and after the implementation of ISPO certification (2016, 2017). Table 2 below shows the data variables, types of data and data sources.

Table 2. Variables, Data, and Data Sources

No	Variables	Data	Data Sources
1.	Management Process for Obtaining ISPO Certification	- The flow of the submission process and management receives ISPO certification - Application of ISPO principles & criteria to obtain ISPO certification	Key Informant Plantation Company
2.	Impact of the Implementation of ISPO Certification on economic aspects	- Production of FFB, CPO and Kernel (Kg) - The Productivity of FFB (Kg / Ha) - Selling Price of CPO and Kernel (Within Rp.) - CPO and Kernel (Kg) sales levels - Revenue from sales (In IDR)	Key Informant Plantation Companies / Trade Offices
3.	Impact of the Implementation of ISPO Certification on social	- Completeness of company documents - Plant maintenance activities - Post Harvest Work	Key Informant Plantation Company

No	Variables	Data	Data Sources
	and environmental aspects	- Total FFB production - Problems with conflict with the community - Relations with related stakeholders, the community, buyers - Number of accidents and fire cases - Environmental management and monitoring - Concern for the environment	

2.3. Data analysis

We applied descriptive approach to describe the process to get ISPO certification for Palm Oil Plantation Companies. Information on stages of the ISPO certification process was collected to find out the process of submitting and managing requirements to obtain ISPO certification. It was identified based on ISPO indicator criteria which must be applied by companies to obtain the certification.

The impact of applying ISPO certification at PT. Perkebunan Nusantara VI Ophir on economic aspects is described by comparing production costs, revenues from sales, and income before and after the implementation of ISPO certification. This can be seen by comparing the average amount of production costs that must be issued by the company, the level of revenue from sales, and the amount of income received by the Company at the time before and after the application of ISPO certification.

To calculate the income of palm oil plantation companies before and after the implementation of ISPO, we used the formula below:

$$\text{Profit} = TR - TC$$

Explanation:
Profit = Plantation profit
TR = *Total revenue*
TC = *Total Cost*

Total revenue (TR) is calculated using the formula:

$$TR = y \times Py$$

Explanation:
TR = Total revenue (Rp)
y = Total sales of CPO (Ton)
Py = Selling price of CPO (IDR/Ton)

To analyze the impact of implementing ISPO certification on the financial aspect of PT. Perkebunan Nusantara VI Ophir, it was conducted by calculating the ratio of revenue to production costs (*R / C ratio*) at the time before and after ISPO certification. As a measure of assessment, if the results of the *cost-revenue ratio* (R / C ratio) are greater than 1 (one), it means the application of ISPO certification is efficient for the costs incurred. If the R / C ratio is equal to 1 (one), it means that the *revenue* obtained is only enough to cover the *cost* so that in terms of financial and economic aspects is not profitable. If the R / C ratio is smaller than 1 (one), it means that the *revenue* obtained is not enough to cover *costs* so that the application of ISPO certification is not efficient from its financial aspects.

To find out the impact of implementing PT. Perkebunan Nusantara VI Ophir's ISPO certification on social and environmental aspects, we describe various data variables resulting from interviews with *key informants* and direct observations in the field related to following items: company documents; plant maintenance activities; the problem of conflict with the community; relationships with stakeholders, the community, and buyers; cases of fire and work accidents, as well as the company's concern for environmental management and monitoring. The thought framework that underlies this research can be seen in figure 1 below.

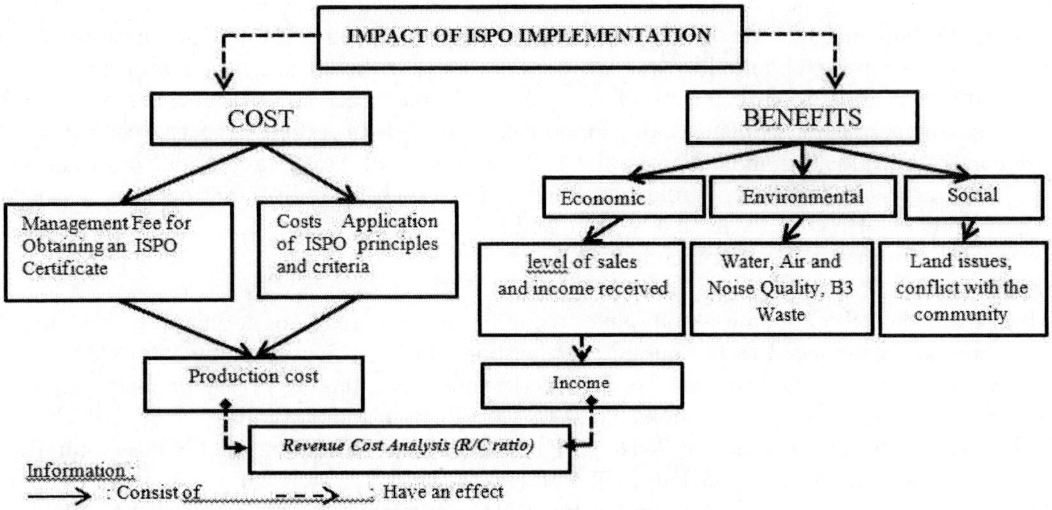

Figure 1. Schematic Framework

3. Results and discussion

The Ophir Business Unit is one of sixteen business units under the management of PT. Perkebunan Nusantara VI. PT. Perkebunan Nusantara VI is a State-Owned Enterprise (SOE) in the form of a Limited Liability Company with the working area of the Province of West Sumatra and Jambi.

PT. Perkebunan Nusantara VI Ophir is located in Kejorongan Sarik Kenagarian Koto Baru, Luhak Nan Duo District, West Pasaman Regency. The area of plantation and palm oil processing of PT. Nusantara VI Ophir Plantation is an area of Cultivation Rights covering an area of 3,549.16 Ha. Consists of a productive area of palm oil (core area) covering an area of 3,256 ha, and a non-productive area (housing, MCC, river/lowland, etc.) covering an area of 293.16 ha.

3.1. ISPO activities of PT Pekebunan Nusantara VI Ophir

3.1.1. Results

Stages of the ISPO certification process at PT. Perkebunan Nusantara VI Ophir began with the submission of an ISPO certification request to PT. AJA Indonesian certification, one of the certification bodies that has received recognition from the ISPO Commission. PT. Perkebunan Nusantara VI Ophir carried out the ISPO certification process starting in 2014. However, the process of preparing the documents completeness and completeness of the field preparation process for the ISPO certification process has been carried out since 2013.

The certification body, after receiving a certification request from a plantation company, conducts research on the completeness and correctness of the document (review document). Based on the completeness and validity of the company's ISPO certificate submission documents, PT. AJA Indonesian Certification made an ISPO certification work contract with PT. Perkebunan Nusantara IV Ophir, as well as conducting an Audit Stage I, Audit Stage II, and surveillance. Based on the completeness and correctness of the document (*document review*) for submitting an application for ISPO certification, PT. AJA Certification Indonesia has entered into an ISPO certification work contract dated June 17, 2014, with PT. Nusantara Plantation VI Ophir Unit, consisting of Ophir Area (Planted 3,256 Ha), and Ophir Palm Oil Mill (capacity 40 Tons FFB / hour).

The next ISPO certification process is the implementation of PT. Nusantara VI Plantation's ISPO certification audit for Ophir MCC and Ophir Unit Area by PT. AJA Certification Indonesia, and improvement of audit findings by the Ophir Business Unit and Head Office of PT. Perkebunan Nusantara VI. The implementation of Audit Stage I was conducted on the 8th of September 13, 2014, and Audit Stage II on the 2nd of February 6, 2015.

The results/report of the Phase II audit that fulfilled the ISPO requirements were submitted to the ISPO Commission through the ISPO Commission Secretariat. The ISPO Commission Secretariat verified the audit report submitted by the certification body. If the document is complete and correct, the audit report is forwarded to the ISPO Assessment Team to get an assessment. Furthermore, the ISPO Assessment Team evaluates all documents along with other aspects relating to ISPO requirements. In evaluating audit reports, the ISPO Assessment Team can collect information from various sources, including several relevant stakeholders, such as indigenous peoples, associations, local government officials, local NGOs, and audited company employees. The next process, the Assessment Team provides recommendations to plantation companies to the ISPO Commission for approval.

After that, the ISPO Commission gave recognition to plantation companies that met ISPO requirements and announced to the public. Furthermore, the certification body PT. AJA Indonesian Certification issues ISPO certificate to PT. Perkebunan Nusantara VI Ophir that has received recognition from the ISPO Commission. PT. AJA Certification Indonesia has issued ISPO certificate No. IN-2-AJA-ISPO on February 5, 2016, to PT. Perkebunan Nusantara VI for the Ophir Business Unit. The certified unit is the Ophir Palm Oil Mill (POM) with a capacity of 40 Tons of FFB/hour and the Ophir Area with a total area of 3,256 Hectares. PT. Perkebunan Nusantara VI (LLC) Ophir Unit's ISPO certificate is valid for 5 (five) years, effective from February 5, 2016, until February 5, 2021.

In general, the process carried out by PT. Perkebunan Nusantara VI The Ophir Unit for obtaining ISPO certificates is explained in figure 2 below.

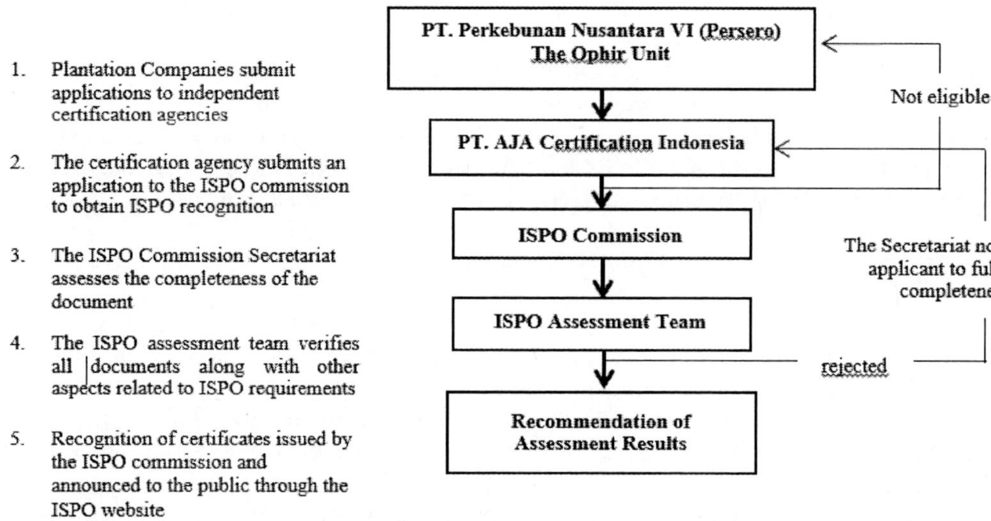

Figure 2. Flowchart of ISPO implementation process by PT. Perkebunan Nusantara VI ophir

3.1.2. Discussion

PT. Nusantara Plantation Ophir VI has carried out the process of preparing the complete document and preparation for completing the ISPO certification process field. Quite a number of requirements must be completed to register until the process of obtaining an ISPO certificate. [6] stated that palm oil plantation businesses that would apply for ISPO certification must complete several documents to fulfil legal aspects. According to [7] ISPO is a stronger system because it is based on legality. The legality aspect in ISPO is one of 7 (seven) principles that are used as references in the application of ISPO certification. ISPO certification has binding legal force and violations of law will be faced with authorities in Indonesia. Violations of the law can result in the cancellation of plantation permits.

Furthermore, [6] stated that the implementation of the ISPO certification system had been considered to be running less optimally because of several factors, including related to the fulfilment of the legality aspects and the issue of financing ISPO certification. In addition to having to prepare a number of legal documents, plantation companies who will propose ISPO certification must also prepare a number of certification costs that have been borne by the businessmen themselves. The reality in the field shows that business people sometimes cannot fulfil the required legal documents. In fact, some businessmen complain about the high cost of ISPO certification. For the issue of legality, the Indonesian government and other stakeholders must work harder to find the main problems and solutions related to the slow or non-fulfilment of legality requirements by plantation businessmen. For financing problems, the government must be able to formulate price standards and look for alternative financing for ISPO certification, especially for plasma planters and/or independent smallholders.

Procedures for recognizing PT. Nusantara Plantation Ophir VI's ISPO certification is in accordance with the mechanism as stipulated in Minister of Agriculture Regulation No. 11/Minister of Agriculture Regulation/OT.140/3/2015. The stages of the ISPO certification process begin with an audit conducted by the certification body. The audit results from the certification body were verified again by the ISPO Commission Secretariat. After that, an assessment was carried out at the level of the ISPO Assessment Team. After the Assessment Team gave recommendations, discussions were then carried out at the ISPO Commission level to obtain recognition. Finally, the certification body issues an ISPO certificate [8].

Reference [7] highlight that the third party ISPO certification procedure is more complicated, the first stage of verification is carried out by an independent auditor, but at a later stage, the ISPO Commission that conducts an assessment before being able to approve the certificate that is issued by the independent certification body. Furthermore, [6] stated that the large role given to the ISPO Commission in the ISPO certification process, namely through the ISPO secretariat which is under the auspices of the Ministry of Agriculture, was considered by some as a practical obstacle for ISPO certification executants and institutions that should be independent. Therefore, there must be improvements to the institutional mechanism in the implementation of ISPO certification. Based on a study conducted by the Coordinating Ministry for the Economy, the fundamental change in the proposed strengthening of the ISPO institutional scheme is that the implementation of ISPO certification will be carried out fully and independently by the ISPO certification body in accordance with a scheme that has been regulated and/or established by the National Accreditation Committee. In addition, there are independent monitors tasked with monitoring the implementation of ISPO certification.

3.2. Impact of PT Perkebunan Nusantara VI ophir's ISPO certification application: economic aspects

3.2.1. Results

The impact of the application of ISPO certification from the economic aspect financially influences the increase in production costs, revenue from sales, and income received by PT. Perkebunan Nusantara VI Ophir. To analyze the impact of ISPO implementation on PT. Perkebunan Nusantara VI Ophir on economic aspects, comparing the average level of revenue from sales, production costs, and income received by the company after ISPO certification compared to before the ISPO certification, with the results of the comparison as follows:

- *Production of Fresh Fruit Bunches (FFB)*

Comparison of FFB production rates and productivity of FFB core area plantations at PT. Perkebunan Nusantara VI Ophir at the time before (2013, 2014, 2015) and after (2016, 2017) ISPO certification, can be seen in table 3 below.

Table 3. Comparison in PT. Perkebunan Nusantara VI Ophir FFB Average Production

No.	Description	Before the application of ISPO	After the application of ISPO
1.	FFB Production (Ton)	48.101.613	75.351.955
2.	FFB Productivity (Ton / Ha)	24.904,34	25.740,29

Source: FFB Production of PT. Perkebunan Nusantara VI Ophir from 2013 till 2017.

Based on the table above, it can be seen that at the time after ISPO certification compared to before ISPO certification, the average FFB production level was 56.65% and the average level of FFB productivity was 3.36%. So that it can be concluded, the application of ISPO certification affects the increase in FFB production of PT. Perkebunan Nusantara VI Ophir.

- *Production of Palm Oil CPO and Palm Kernel*

Comparison of the total production of *CPO* and *Kernel* from the Palm Oil Mill (POM) processing of PT. Perkebunan Nusantara VI Ophir at the time before (2013, 2014, 2015) and after (2016, 2017) ISPO certification, can be seen in table 4 below.

Table 4. Comparison in PTPN VI Ophir *CPO* and *Kernel* Average Production

No.	Description	Before the application of ISPO	After the application of ISPO
1.	CPO Production (Ton)	11.326.673	16.803.935
2.	Kernel Production (Ton)	2.046.864	2.087.363
3.	**CPO & Kernel Production**	**13.373.537**	**18.891.298**
4.	**CPO Productivity (Ton / Ha)**	**5.917,76**	**5.219,46**
5.	**Palm Kernel Productivity (Ton / Ha)**	**1.076,20**	**649,89**
6.	**CPO & Kernel Productivity (Ton / Ha)**	**6.993,959**	**5.869,348**

Source: CPO and Palm Kernel Production of PT. Perkebunan Nusantara VI Ophir from 2013 till 2017

Based on the table above, it can be seen that the average level of *CPO* and *Palm Kernel* production at the time after ISPO certification was 41.26% higher than before ISPO certification. However, the average productivity of *CPO* and *Palm Kernel* of PT. Perkebunan Nusantara VI Ophir at the time after ISPO certification was 16.08% lower than before ISPO certification. So it can be concluded, the application of ISPO certification does not affect the increase in the production of *CPO* and *Palm Kernel* from the POM processing of PT. Perkebunan Nusantara VI Ophir.

- *The yield of Crude Palm Oil (CPO) and Palm Kernel*

Comparison of the yield of CPO and Palm Kernel of PT. Perkebunan Nusantara VI Ophir at the time before (2013, 2014, 2015) and after (2016, 2017) ISPO certification, can be seen in table 5 below.

Table 5. Comparison in PT. Perkebunan Nusantara VI Ophir *CPO* and *Kernel* Average Yield

No.	Description	Before the application of ISPO	After the application of ISPO
1.	CPO Yield (%)	23,69	22,23
2.	*Palm Kernel* Yield (%)	4,31	2,83

Source: The yield of CPO and Palm Kernek of PT. Perkebunan Nusantara VI Ophir from 2013 till 2017

Based on the table above, it can be seen that after ISPO certification compared to before ISPO certification, the average CPO yield was 2.05% lower and the average Palm kernel yield was 11.48% lower. So it can be concluded, the application of ISPO certification does not affect the increase in the yield of *CPO* and the yield of *Palm Kernel* from the MCC processing of PT. Perkebunan Nusantara VI Ophir.

- *Selling Price of Crude Palm Oil (CPO) and Palm Kernel*

Comparison of the selling prices of *CPO* and *Palm Kernel* of PT. Perkebunan Nusantara VI Ophir at the time before (2013, 2014, 2015) and after (2016, 2017) ISPO certification, can be seen in table 6.

Table 6. Comparison in PT. Perkebunan Nusantara VI Ophir *CPO* and *Kernel* Average Selling Price

No.	Description	Before the application of ISPO	After the application of ISPO
1.	CPO Selling Price	7381,96	8154,17
2.	Kernel Selling Price	4336,46	6606,90

Source: Selling Price of CPO and Palm Kernel of PT. Perkebunan Nusantara VI Ophir from 2013 till 2017

Based on the table above, it can be seen that at the time after ISPO certification compared to before ISPO certification, the average selling price of CPO was 10.46% higher and the average palm kernel selling price was 52.36% higher. So it can be concluded, the application of ISPO certification affects the selling price of PT. Perkebunan Nusantara VI Ophir's *CPO* and *Palm Kernel*.

- *Sales of PT. Perkebunan Nusantara VI Ophir*

Comparison of sales volume and revenue from sales of PT. Perkebunan Nusantara VI Ophir's *CPO* and *Palm Kernel* at the time before (2013, 2014, 2015) and after (2016, 2017) ISPO certification, can be seen in table 7 below

Table 7. Comparison in PTPN VI ophir average sales rate at the time before and after ISPO

No.	Description	Before the application of ISPO	After the application of ISPO
1.	Sales Volume of CPO and Kernel (Ton)	13.373.537	18.891.298
2.	Acceptance of CPO and Kernel Sales (Rp)	Rp. 92.650.901.570,-	Rp.150.892.397.565,-
3.	**Receiving Sales of Rp. / Ton**	Rp. 6927,928,-	Rp. 7987,4023,-

Source: Income Statement of PT. Perkebunan Nusantara VI Ophir from 2013 till 2017

From the table above, it can be seen that at the time after ISPO certification compared at the time before ISPO certification, the average level of acceptance of PT. Perkebunan Nusantara VI Ophir from the sale of CPO and Palm Kernel is 62.86% larger and the average revenue per Ton is greater by 15.29%. This shows that there is an increase in the amount of revenue from the sale of CPO and Kernel of PT. Perkebunan Nusantara VI Ophir at the time after the adoption of ISPO certification. This is due to an increase in the volume of CPO and Kernel production at the time after ISPO certification. Comparison of the average volume of production of CPO and Kernel at the time after ISPO was 41.26% greater than at the time before ISPO. So it can be concluded, the application of ISPO certification affects the increase in revenue from the sale of PT. Perkebunan Nusantara VI Ophir's *CPO* and *Palm Kernel*.

- *Production Costs of PT. Perkebunan Nusantara VI Ophir*

Comparison of the average production cost of PT. Perkebunan Nusantara VI Ophir at the time before (2013, 2014, 2015) and after (2016, 2017) ISPO certification can be seen in table 8 below

Table 8. Comparison in PTPN VI ophir average production costs at the time before and after ISPO

No.	Description	Before the application of ISPO	After the application of ISPO
1.	Garden Production Costs	Rp. 24.024.819.720	Rp. 43.757.968.086
2.	Garden Processing Fees	Rp. 3.572.997.671	Rp. 10.034.514.695
3.	Depreciation Costs	Rp. 8.500.245.946	Rp. 14.095.522.869
	Total Production Costs of Areas	**Rp. 27.597.817.392**	**Rp. 53.792.482.781**
	Plant Cost per Hectare	Rp. 10.011.421,93	Rp. 10.428.388,30
	Production Costs per Hectare	Rp. 13.528.774,39	Rp. 16.688.669,49
	Production Costs per Kg	Rp. 739,88	Rp. 889,00

Source: Production cost in PT. Perkebunan Nusantara VI Ophir, from 2013 till 2017

Based on the table above, it can be seen that at the time after ISPO certification compared to before ISPO certification, the average production cost was 94.92% higher, the average production cost per hectare was 23.36% higher, the average production cost per Kg was 94.92% higher.

This shows that there are additional costs for the production costs of PT. Perkebunan Nusantara VI Ophir at the time after the adoption of ISPO certification. This is due to an increase in the cost of the components of garden production costs by an average of 82.14%, the average cost of gardening is 180.84%, and the average depreciation expense is 65.82% after ISPO certification, compared to before ISPO certification.

To see a comparison of garden production costs that greatly affect the production costs of PT. Perkebunan Nusantara VI Ophir Unit at the time before and after ISPO certification was analyzed by comparing the average cost of production at the time before (2013, 2014, 2015) and after (2016, 2017) ISPO certification, with the results of the analysis that could be seen in table 9 below.

Table 9. Comparison of the average of PTPN VI ophir's production costs before and after ISPO

No.	Analysis	Before Implementation of ISPO	After Application of ISPO
1.	General Costs	Rp. 3.678.097.420	Rp. 10.159.714.136
2.	Generating Plant Costs	Rp. **12.542.778.790**	Rp. **21.347.833.118**
3.	Harvest	Rp. **5.240.408.215**	Rp. **8.163.145.540**
4.	Transportation to the factory	Rp. **2.563.535.295**	Rp. **4.087.275.293**
	Gardens Total Production Costs	**Rp. 24.024.819.720**	**Rp. 43.757.968.086**
	Gardens Production costs per hectare	**Rp. 11.861.110**	**Rp. 13.577.421**

Source: Production Costs of PT. Perkebunan Nusantara VI Ophir from 2013 till 2017

From table 9 above, it can be seen that the average cost of garden production at the time after ISPO certification was 82.14% greater than at the time before ISPO certification, and the average cost of garden production per hectare was 14.47% higher. This is due to an increase in the components of plantation production costs at the time after ISPO certification compared to before the ISPO certification, including general costs 176.22%; plant maintenance costs produce 70.20%; harvest costs 55.77%; factory transportation costs 59.44%.

To see a comparison of the cost of maintaining a plant that greatly influences the cost of producing PT. Perkebunan Nusantara VI Ophir Unit at the time before and after the ISPO certification was analyzed by comparing the average maintenance costs of the plants before (2013, 2014, 2015) and after (2016, 2017) ISPO certification, with the results of the analysis can be seen in table 10 below.

Table 10. Comparison of average PTPN VI Ophir maintenance costs before and after ISPO

No.	Analysis	Before Implementation of ISPO	After Application of ISPO
1.	Staff Salary	Rp. 878.365.908,-	Rp. 1.157.067.743,-
2.	TM Maintenance	Rp. 949.839.464,-	Rp. 1.543.628.201,-
3.	Road, Water Channels, Terrace Maintenance	Rp. 691.508.993,-	Rp. 1.041.766.324,-
4.	Weeding	Rp. 2.358.151.764,-	Rp. 4.628.455.962,-
5.	Pests & Diseases Eradication	Rp. 101.635.043,-	Rp. 135.686.145,-
6.	Fertilization	Rp. 7.281.136.864,-	Rp. 12.374.897.740,-
7.	Etc	Rp. 282.140.753,-	Rp. 466.331.005,-
	Total Cost of Maintenance	**Rp. 12.542.778.790,-**	**Rp. 21.347.833.118,-**
	Maintenance costs per hectare	**Rp. 6.021.577,-**	**Rp. 6.621.006,-**

Source: Costs for Garden Maintenance for PT. Perkebunan Nusantara VI Ophir from 2013 till 2017

From table 10 above, it can be seen that the average maintenance costs of generating plants for PT. Perkebunan Nusantara VI Ophir at the time after ISPO certification was 70.20% greater than before the ISPO certification. This is due to an increase in the components of plant maintenance costs at the time after ISPO certification compared to before the ISPO certification, including weeding costs 96.27%; fertilizer costs 69.96%; maintenance costs for plants produced 62.51%; the cost of maintaining roads, waterways and drainage is 50.65%; cost of eradicating pests and diseases 33.50%; the cost of staff salaries is 31.73%; other costs of 10.19%.

Based on the description above, it can be seen that there are additional costs for the cost component: weeding; fertilization; maintenance of produce; maintenance of roads, waterways and drainage; and the cost of eradicating pests and diseases, after the ISPO certification. This causes an increase in costs for the cost of garden production at the time after ISPO certification, resulting in production costs of PT. Perkebunan Nusantara VI Ophir at the time after the ISPO certification was greater than before the ISPO certification. This shows that there is an increase in the cost of garden production at the time after the adoption of ISPO certification, which will greatly affect the production costs of PT. Perkebunan Nusantara VI Ophir at the time after the adoption of ISPO certification. So that it can be concluded, the application of ISPO certification affects the increase in production costs of PT. Perkebunan Nusantara VI Ophir at the time after ISPO certification.

- *Revenue of PT. Perkebunan Nusantara VI Ophir*

Comparison of the average revenue of PT. Perkebunan Nusantara VI Ophir at the time before (2013, 2014, 2015) and after (2016, 2017) ISPO certification, can be seen in table 11 below.

Table 11. Comparison in PTPN VI ophir average revenue level at the time before and after ISPO

No.	Description	Before the application of ISPO	After the application of ISPO
1.	Total Receipt	Rp. 92.650.901.570,-	Rp. 150.892.397.564,-
2.	Total Cost	Rp. 36.098.063.338,-	Rp. 67.888.005.649,-
3.	**Revenue (Rp)**	**Rp. 56.552.838.232,-**	**Rp. 83.004.391.915,-**
4.	**Revenue Rp/Ton**	**Rp. 4.256,-**	**Rp. 4.422,-**

Source: Income Statement of PT. Perkebunan Nusantara VI Ophir from 2013 till 2017

From the table above, it can be seen that the average revenue level of PT. Perkebunan Nusantara VI Ophir at the time after ISPO certification was greater than before ISPO certification. Although the revenue from the sale of *CPO* and *Palm Kernel* at the time after ISPO certification increased by 62.86%, the production costs incurred by the company in producing *CPO* and *Palm Kernel* also increased by 88.07%, resulting in revenue from *CPO* and *Palm Kernel* sales received by the company only increased by 46.77%.

Likewise, the average income level per Ton received by PT. Nusantara Plantation at 3.92% higher after ISPO certification compared to before ISPO certification. So that it can be concluded, the application of ISPO certification has an effect on increasing the revenue of PT. Perkebunan Nusantara VI Ophir.

- *Analysis of Acceptance Fees Ratios (R / C ratio)*

To find out the financial benefits at the time after the adoption of ISPO certification at PT. Perkebunan Nusantara VI Ophir, explained by conducting a comparison of the average ratio of revenue to costs (R / C ratio) at the time before (2013, 2014, 2015) and after (2016, 2017) the application of ISPO certification, with results that can be seen in table 12 below.

Table 12. Comparison in PTPN VI ophir average acceptance fees at the time before and after ISPO

No	Description	(R/C ratio)
1.	Before the application of ISPO	2,64
2.	After the application of ISPO	2,29

Source: Ratio of revenue to cost (R / C ratio) of PTPN VI Ophir from 2013 till 2017

From the table above, it can be seen that the average comparison of the calculation of the acceptance fees ratio is seen by the R / C ratio at the time after the ISPO certification is smaller by 13.45% compared to before ISPO certification. Even so, the R / C ratio is greater than one (R / C> 1), meaning that the *revenue* obtained is greater than the *cost*. So that it can be concluded, although there is no increase in the ratio of calculation of acceptance fees at the time after the application of ISPO certification, the application of ISPO certification to PT. Perkebunan Nusantara VI Ophir is financially efficient over the costs incurred.

3.2.2. Discussion

Judging from the comparison of the selling prices of CPO and Palm kernel of PT. Perkebunan Nusantara VI Ophir, the average selling price of CPO and Palm Kernel at the time after ISPO certification is higher than before the ISPO certification. However, this is not influenced by ISPO certification, but the quality of CPO is good, the level of FFA (free fatty acids) is low, and the increase in world CPO prices. The application of ISPO certification does not appear to influence the price of the selling price of Palm Oil (CPO) and Palm Kernel of PT. Perkebunan Nusantara VI Ophir. Thus, it can be concluded that there is no difference in the selling price of CPO between ISPO-certified plantations and non-ISPO certified plantations.

This is in line with the results of research by [9], that there was no difference in the selling price of CPO between RSPO-certified plantations and non-RSPO certified plantations. The results of the research by [9] related to "*Comparative Analysis of Revenue between Certified Plantations and the Non-Certified Plantation Roundtable On Sustainable Palm Oil (RSPO)*" shows that before and after PT. Nusantara Plantation III RSPO certified, there is no difference in price both nominal price and real price, sales volume, production costs and income.

The application of ISPO certification at PT. Nusantara Plantation Ophir has provided benefits that have an effect on revenue from sales had increased. Whereas [10] in him research entlited "*Analysis of Costs and Benefits of RSPO Certification for Palm oil Companies*" shows that the direct benefits received after RSPO certification has not affected the increase in selling prices, sales volume and market reach of CPO. Likewise with the results of the research by [11] in a study entitled "*The Impact of RSPO Implementation on CPO Export Sales Volume and Revenue in State Plantation Companies*" shows that there were no significant differences in both the CPO export sales volume differences, production costs, prices and revenues before and after the RSPO implementation.

PT. Nusantara Plantation Ophir VI in obtaining ISPO certification has issued a number of costs for the process of fulfilling the requirements for processing the certificate, as well as the cost of applying 7 principles of 41 criteria and 128 indicators as requirements for implementing ISPO certification. Judging from the number of costs to obtain ISPO certification at PT. Nusantara Plantation Ophir VI, it does not significantly affect production costs because most of the ISPO principles and criteria have been met before obtaining ISPO certification. Besides that, the application of ISPO certification at PT. Nusantara Plantation Ophir VI does not burden the company in terms of finance, because most of the ISPO certification costs are not charged to the Ophir Business Unit but are the cost to the head office of PT. Nusantara Plantation VI. This is in line with the research results of [10], that the production costs before and after RSPO certification are not significantly different because most principles and criteria have been applied before RSPO certification.

From the economic aspect, the application of ISPO certification at PT. Nusantara Plantation Ophir has provided benefits that have an effect on increasing income. Although the average amount of production and the yield percentage of *CPO* and Palm Kernel decreased, while the production cost increased, revenue from sales had increased so that it had an impact on the increase in income of PT. Nusantara Plantation Ophir VI. Whereas [12] in her research entitled "*Impact of RSPO Implementation on Revenue at PT. Bakrie Sumatera Plantation, Inc*" shows that the application of the RSPO has no significant effect on profits at PT. Bakrie Sumatera Plantation. Even though the price of CPO and Palm Oil products is increasing, but because the production of CPO and Palm kernel is very declining, so it does not affect the profits of CPO and Palm Oil products of PT. Bakrie Sumatera Plantation, Inc.

The results of the calculation of the acceptance ratio to the costs of implementing ISPO certification at PT. Nusantara Plantation Ophir VI, the value of R/C ratio is greater than 1 (one). Means, the implementation of ISPO certification at PT. Nusantara Plantation Ophir VI is financially efficient for the costs incurred. While the results of the [13] study that identified the benefits and costs of the current scenario of palm oil plantation operations or if the company registered as an RSPO member, indicated that the overall financial benefits obtained from the sale of certified CPO and PKO were not greater than the costs incurred.

3.3. Impact of ISPO certification implementation: non-financial aspects

3.3.1. Results

The impact of implementing ISPO certification aside from the financial aspects that do not directly affect the level of sales and revenues received by PT. Perkebunan Nusantara VI Ophir after obtaining the ISPO certification, it can be seen from the aspect of the operational, environmental and social relations, and trade of *CPO* and *Palm Kernel* as follows:

- *Operational Aspects*

Benefits of implementing ISPO certification in terms of operations received by PT. Perkebunan Nusantara VI Ophir after the implementation of ISPO is as follows:

a) Repairing and completing existing documents in plantation companies and adjusting and uniformizing operational activities and plantation company documents.

Indicators of fulfilling the ISPO principles and criteria in each article, in large part, are in the form of availability of documents for the implementation of each of the ISPO principles and criteria. In principle, the majority are in the form of SOP documents/work instructions, evidence of documentation of activities, and records of the realization of the implementation of each of the ISPO criteria. Indicators of the availability of documents that fulfil the ISPO criteria, most of which have been met by PT. Perkebunan Nusantara VI Ophir at the time before the adoption of the ISPO.

b) Plant maintenance work is well managed.
Plant maintenance activities by applying Palm Oil *Good Agriculture Practice* (GAP) according to the Standard Operating Procedure (SOP) have been carried out by PT. Perkebunan Nusantara VI Ophir intensively includes goal maintenance; dish maintenance; fertilization; eradication of pests and diseases; and repayment.

c) Post-harvest work is well managed.
Plantation companies have harvested on time in a good and correct manner and recorded FFB production. The plantation company ensures that harvested FFB is then collected at the harvest collection site then brought to the processing plant as soon as possible to avoid deterioration in quality. Regarding FFB receipts in the Palm Oil Processing Unit, the plantation company ensures that the FFB received is in accordance with the requirements. The company has provided TBS acceptance and inspection/sorting SOPs, records of TBS receipt documentation that are appropriate and not in accordance with the requirements, and documentation of FFB prices.

d) A well managed Manpower Acceptance System.
Workforce recruitment required by PT. Perkebunan Nusantara VI Ophir is carried out by recruiting workers who do not need special expertise based on the needs of the company with the approval of the board of directors, and the recruitment of workers with special expertise based on needs is carried out by the head office of PT. Perkebunan Nusantara VI Jambi. The labour recruitment mechanism set by PT. Perkebunan Nusantara VI Ophir has given local communities the opportunity to seize the available employment opportunities in the company according to the requirements needed. Labour needed by companies can be recruited from communities in the surrounding environment, both casual daily workers and permanent workers.
The wage/payroll system of workers is grouped into two large groups, namely permanent and non-permanent workforce. Permanent Labour, with monthly salaries varying per month. Honorary staff, with salary according to the Regional Minimum Wage. Non-permanent labour, with salary per day. Every overtime job is paid overtime in accordance with applicable regulations.
In managing comfort, safety and health, PT. Perkebunan Nusantara VI Ophir has prepared: polyclinic facilities and infrastructure; cooperation; religious facilities (mosques, prayer rooms, and churches); sports and arts facilities; housing; SPS work unit; kindergarten; an elementary school and car transfer for school students; scholarship for company employees' children.

e) Fire cases can be minimized.
During the past few years, there have been no incidents of land fires in the PT. Perkebunan Nusantara VI Ophir Cultivation Rights area caused by plantation activities. This is because the effectiveness of prevention efforts carried out by officers both through coaching and counselling to workers and the community as well as preventive measures against the sources of land fires. The fire department coordination team is very active not only in the dry season.

f) Work Accident Cases can be minimized.
Based on Reports of Common Problems of PT. Perkebunan Nusantara VI Ophir Unit which has been carried out regularly, it is known that there are no work accidents. Occupational health and safety program at PT. Perkebunan Nusantara VI Ophir has been implemented in accordance with applicable regulations.

g) FFB theft cases can be minimized.
In the core plantation area, there were no incidents of FFB theft, thanks to the readiness of the company security officers and good cooperation with community leaders.

- *Environment Aspects*

PT. Perkebunan Nusantara VI Ophir unit has been awarded a blue rank in the assessment of the company's performance in the environmental field. During this time, the company's management has carried out environmental monitoring, especially on environmental management that has been carried out so far, and the results of its implementation have been analyzed based on the standards of environmental quality that have been established so as to obtain a level of decline and level of improvement in the quality of life carried out.

The commitment of the company has been able to provide good results for the environment so that there are not many complaints from the surrounding community. The application of ISPO which regulates the environmental management of plantations and minimizes existing environmental problems, including the following:

a) Managing the Impact of Water Quality Reduction.
   Monitoring the quality of wastewater, most of the results of the analysis have met the quality standards of liquid waste in accordance with the decision of the Minister of the Environment number 51 of 1995 concerning the quality standards of wastewater for the palm oil industry, the average value has met quality standards. The results of monitoring the quality of river water are good before and after being mixed with wastewater, almost all parameters have met the water quality standards of Class II rivers in accordance with government regulation number 82 of 2001.
b) *Managing Air and Noise Quality.*
   The results of monitoring the quality of gas emissions and total particles in the boiler chimney of PT. Perkebunan Nusantara VI factory Ophir still meet the air quality standards of emissions in accordance with the regulation of the minister of environment number 7 of 2007 concerning emission quality standards of immovable sources. The results of the analysis of ambient air quality and noise levels around the PT. Perkebunan Nusantara VI Ophir factory both inside the factory and employee housing are still below the threshold value as stipulated in Government Regulation number 41 of 1999 concerning the standard level of noise, vibration and noise.
c) Managing B3 Waste Management.
   To avoid the impact of environmental pollution, the used lubricants produced by the palm oil mill activities are put into drums that have been provided and stored in a particular warehouse that is protected from sunlight and waterproof floors. So that spillage of used lubricants does not pollute the environment in temporary storage warehouses equipped with oil traps. Some of these used lubricants are used as engine chain lubricants, but if there is anything left, it will be sold to collectors who already have permits.
d) Managing Land Use Throughout the Watershed.
   Land use along the watershed is residential, community fields, religious facilities, planting mahogany trees in the watershed. While the use of river water is for fish prohibition, management of liquid waste that has high potential as a nutrient for plant growth that will provide benefits as nutrients for fish cultivation, as well as maintaining moisture and biogas energy sources.

- *Social Community Aspects*

The benefits of applying ISPO certification based on social community relations are as follows:

a) The company does not experience land problems or land conflicts.
   PT. Perkebunan Nusantara VI Ophir Unit does not experience the problem of conflicts with the community. The problem of conflict with the community such as problems about land and so on can be minimized and can be overcome due to a system of handling land conflicts that are good, in accordance with the principles and criteria of ISPO which regulates the mechanism for resolving land disputes.
b) The problem of conflict with community due to pollution can be controlled.
   In an effort to prevent health problems from local communities due to factory activities, PT. Perkebunan Nusantara VI Ophir has made various efforts by directly contributing funds to the community for the construction of clean water facilities, as well as equipping adequate health

facilities and infrastructure with medical and paramedical personnel so that factory activities do not cause public unrest.

For comfort, safety and health, PT. Perkebunan Nusantara VI Ophir has prepared tools and facilities for polyclinics, cooperatives, religious facilities (Mosques, Musholla, and Churches), sports and arts facilities, housing, Indonesian Workers Union work units, kindergartens, elementary schools and cars for picking up students, scholarships son of a company employee. And to anticipate and overcome public health problems and labour due to company activities, 1 (one) health clinic has been built complete with the provision of doctors, medical/medical personnel and medicine. All health costs are borne by the Social Security Agency of Health.

c) Problems of public unrest as a result of changes in land function to plantation land can be overcome.

In anticipating the impact of changes in the functions of land use on the decline in community welfare income, PT. Perkebunan Nusantara VI Ophir has made the following management efforts:

1) Providing opportunities for local people (local residents) to become workers in companies, both non-permanent workers and permanent workers.
2) Developing a plasma plantation covering 4,800 hectares with a primary cooperative credit pattern for Primary Cooperative Credit Members which has an impact on increasing community income/income (2400 households).
3) Distributing aid funds to 103 fostered partners around the company, related to the development of Small Businesses and SME Development and Cooperation Cooperatives through the utilization of a portion of SOE profits.

In addition to impacting the decline in income and community welfare, the existence of the MCC also has several impacts that can benefit the community around the company's environment. Ophir MCC has helped accommodate community harvests that will support the company's productivity. In addition, it opens up business opportunities for the local community, especially in the transportation services business, namely the provision of FFB transportation equipment from the plantation to the factory, and the loading and unloading of FFB labourers and goods that have implications for people's income.

d) Improve relations with local stakeholders, including government, labour, civil society and buyers.

ISPO certification is a rule and principle that has been made by the Indonesian government by taking into account 3 aspects (3P), namely *profit* (economy), *people* (social), *planet* (environment). Social responsibility is related to the company's commitment to social and environmental problems, enhancing relationships with local stakeholders, including government, labour, civil society and buyers. ISPO criteria are based on social community relations, ISPO requires companies to provide data and information to stakeholders, social and community responsibilities, empowerment of indigenous peoples, addressing problems of public unrest as a result of changes in land functions to plantation land, and problems with conflict with communities as a result of environmental pollution. The socioeconomic aspects contained in the ISPO principles and criteria include the process of facilitating the development of gardens for the surrounding community, land tenure, responsibility for workers in occupational safety and health and in improving the welfare of workers 'ability, the formation of trade unions and workers' cooperatives, economic empowerment of the community.

- *Trading of CPO & Palm Kernel Products*

Based on the results of the study, it can be seen that the application of ISPO certification at PT. Nusantara Plantation Ophir has provided benefits that have an effect on increasing income. Although the average amount of production and the yield percentage of *CPO* and Palm Kernel decreased, while the production cost increased, revenue from sales had increased so that it had an impact on the increase in income of PT. Nusantara Plantation Ophir VI.

With the implementation of ISPO, it is expected that the CPO products produced have superior quality and grade as well are environmentally oriented so as to be able to ward off the negative issues of managing palm oil plantations and meet market demand for CSPO (*Certified Sustainable Palm Oil*). PT. Nusantara Plantation Ophir VI, whose main target market is the domestic (local) market, is developing not only in the local market but also in the export market.

3.3.2. Discussion

The results of research showing that there are still other positive impacts from the application of ISPO certification in addition to the financial aspects of PT. Nusantara Plantation Ophir VI. These benefits can be seen from the operational aspects, social community relations, and the company's concern for the environment. In terms of operations, it affects the completeness of documents and conformity with operational activities; managed plant maintenance activities and labour recruitment systems; minimize fire cases, work accident rates, the incidence of FFB theft. Based on the relationship with the social community, it influences the handling of conflict issues with the community; relations with government, labour, society and buyers. From the environmental aspect, it affects the management of the impact of decreasing water quality, air quality and noise, hazardous and toxic materials waste treatment; and the company received a blue rating in the assessment of the company's performance in the environmental field. The results of the research by [10], shows that the indirect benefits received after the RSPO implementation were decreasing the number of workplace accidents, decreasing cases of land conflicts and decreasing fire cases.

Based on the results of the study, it can be seen that the application of ISPO at PT. Nusantara Plantation Ophir VI has been able to increase the concern of palm oil plantation businesses for the importance of producing sustainable palm oil and ultimately increase the concern of plantation companies in preserving natural resources and the environment. This is in line with the research result of [11] highlighted that although there were no significant differences in both the CPO export sales volume differences, production costs, prices and revenues before and after the RSPO implementation, the application of RSPO at PT. Nusantara Plantation III has a positive impact because it cares more about the environment.

ISPO is a policy taken by the Government of Indonesia in this case the Ministry of Agriculture in positioning palm oil development as an integral part of the Indonesian economy, which aims to meet global demands and improve the competitiveness of Indonesian palm oil in the global market, and support Indonesia's commitment to reduce the contribution of greenhouse gases and support Indonesia's commitment to the preservation of natural resources and the environment. With the ISPO arrangement, it is expected that all palm oil plantation businessmen in Indonesia can raise awareness of the importance of producing sustainable palm oil which can influence the increase in the competitiveness of Indonesian palm oil on the world market [8].

According to [7], with the existence of ISPO which is a principle and rule originating from Indonesia itself, it is expected that in the future Indonesian CPO can be accepted by the world because it has followed the ISPO principles and rules, including compliance with aspects/legal aspects; economy; environment; and social as applicable laws and regulations in Indonesia. With the implementation of ISPO, it is expected that the CPO products produced have superior quality and grade as well are environmentally oriented so as to be able to ward off the negative issues of managing palm oil plantations and meet market demand for CSPO (*Certified Sustainable Palm Oil*). [14] in her research entitled "*Interaction of Renewable Energy Directive Policy and Indonesian Sustainable Palm Oil Policy Against Indonesian Palm Oil Exports to the European Union*" shows that's application of the ISPO is a tool for Indonesia to penetrate trade protection issued by the European Union through the Renewable Energy Directive (RED).

However, the study of [15] in the "*Six Year ISPO*" found that the adoption of ISPO had not been able to respond to the negative impacts caused by palm oil development so far, especially on environmental and social aspects. The ISPO certification system, which is expected to be the entry point for improving plantation and land governance, is felt to be limited to an instrument to gain recognition in the international market and dismiss negative issues related to the management of palm oil plantations in Indonesia.

Although the application of the ISPO certification obligation at PT. Nusantara Plantation Ophir VI has not been able to meet global demands and improve the competitiveness of Indonesian palm oil in the global market but in principle the fulfilment of ISPO certification obligations at PT. Nusantara Plantation Ophir VI has increased the awareness of businessmen about the importance of fulfilling the law, in an effort to encourage palm oil plantation companies to fulfil their obligations in accordance with laws and regulations. This is in line with the results of [16] research which shows that fulfilling ISPO certification obligations despite burdening the Company in terms of finance and fulfilment of bureaucracy, in the end, encourages the quality and productivity of *CPO* products produced and increases awareness of the importance of legal compliance.

4. Conclusions and suggestions

4.1. Conclusions

The results "Impact of ISPO Implementation on Economic, Social and Environmental Aspect of Palm Oil Plantation" for palm oil companies at PT. Perkebunan Nusantara VI Ophir Business Unit in West Pasaman Regency can be concluded as follows:

1. Stages of procedures for PT. Perkebunan Nusantara VI Ophir to obtain ISPO certification begins with the submission of an ISPO certification application to PT. AJA Certification Indonesia, then audit conducted by the certification body. After that, an assessment was conducted at the level of the ISPO Assessment Team to get recognition to plantation companies that met ISPO requirements and were announced to the public. Finally, PT. AJA Certification Indonesia issued an ISPO certificate to PT. Perkebunan Nusantara VI Ophir.
2. The impact of the application of ISPO certification on economic aspects is proved by an increase in production costs, revenues, and profit received by PT. Perkebunan Nusantara VI Ophir. Application of ISPO certification at PT. Perkebunan Nusantara VI Ophir has been financially efficient at the costs incurred.
3. ISPO certification received by PT. Perkebunan Nusantara VI Ophir affect operational aspects, relationships with the social community, the company's concern for the environment, as well as trading CPO and Kernel. This can be seen from affect the completeness of documents and conformity with operational activities; managing plant maintenance activities and labour recruitment systems; work accident rates; influencing the handling problems of conflict with the community; labour, society and buyers; the management of the impacts of water quality degradation, air and noise quality, B3 waste processing; and the company was awarded a blue rank in the assessment of the company's performance in the environmental field.

4.2. Suggestions

For plantation companies, as businessmen, it is expected to maintain environmental sustainability and preservation in producing palm oil in order to produce sustainable and environmentally friendly products. In addition, the company is also expected to make efforts such as separating certified CPO products from non-ISPO certification, improving marketing management systems, and conducting sales promotions so that the range of market interest is wider.

For the government, as the policymaker, must work harder to find issues and solutions related to the slow implementation of ISPO certification by plantation companies. The government is expected to make improvements to the institutional mechanism in the management of ISPO certification, must be able to formulate and find alternative financing for ISPO certification so that the costs for obtaining ISPO certification are not too large, and to be able to increase the acceptance of ISPO certification in the global market.

For further researchers, further research is needed on the impact of applicating ISPO certification, both in palm oil plantation companies and in non-governmental organizations.

Acknowledgments
On this occasion, the author expressed her sincere thanks to Mr. Dr. Ir. Ifdal, M.Sc; Mrs. Dr. Ir. Ira Wahyuni Syarfi, M.Si; and Mrs. Hasnah, SP, MEc, PhD as mentors who have provided advice, direction and guidance to the author in preparing this published article. The author also expressed her gratitude and utmost appreciation to Mr. Sumardi and Mr. Muhardiansyah, as key informants at PT. Perkebunan Nusantara VI Jambi which has helped a lot in providing information regarding the implementation of the Ophir Business Unit's ISPO certification. The author wishes to express her gratitude to mother Syamsiah, brothers and sisters, friends of IEP 2016, as well as other parties who have participated in helping who the author cannot mention one by one. The author realizes that this publication article is still far from perfection. Therefore the authors expect constructive criticism and suggestions from all parties for improvement in the future. Finally, the authors hope that this publication article can contribute and benefit the reader.

References

[1] Central Bureau of Statistics. 2016. Indonesian Palm Oil Statistics. Jakarta: The Central Bureau of Statistics of the Republic of Indonesia. 84 p

[2] Putri, M. D. 2013. Impact Analysis of Palm Oil (CPO) *Black Campaign* on Indonesian CPO Export Volume. [Essay]. Bogor. Bogor Agricultural Institute

[3] Trisna, A. 2012. Policy Analysis of Increasing Sustainable Palm Oil Production. [Thesis]. Bogor. Agribusiness Management Study Program. Bogor Agricultural Institute

[4] Casson A. 2000. The Hesitant Boom: Indonesia's Oil Palm Sub-sector in An Era of Economic Crisis and Political Change. Bogor (ID): CIFOR Occasional paper

[5] Marti S. 2008. Losing ground The Human Rights Impacts of Oil Palm Plantation Expansion in Indonesia. Bogor (ID): Sawit Watch

[6] Fahamsyah, E. 2017. ISPO System to Address the Challenges in Indonesia's Sustainable Palm Oil Development. Indonesian Society Journal. **43** (1), 1-15

[7] Wijaya, A. and Glasbergen, P. 2016. Toward a New Scenario in Agricultural Sustainability Certification? The Response of the Indonesian National Government to Private Certification. Journal of Environment & Development **25** (2) : 219–246

[8] Ministry of Agriculture. Ministry of Agriculture of the Republic of Indonesia. 2011. Regulation of the Minister of Agriculture of the Republic of Indonesia Number: 19/Permentan/OT.140/3/2011 about the Guidelines for Indonesia's Sustainable Palm Oil Plantations *(Indonesian Sustainable Palm Oil/ISPO)*. Jakarta. Ministry of Agriculture of the Republic of Indonesia

[9] Ginting, S.M. 2011. Comparative Analysis of Revenue between Certified Plantations and the Non-Certified Plantation Roundtable On Sustainable Palm Oil (RSPO)" [Essay]. Medan. The University of North Sumatra. 80 p

[10] Nasution, M. R. 2013. Analysis of Costs and Benefits of RSPO Certification for Palm oil Companies". [Thesis]. Medan. The University of North Sumatra. 70 p

[11] Kurniawan, Y. 2013. The Impact of RSPO Implementation on CPO Export Sales Volume and Revenue in State Plantation Companies (Study at PT. Perkebunan Nusantara III). [Essay]. Medan. The University of North Sumatra. 48 p

[12] Febrini, N. 2014. The Impact of RSPO Implementation on Revenue at PT. Bakrie Sumatera Plantation, Tbk. [Essay]. Medan. The University of North Sumatra. 63 p

[13] Salam, F. 2017. Cost And Benefit Analysis Of RSPO Certification (Case Study in PT BCA Oil Palm Plantation in Papua. Indonesian Journal of Business and Entrepreneurship. **3** (3) : 1-10

[14] Kartika, Intan Tiara. 2016. Interaction of Renewable Energy Directive Policy and Indonesian Sustainable Palm Oil Policy Against Indonesian Palm Oil Exports to the European Union. [Essay]. Makasar. Hasanuddin University. 58 p

[15] Forest Watch Indonesia. 2017. Six years of ISPO "Study Related to Strengthening ISPO Instruments in Responding to Negative Impacts such as Deforestation, Peat Ecosystem

Damage, Forest and Land Fires, and Tenurial Conflicts". May edition. Bogor : *Forest Watch Indonesia*. 80 p
[16] Panjaitan, M. 2013. Legal Analysis of ISPO Certification Obligations in Relation to Investment Growth in Oil Palm Plantations in Indonesia (Study at PT. REA KALTIM PLANTATION - Jakarta). [Thesis]. Medan. Faculty of Law. The University of North Sumatra. 47 p

IOP Conf. Series: Earth and Environmental Science **336** (2019) 012014 doi:10.1088/1755-1315/336/1/012014

Do oil palm impacts on species richness negatively affect the local livelihoods?

Arzyana Sunkar[1], Stephany Debby Eka Putri[1], Shinta Nur Rahmasari[1], Ade Saraswati[1], Farras Imantoko[1], Lutfy Alif Putra[1]

[1]Department of Forest Resources Conservation and Ecotourism, Faculty of Forestry, Bogor Agricultural University, P.O Box 168 Bogor, Indonesia

*E-mail: arzyanas@gmail.com

Abstract. Land use conversion is the leading cause of biodiversity changes, and biodiversity is often used as an indicator of sustainability. A great number of scholars and public, in general, have often associated biodiversity loss, such as species richness, with unsustainable oil palm plantations. Similarly, oil palm expansions were also associated with negative impacts on the local people's livelihoods. Seldom in discussions on oil palms sustainability, do they take into consideration the perceived biodiversity changes from social points of view. This study seeks to determine the local people's perceptions of the changes in species richness due to land conversion into oil palm plantations and how they have affected the local livelihoods. Household surveys were conducted using direct observation, literature study, and in-depth structured interview to 373 respondents from various ethnicities including indigenous people in 15 villages around 6 oil palm estates in North Sumatera, Riau, and West Kalimantan, whom were selected based on convenience sampling. Respondents were stratified into indigenous people, migrants, and estate's workers. A five-point Likert Scale was employed to identify people's perceptions while recollection technique was used to identify biodiversity loss. Results of the study revealed that there were varied perceptions related to the impacts of oil palms expansions on species richness from social perspectives. The species had different ecological and social values, depended on the benefits that people attach to the species. Although, as many as 16 wildlife and 15 plant species loss were recalled following the plantations' development, these losses provided little contributions to the current livelihood strategies. The locals appeared to be responsive to better economic opportunities and willingness to shift the livelihood system for improved incomes. These results suggested that species richness might not always be the best indicator of plantation sustainability, as sustainability other than approached from ecology, must also have social and economic dimensions.

1. Background

Oil palm plantations, one of the several important human-induced ecosystem change thorugh land conversion [1], have drawn much divergence from various stakeholders, covering a range of issues related to their impacts on biodiversity loss, as well as on rural communities, rural economics and household economics [2], [3], [4], [5]. The emergence of oil palm villages has also been associated with changes in the local cultivation practices that was blamed for altering the biodiversity composition (richness) as well as leading to local food insecurity [6], [7]. Yet, if managed properly, the socio-economic benefits of the plantations can improve livelihood and alleviate poverty [8], [9], [10], [11], [12].

Management of sustainable environment has so far been based on biological/ecological data [13], including sustainable oil palm plantations, which often ignores biodiversity relationships with humans [14]. Many international debates and discussions use biodiversity as their main instrument to value environmental management effectiveness [13] since a healthy environment is often synonymous with high biodiversity [15], [16]. Biodiversity is said to influence the functioning of ecosystems [17], [18], a term that defines the capacity of natural processes and components, in providing goods and services that will directly or indirectly satisfy human needs [19]. It provides a mixture of ecosystem services, which directly and indirectly contribute to human well-being [20].

People's level of support to the local biodiversity would depend on several factors. In their research [21], [22] have found that preferences over wildlife are diverse and are closely related to the differences in socio-demographic background, use of the resources for consumption purposes and the effect on their welfare, years of living in the area, frequency of interaction with wildlife, and attitudes toward wildlife in general. Thus, biodiversity is part of a social construct [23], since the social values attached to biodiversity would be attributable to the benefits that such wildlife or plants provides, implying that biodiversity have various potential uses with different values depending on the beneficiaries [24].

As the direct beneficiary of oil palm plantations, the social opinions of the local communities became increasingly important to be acknowledged and understood [14], because community's perceptions will be positively correlated with the development of the plantations [25]. Perception reflects the attitude and understanding of the people's expectations of oil palm plantations, which will also differ depending on how the environment and resources are aligned with their respective livelihood strategies [25], [26]. It is important to have understandings on how the local people perceive biodiversity loss, because in the presence of better economic option, does the willingness to shift livelihood options, suggest that the loss of biodiversity in the ecological term, also mean a loss of livelihood?

2. Methodology

2.1. Study sites

The study was conducted from February – March 2018 in the periphery villages of six private oil palm estates in Indonesia (names abbreviated): PHLE (2° 19' 09" N; 99° 50' 22" E) & LPYE (3° 26' 40.41" N; 98° 49' 12" E) in North Sumatera Province; KJNP (0° 46' 59.58" N; 101° 03' 01.03" E) & BPME (0° 36' 14.58" S; 102° 59' 14.58" E) in Riau Province; and PKWE (1° 31' 07" S; 110° 28' 18" E) & NTYE (1° 34' 0" S; 110° 32' 0" E) in West Kalimantan Province. The villages have experienced economic growth due to the existence of oil palm companies; thus in this study, these villages will be termed ***oil palm villages***.

2.2. Survey methodology

2.2.1 Data collection techniques

This study is historical-comparative research using the method of historiography, in gathering oral history evidence [27]. In order to identify the species that were lost due to the establishment of oil palm plantations, historical evidence was drawn based on recollection technique that is using the statements of individuals about their experiences in sightings a species based on memory. These were done using open-ended interviews. Attitudes and responses of the local people related to the existing biodiversity were gathered using five-point Likert Scales.

2.2.2 Selection of respondents

The survey design involved a multi-stratified sampling. The household was the basic unit of analysis and respondents were heads of households. In case of absence, a member of the household was identified as a replacement. The origin/culture influences social perceptions of private oil palm plantations; while environmental perceptions are affected by the length of time, a person lives in the area, whereas economic perceptions are not influenced by any socio-economic variables [28]. Considering these factors, the population studied were stratified into 3 groups: (a) indigenous people((I) – a group of individuals attached to a specific group of ethnicity, who were born and resided in an area, and whose family have been inhabiting the place for generations and do not work in the estates; (b) migrants(M) – an individual who came and live in the area following the government transmigration scheme or an individual who migrated to another area using self-supporting scheme and does not work in the estates; (c) estate workers(W) – an individual who works at an oil palm plantation estate and is not categorized under (a) and (b) strata. The total number of respondents (T) and stratification within each estate varied as shown in Table1. Where possible, the indigenous community and migrants strata were further stratified into households with oil-palm related livelihoods and non-oil palm related livelihoods.

Table 1. A total number of respondents in each oil palm estate.

No.	Estate	Province	Villages	Total Respondents			
				I	M	W	T
1.	PHLE	North Sumatera	Aek Korsik, Padang Halaban, Karang Anyar	18	18	9	45
2.	LPYE	North Sumatera	Huta Baru Nangka, Siopuk Baru	15	15	15	45
3.	KJNP	Riau	Sekijang, KijangMakmur	17	25	4	46
4.	BPME	Riau	Suhada, Bagan Jaya	24	16	6	46
5.	PKWE	West Kalimantan	SP3 Semblangaan, Tanjung Medan, Lembah Hijau 1, Lembah Hijau 2	20	n/a	40	60
6.	NTYE	West Kalimantan	Nanga Tayap, Tajok Kayong	110	21	n/a	131

2.2.3 Data analysis

Names of the lost species and their social values were tabularized. Each Likert item was coded: 0 = do not know, 1= strongly disagree, 2= disagree, 3= neutral, 4= agree, 5=strongly agree. To enable interpretation of the data, the numbers of response for each code in each item were analysed using frequencies, i.e., percentages and placed in tables. Distributions of the most responses (% that agree, disagree, etc.) would enable the comparisons between the respondents' groups.

3. Result and discussion

3.1. Biodiversity

This result showed that there might be a relationship between the organic material and soil macro fauna abundance, this result expected that the organic material provides food for the soil macro fauna, for example, earthworm. The earthworm is one of the examples of soil macro fauna, which affects the dynamics of changes of organic matter in the soil, as they decomposed the bigger fragments of organic material and turned in to cast (smaller fragments). Soil fauna feeding activity has a positive correlation with the earthworm and soil insect abundance. This result was linked with the observations made by reference [17] that bait perforation was not found when only meso fauna (such as collembolans) was present but seemed to be dependent on medium and large fauna like earthworms.

As a source of economic options, the daily life of a rural community cannot be separated from nature and its environment. As a neighbour of oil palm plantations, one or more of these interactions affected the daily life of most of the people living in these oil palm villages. However, the actual level of interactions between the communities and wildlife around the oil palm plantations varied greatly depending on the location of the village (distance to the plantation), species, and daily human activities (farming, livestock grazing, trading, working in government, etc.). Table 2 lists wildlife and plant species, recalled by the local communities as no longer existed in the area after the establishment of the plantations.

Table 2. Species loss as perceived by the local people after oil palms development

No	Wildlife Loss	Estates						Plants Loss	Estates					
		P	L	K	B	W	N		P	L	K	B	W	N
1.	*Manis javanica*[1,3,4]	√	√	√			√	*Cassia* sp.[6]	√					
2.	*Hystrix* sp.[1,3]	√	√	√				Langkat wood[7]	√					
3.	*Cervus* sp.[1,3,5]	√		√				*Durio zibethinus*[3,5]	√		√			
4.	*Panthera tigris sumatrae*[1]	√	√	√				Karamonting		√				
5.	*Elephas maximus sumatrensis*[1]		√	√				*Nephelium lappaceum*[5]		√				
6.	*Muntiacus muntjak*[1]			√				*Lansium domesticum*[5]		√				
7.	*Pongo abelii*[1]			√			√	*Calamus sp.*[4]			√			
8.	*Tapirus indicus*[1]			√				*Shorea sp.*[3,7]				√		√
9.	*Sus scrofa*[1,3,5]			√				*Peronema cenescens*						
10.	*Helarctos malayanus*[1]			√	√	√	√	*Nepenthes sp.*[3,4]				√		
11.	*Tragulus kanchil*[1]			√				*Neolamarckia cadamba*[3,7]						√
12.	*Presbytis sp.*[1]			√	√		√	*Durio kutejensis*[3,5]						√
13.	*Macaca fasicularis*[1]				√			Local durian[3,5]						√
14.	*Amyda cartilaginea*[2]				√			*Momordica* sp.[4,5]						√
15.	*Paradoxurus hermaphroditus*[1]				√			*Artocarpus anisophyllus*[5]						√
16.	*Hylobates muellerii*[1]					√								

Note: P (PHLE); L (LPYE); K (KJNP); B (BPME); W (PKWE); N (NTYE); [1]mammal; [2]reptile; [3]source of income; [4]medicine; [5]food;[6]honey bee;[7]-building materials

Given the proximity to oil palm plantations, most communities, especially those with a source of livelihood from oil palm, were often in contact with wildlife and plants thrived in the plantations and surrounding areas. Thus, they were more observant of their environment and knew when the local resources have changed. People use their local knowledge of plants and animals for either food or medicines to fulfil their household living needs, including using them for sources of income. In fact, 80% depends on this traditional knowledge to obtain their medicinal needs [29].

Table 2 indicated that 93.75% of the wildlife loss was mammals, ranging from small to large, of which 73.3% did not provide any contributions to the livelihood systems of the local communities. These results were in line with other research on biodiversity in oil palm plantations, that suggested mammals were one animal kingdom that was mostly lost after an oil palm plantation's establishment [30], [31].

Most of the listed wildlife species loss were those inhabiting dense forest of which were mostly occurred in KJNP and BPME of Riau Province. Sumatra is home to big mammals. The communities living around KJNP and BPME could recall of the time when they were still living close to forests and were faced by threats from the big predatory mammals (table 2). Forests were cleared by the local people and plantation companies before the oil palm establishment [31]. However, according to the local communities and regulations studied [31], the stated forest area was not area appointed by the government as forest.

In addition, based on the research in the same areas, several butterflies and few birds species were noted to be ecologically loss [31], but these species were not recalled by the communities as loss. This suggested that people tend to recall wildlife based on size and their contributions to the livelihood and safety. Butterflies and most birds were small in size. They were neither used as a source of income, medicine nor food, hence were ignored in the memories. The small size and diversities of their colours added to the difficulties in recognizing different species. Aside from butterflies and birds, compared to the wildlife loss from an ecological perspective in the same study sites [31], the number of mammal species that were lost based on the recollection of the local people's memories was higher, since the land use baseline that was used to identify ecologically wildlife loss was not forest, since remnants of forest were not found within the area [30]. This highlighted the importance of incorporating local knowledge in historical, ecological studies.

Contrary to wildlife, plants species that were lost had significant importance for the local livelihood with 46.67% (including honeybee) used as sources of income, 20% as medicines, 46.67% as food plants and 20% as building materials. The local people have a better recollection of memories with regards to plants that supported their livelihoods. Table 2 indicated that NTYE experienced the most plant loss in terms of a number of species, of which the plants functioned as a wild food. The people living in NTYE were mostly the indigenous Dayaks who used plants to fulfil their daily food needs [32].

3.2. *Local perceptions on wildlife loss*

Biodiversity is part of social construct, where people showed preferences over them, of which can be liked or disliked for various reasons. Attitudes towards wildlife are influenced by factors such as interactions with wildlife regarding space and heterogeneity [33], [34]. Some studies conducted on local attitudes towards carnivores in Asia found that in general, crop damage and animal predation by wild animals had a negative effect on attitudes of local communities, while the benefits that can be gained by the community have given rise to their positive attitudes [33], [35], [36]. Furthermore, the various preferences over certain wildlife were very much related to varied socio-demographic backgrounds, use of wildlife as ingredients of consumption and better well-being, length of stay in the area, interaction with wildlife, and attitudes towards wild animals in general [37], [38].

Indigenous people were often associated with their local traditional knowledge of the living biodiversities since they are environment-dependent. It is assumed that because these people were indigenous, they have been living in the areas for generations. Table 3 below showed that the indigenous people of PHLE showed the most contrasting responses, where most of them agreed that oil palm plantations of PHLE Estate had caused increased wildlife diversity as well as increased sightings. These responses were complemented by the fact that PHLE and LPYE, both located in North Sumatra, experienced the least loss. North Sumatra was one of the first areas in Indonesia to be cleared for plantations, where according to historical records, the area was largely cleared from forest since 1917 [39].

Before the development of the first planting period of PHLE in 1983, the area was already planted with oil palms. Most of the indigenous people, both that were dependent on oil palm and non oil palm for their livelihood in other study sites, on the other hand, have disagreed that conversion to oil palm plantations has increased wildlife diversity. In fact, they stated that there were fewer animal sightings which impacted on the number of animals that could be used as sources of income. Table 2 above indicated that KJNP and BPME showed the most wildlife loss with 80% in KJNP, since KJNP were used to be the forested area.

Table 3. Local perceptions on the existing wildlife species richness after the establishment of oil palm plantations (highest percentage)

Estate		Likert Statement (%)											
		Source of income (+)			Total species (+)			Species density (+)			Sighting frequency (+)		
		I	M	W	I	M	W	I	M	W	I	M	W
PHLE	1	A 66.67	A 44.44	A 77.78	N 55.56	D 44.44	N 44.44	A 55.56	D 55.56	D,N 33.33	A 88.89	N 44.44	A 77.78
	2	D 66.67	D 44.44	-	A 55.56	N 66.67	-	A,N 55.56	D 55.56	-	A 55.56	A 55.56	-
LPYE	1	D 60.00	N 73.33	A 46.67	A 66.67	A 66.67	A 60.00	A 80.00	A 53.33	A 53.33	A 73.33	A 53.33	A 80.00
KJNP	1	D 81.82	D 37.50	D 75.00	D 63.64	D 68.75	D 75.00	D 72.73	D 37.50	D 50.00	D 72.73	A 43.75	D 50.00
	2	D 66.67	D 66.67	-	A 66.67	A 33.33	-	D 66.67	D 66.67	-	D 66.67	D 77.78	-
BPME	1	D 35.00	D 91.67	A,D 33.33	D 85.00	D 75.00	D 83.33	D 40.00	D 33.33	D 67.77	D 50.00	K 33.33	D 50.00

Estate		Likert Statement (%)											
		Source of income (+)			Total species (+)			Species density (+)			Sighting frequency (+)		
		I	M	W	I	M	W	I	M	W	I	M	W
	2	D 50.00	A,D,K,N 25.00	-	D 50.00	A,D 50.00	-	D 75.00	K 50.00	-	D, K 50.00	-	-
PKWE	1	D 30.00	-	D,K 35.00	D 100.00	-	D,K 30.00	D 90.00	-	D 62.50	D 100.00	-	D 662.50
NTYE	1	D 60.20	-	D 80.95	D 96.50	-	D85.71	D 61.50	-	A 57.14	D 61.50	-	A 57.14
	2	D 87.00	-	-	D 100.00	-	-	D 92.50	-	-	-	-	-

Note: 1- income from oil palm; 2- income from non-oil palm; A- agree; D- disagree; K- don't know; N- neutral/doubtful

Table 4. Local perceptions on the existing plant species richness after the establishment of oil palm plantations (highest percentage)

Site		Likert Statement (%)											
		Source of income (+)			Medicinal species (+)			Food plants (+)			Diversity (+)		
		I	M	W	I	M	W	I	M	W	I	M	W
PHLE	1	A 77.78	A 88.89	A 88.89	A 77.78	A 55.56	A 100.00	A 66.67	A 77.78	A 100.00	A 44.44	N 33.33	A 88.89
	2	A 88.89	A 66.67	-	A,K 44.44	A 55.56	-	A 10.00	A 77.78	-	A,D 44.44	D 66.67	-
LPYE	1	A 80.00	N 80.00	A 93.33	N 40.00	A 33.33	A 66.67	A 66.67	N 46.67	A 60.00	N 40.00	N 46.67	A 73.33
KJNP	1	D 63.64	D 62.50	D 75.00	D 55.54	D 50.00	D 100.00	D 55.54	D 56.25	D 75.00	D 63.64	D 56.25	D 75.00
	2	A,D 33.33	D 88.89	-	D 50	A,D 44.44	-	A,D 50.00	D 66.67	-	D 50	A,D 44.44	-
BPME	1	D 45.00	D 75.00	D 66.67	K 35.00	D 66.67	A,D 50.00	A 55.00	D 66.67	D 83.33	D 55	D 58,33	D 83.33
	2	N 50.00	A,D 50.00	-	K 50.00	D 50.00	-	K 50.00	A 75.00	-	D 75.00	D 25.00	-
PKWE	1	K 50.00	-	D 42.50	D 50.00	-	K 47.50	D 55.00	-	D 32.50	D 60.00	-	D 50.00
NTYE	1	D 87.04	D 76.19	-	D 74.31	D 52.38		D 75.93	D 63.64	-	D 52.78	D 61.50	-
	2	D 87.40	-	-	D 74.30	-	-	D 75.93	-	-	D 52.70	-	-

Note: 1- income from oil palm; 2- income from non-oil palm; A- agree; D- disagree; K- don't know; N- neutral/doubtful

Table 3 suggested that both migrants and the indigenous people not show many variations in their responses. Specifically, in PHLE and LPYE, it was noted that migrants had an opposite response with regards to the sightings of individuals per species, since in this area, migrants have lived around the oil palm plantation longer than the indigenous people, and thus they were similarly observant of their surrounding areas.

Estate workers mostly spent their times in the company's plantation, thus showed the most varied responses towards their perceptions of species diversity, and indicated by the "neutral" responses and "don't know" responses among the agree' and "disagree" responses These suggested that the workers have the least knowledge on species diversity, due to the relatively homogenous environment of the plantations.

Concerning wildlife, table 3 indicated that the majority of the respondent's groups showed similar responses on the local species loss, that oil palm plantations development had caused loss of economically important species (see table 2). Similar patterns of response were also showed towards perceptions on species density and sightings frequency. People who were dependent on oil palm for their livelihood and those who were not, had tendencies to show a different pattern of responses, except for indigenous people, where both groups showed similar responses (except in PHLE as the only estate which showed varied responses from each stratum and oil palm dependent/not dependent). Such similar and varied responses were influenced by how much values were placed on the species by the people about their functions in fulfilling the economy or livelihood systems [40] as well as their interactions with the species [34].

PKWE and NTYE of West Kalimantan Province demonstrated the most homogenous responses among all strata and livelihood dependence groups, although different responses were found among the estate workers about species density and sightings. West Kalimantan indigenous people were the Dayak people and Melayu Kalimantan. It was observed, that there was similarity of the local knowledge on the resources. This was possible since the oil palm plantations are the youngest among the six plantations; hence most of the local communities observed and experienced similar surroundings, in other words, experienced similar interactions with the wildlife [34]. The only difference between the two community groups was culture, where Melayu is Muslim, and Dayak is Christian, thus influenced their perceptions toward wild boar and certain local plants.

In all six study sites, although the local people could recall the names of species that were lost, however, according to them, they have no significant impacts of their well-being, since they could compensate the loss of economically valued wildlife with other species or since at present, they have a better livelihood in oil palms. Whereas for the Dayak people whom most had not to shift to oil palm plantations means of living, they stated that the wildlife loss related to construction woods for building materials could be compensated with buying another type of woods as materials. With regards to wild boar, they also stated that they were not much affected since they manage their wild boar in the garden of their houses.

3.3. *Local perceptions on plant loss*

The plant categories were divided based on their purposes as a source of income, medicine, consumption, and diversity (table 4). Of the medicinal plants that have disappeared, the most rapid occurred in KJNP and BPME of Riau Province and least rapidly in NTYE. Parts of the study area in Riau were used to be peatland, where for many years prior to the establishment of the oil palm plantations, the area experienced selective felling of valuable tree species, such as *Shorea* spp (see table 2) [41], while *Nepenthes* sp. was exploited due to its increased ornamental values/commercial values [42]. The plants that disappeared were mostly plants that had economic values, in which the people used as sources of income and these use as consumptions.

On the contrary to the local people perceptions on wildlife loss, table 4 indicated that perceptions of the various groups of people and oil-palm dependent/not dependent were varied in each study site, except in NTYE. The reason for this might be because the oil palm company in NTYE was the youngest among other oil palm plantations in another study area, where the planting was just started in 2010, where it was 1983 in PHLE, 1985 in LPYE, 1991 in KJNP, 1998 in BPME and 2006 in PKWE.

Similar to the perceptions of wildlife as sources of income, the people in PHLE positively perceived the local perceptions on plants loss that had economic value. In the same line, the local people of NTYE showed no different perceptions among all respondents groups and the oil-palm dependent/not dependent.

In the case of other study sites, especially in KJNP and BPME, estates, it was noted from table 4 that the perceptions of the local people who were oil palm dependent were mainly negative towards oil palm impacts on increasing plants diversity, while those who were not dependent on oil palms, tent to show varied responses, where some stated “don’t know” and some remained neutral.

The most varied responses were shown towards perceptions on medicinal plants. Before the oil palm establishments, most of them were subsistent oriented, where they had a high dependence on the local resources, especially on plants. Before the establishment of oil palm companies, most of these oil palm villages have experienced a touch of modernization, where they no longer use medicinal plants but go to the local community health centres for medicines. Furthermore, markets were slowly developed, where the people could buy the food plants they needed.

Overall, there was an indication of high anthropocentric views of the local people towards wildlife for both the oil palm dependents and those who were not, indicated by the high negative responses (disagree) towards the impacts of oil palm plantations on wildlife as sources of income from all respondents strata. Despite biodiversity loss regarding ecology and economy, nevertheless, the local people perceived the ecological biodiversity loss as not a loss since they have adjusted their standards of living or even shift livelihood to continue survival, in the presence of better economic opportunities. The local people now have a higher economic living from oil palms. Thus they are becoming more market-oriented.

4. Conclusion

The local people's perceptions towards biodiversity indicated that there had been some declines in the frequencies of biodiversity sightings across the study sites, with KJNP and BPME in Riau Province showing the most species loss. The varying perceptions among indigenous people, migrants and workers simply indicated that perceptions on biodiversity were mostly influenced by their economic value in fulfilling household needs. Among all study sites, NTYE showed the least varied perceptions among the respondents and between the oil-palm dependent and those who were not. The establishment of oil palm plantations was a part of the modernization process, which embraced transformation of household economic pattern that led to a change in the household means of living. Thus, although ecologically the areas have experienced declines in species richness, however, from a social perspective, they were not considered as losses, since the people tent to show preferences over better economic opportunities.

Although the people acknowledged that changes in gains and loss of both plants and wildlife species richness were due to land conversion to oil palms, nevertheless, the majority agreed that their impacts upon their livelihoods were insignificant, since, in the absence of the species, they still enjoyed the welfare gains from the plantations

Acknowledgement

This work was supported by PT. SMART Tbk. Thanks are also forwarded to all the Directors and Staffs in the six oil palm estates for the permissions and facilities provided during the study period.

References

[1] N. Titeux, K. Henle, J. Mihoub, A. Regos, I.R. Geijzendorffer, W. Cramer, P.H. Verburg, L.L. Brotons, Glob. Change Biol, 1-11 (2016)13P.A. Harrison, P.M. Berry, G. Simpson, J.R. Haslett, M. Blicharska, M. Bucur, R. Dunford, B. Egoh, M. Garcia-Llorente, N. Geamănă, W. Geertsema, E. Lommelen, L. Meiresonne, F. Turkelboom, EcosystServ, **9**, 191-203 (2014)

[2] M. Colchester, S. Chao, J. Dallinger, H.E.P. Sokhannaro, V.T. Dan, J. Villanueva, *Oil palm expansion in South east Asia: trends and implications for local communities and indigenous peoples*, (Forest People Programme and Perkumpulan Sawit Watch, 2011)

[3] D.Sheil, A.Casson, E. Meijaard, M. Van Noordwijk, J. Gaskell, J. Sunderland-Groves, K. Wertz, M. Kanninen,*The impacts and opportunities of oil palm in Southeast Asia: What do we know and what do we need to know?*(CIFOR, Bogor, 2009)

[4] K.M. Carlson, L.M. Curran, D. Ratnasari, A.M. Pittman, B.S.S Filho, G.P. Asner, S.N. Trigg, D. Gaveau, D. Lawrence, H.O. Rodrigues, Proceedings of the National Academy of Sciences **109**, 7559-64 (2012)

[5] L. Hein, P. van der Meer,CurrOpin Environ Sustain,**4**, 604–11(2012)
[6] G. Koczberski, G.N. Curry, V. Bue, Asia Pacific Viewpoint, **53**, 288-299 (2012)
[7] K. Obidzinski, R. Andriani, H. Komarudin, A. Andrianto. Ecol Soc,**17**, 25 (2012)
[8] R. Myers, A. Ravikumar, A.M. Larson, Benefit sharing in context: a comparative analysis of 10 land-use change case studies in Indonesia, (CIFOR, Bogor, 2015)
[9] P. Semedi. Forum for Development Studies, **41**, 233-52(2014)
[10] M.A Shahputra, Z. Zen, *IOP Conference Series: Earth and Environmental Science* (Vol. 122, IOP Publishing, 2018)15S.K. Yadav, G.C. Mishra, IJESDM, **4**, 68-72 (2013)
[11] N. Sharma, B. Bohra, N. Pragya, R. Ciannella, P. Dobie, S. Lehmann, Food Energy Secur, **5**, 165-183 (2016)
[12] J.L. Pfund, J.D. Watts, M. Boissiere, A. Boucard, R.M. Bullock, A. Ekadinata, S. Dewi, L. Feintrenie, P. Levang, S. Rantala, D.Sheil, Environ Manage, **48**, 334-49 (2011).16G. Carrus, M. Scopelliti, R. Lafortezza, G. Colangelo, F. Ferrini, F. Salbitano, M. Agrimi, L. Portoghesi, P. Semenzato, G. Sanesi, Lands. Urban Plan, **134,** 221-228 (2015)
[13] P.A. Harrison, P.M. Berry, G. Simpson, J.R. Haslett, M. Blicharska, M. Bucur, R. Dunford, B. Egoh, M. Garcia-Llorente, N. Geamănă, W. Geertsema, E. Lommelen, L. Meiresonne, F. Turkelboom, EcosystServ**,** **9**, 191-203 (2014)
[14] S. Aikanathan, The oil palm plantation industry in Malaysia: a study of sustainable management (University of Malaysia,Kuala Lumpur, 2013)
[15] S.K. Yadav, G.C. Mishra, IJESDM, **4**, 68-72 (2013)
[16] G. Carrus, M. Scopelliti, R. Lafortezza, G. Colangelo, F. Ferrini, F. Salbitano, M. Agrimi, L. Portoghesi, P. Semenzato, G. Sanesi, Lands. Urban Plan, **134,** 221-228 (2015)
[17] D. Tilman, F. Isbell, J.N. Cowles, Annu Rev Ecol Evol Syst, ***45***, 471-493 (2014)
[18] B.J. Cardinale, J.E. Duffy, A. Gonzalez, D.U. Hooper, C. Perrings, P. Venail, A. Narwani, G.M. Mace, D. Tilman, D.A.Wardle, A.P. Kinzig, G.C. Daily, M. Loreau, J.B. Grace, A. Larigauderie, D.S. Srivastava, S. Naeem, Nature, **486**, 59-67 (2012)
[19] R.S. de Groot, M.A. Wilson, R.M.J. Boumans, Ecol Econ, **41**, 393-408 (2002)
[20] T.M. Daw, C.C. Hicks, K.Brown, T. Chaigneau, F.A. Januchowski-Hartley, W.W. L. Cheung, S. Rosendo, B. Crona, S. Coulthard, C. Sandbrook, C. Perry, S. Bandeira, N.A. Muthiga, B. Schulte-Herbrüggen, J. Bosire, T.R. McClanahan, Ecol Soc **21**, 11 (2016)
[21] S.R. Kellert, S.R, E.O. Wilson, The Biophilia Hypothesis, (Island Press, Washington DC, 1993)
[22] S.R. Kellert, M.O. Westervelt, Children's attitudes, knowledge and behaviors toward animals (Phase S) (U.S. Fish 8c Wildlife Service, Washington, DC, 1983)
[23] G.E.Machlis, J. Biodiverse Conserv., **62**, 161-70(1992)
[24] N. Small, M. Munday, I. Durance, IJGEnvl, **44**, 57-67(2017)
[25] L. Feintrenie, S. Schwarze, P. Levang, Ecol Soc 15, 47 (2010)
[26] M.J. White, L.M. Hunter LM. 2009. Social Sc*ience Quarterly **9**, 960-82 [2009]*
[27] W.L. Neuman, Social research methods: qualitative and quantitative approaches (Pearson, Boston, 2006)
[28] E, Sari, Persepsi masyarakat sekitar tentang dampak perkebunan sawit PT. Surya AgrolikaReksa dan PT AdimulyaAgrolestari (IPB, Bogor, 2017)
[29] R.K. Nanyunja, Human perceptions of biodiversity loss: case studies of Sango Bay, Lake Mburo National Park and Rubaale Grasslands, S.W. Uganda, (LUCID Working Paper Series Number 29, Uganda, 2002)
[30] Y. Santosa, A. Sunkar, Erniwati, I. Purnamasari, *Sejarah Perkembangan Status, Penggunaan Lahan, dan Keanekaragaman Hayati Kebun Sawit Indonesia* (Badan Pengelola Dana Perkebunan Sawit dan Lembaga Penelitian Pengabdian Kepada Masyarakat IPB, Bogor, 2016)
[31] Y. Santosa, A. Sunkar, I. Hilwan, L.N. Ginoga, I. Purnamasari, *Biodiversity loss dan gain di perkebunan sawit PT. Smart Tbk* (IPB Press, Bogor, to be published)
[32] J. Sudrajat, Paramita, **26**, 2 (2016)

[33] A. Treves, J. Bruskotter, Science, **344**, 476 (2014)

[34] L. Naughton-Treves, A. Treves, *Socio-ecological factors shaping local support for wildlife: Crop-raiding by elephants and other wildlife in Africa*. In: Woodroffe R, Thirgood S, Rabinowitz A(eds), *People and Wildlife: Conflict or Coexistence?* (Cambridge University Press, New York, 2005)

[35] J.S. Kahler, M.L. Gore, bio Conserv, **189**, 49-58 (2015)

[36] Z.R. Mir,A. Noor, B. Habib, G.G. Veeraswami, Mt Res Dev,**35**, 392-400, (2015)

[37] R. Kansky, A.T. Knight, Biol Conserv, **179**, 93-105 (2014)

[38] J.J. Vaske, J. Roemer, J. Taylor, Wildl. Soc. Bul., **37**, 122-128 (2013)

[39] K.J. Pelzer,Toean kebouen dan petani: Politik kolonial dan perjuangan agraria di Sumatera Timur 1863-1947, (Sinar Harapan, Jakarta, 1985)

[40] M.J. White, L.M. Hunter, Soc Sci Quarterly, **9**, 960-82 (2009)

[41] A. Schrevel, Chapter 6: Oil-palm estate development in Southeast Asia: consequences for peat swamp forests and livelihoods in Indonesia, (FAO, Rome, 2008)

[42] T. Susanti, Edu-Bio, **3**, 14-28 (2012)

Comparative analysis between smallholder and partnership oil palm farming system in Jambi Province

Dompak Napitupulu*, Ernawati HD, Elwamendri, Mirawati Yanita, Gina Fauzia

[1]Lecturer at Agriculture Faculty, Jambi Province

*E-mail: dompak@unja.ac.id

Abstract. Indonesia is becoming the largest palm oil producer in the world. This commodity estate farm is carried out by smallholder, government, as well as big private companies. As the law rules it, both government and big private companies are urged to run the oil palm plantation in partnership system. This study is aimed to compare the oil palm estate revenue between smallholder and partnership farmer in Jambi Province. Data is analyzed in quantitative descriptive model analysis. The results showed that the average farm income of smallholder is significantly higher than the oil palm partnership farmer. Statistical Analysis also showed that the fresh fruit bunches price (TBS), fertilizer, and labour factors used were significantly affected the oil palm estate benefit.

1. Introduction

Indonesia was still known as an agricultural country. Indonesia's economy is still mostly dependent on the agricultural sector. Compared to eight other economic sectors, the agricultural sector played as a major contributor to Indonesia's GDP. The contribution of agricultural sector to national income (GRDP) in 2014 was 15.40%. Aside from being a major contributor to national income, the agricultural sector also acts as the largest employment provider, with around 40% of Indonesia's population still working as farmers. Plantation sub-sector as one of the sub-sectors of the agricultural sector is also noted as the main contributor to the country's foreign exchange earnings. The Indonesian plantation sub-sector in its management can be differentiated into the Smallholder Plantation, Large Private Plantation (PBS) and the Large State Plantation (PBN). State and Private Plantations are generally managed in the form of Nucleus Estate Smallholder Crops (NES), a pattern of plantation development that involves the smallholder as plasm partners and big plantations on the other side as the nucleus that plays a patronage role. Large plantations play a role in helping and leading the smallholder plantations as plasma in a mutually beneficial, complete and sustainable system of cooperation [1].

The most famous crops grow in nucleus estate crops system lately in Indonesia is Oil Palm. Oil palm plantation is growing fast in Indonesia either managed by smallholder farmer, big state plantation or large-scale private corporation. Oil palm is now become a very important crop to Indonesia as well as Jambi Province economy through its large contribution to Indonesia foreign exchange earning [2]. Besides that, oil palm plantations can be used as basic livelihoods to farmers especially in Jambi Province. Data shows that since it was introduced to the community through the transmigration program in 1993 in Jambi Province, the oil palm plantations has multiplied and can be found in almost all districts except Kerinci Regency in Jambi Province. Jambi Province Plantation

Office shows that in 2014 there was 662864 Ha of oil palm plantation in Jambi Province. Not less than a half of it managed by 200,991 households in smallholder scheme farm.

Comparing both types of palm oil farming scheme become more interesting to be studied further. As it was commonly known, smallholder farmers are mostly own limited capital for they most of the time looked for cheaper and lower quality inputs and has limited access to palm oil mill that made them receive less price than what their colleagues made in partnership scheme. However, on the other hand, partnership schemed farmer even though received higher price, they have to share some part of their oil palm production to pay inputs cost and various caring expenses paid by patron during immature oil palm periods. This study is aimed to compare the oil palm estate revenue between smallholder and partnership farmer in Jambi Province and to analyze factors affect the the oil palm farming income.

2. Research methods

This research was carried out in survey method at Tanjung Jabung Barat and Muaro Jambi Regencies, Jambi Province. The selection of the two districts location was made purposively under the consideration that the two districts were the centres of oil palm production in Jambi Province.
Furthermore, in each district was selected one sample sub-district under consideration of having the largest oil palm plantation area, productivity and the ratio of the number of smallholders per ha. The sample districts selected in Tanjung Jabung Barat Regency was Renah Mendaluh Sub-District, while in Muaro Jambi Regency, Maro Sebo Sub-District. At the village level, two villages Lampisi and Lubuk Kambing villages were chosen in Renah Mendaluh Sub-District, while in the Maro Sebo, Niaso and Jambi Kecil Villages were selected.

Respondents in this study were PIR plasm oil palm farmers and smallholder oil palm farmers who manage oil palm with plantation age at least 10 years old and have a land of 2 ha or more. The focus of this research is on oil palm farming activities carried out by farmers including crop maintenance, costs incurred, prices, income received, as well as other data relating to the decisions of oil palm farmers in managing their oil palm plantations.

Data collected in this study consist of primary and secondary data. Primary data is collected on direct interviews by using structured listed questioner with farmers whose data is taken from the year 2017 oil palm plantation period. While secondary data were obtained from literature studies either by book survey, journals and scientific papers evaluation that have been recorded and published.

Sampling is taken by using the snowball method (nonprobability sample) which is started by an initial respondent decided based on village head information. The next respondent was determined based on information from the sample farmers who had been interviewed before. The number of samples in each village was determined by 10 farmers representing plasma farmers and 10 independent smallholders in each village so that 40 farmers were represented representing the plasma farmer population and 40 farmers representing independent smallholder populations. The snowball method (nonprobability sample) is used due to the limited official information about sub-population intended in this research.

The data collected in this research was analyzed in income and multiple linear regression analysis. The first purpose was analyzed by using income analysis which consists of total revenues minus the total production costs. The cost structure in this study referred to fixed costs and variable costs. Mathematically it can be written as follows:

$$Pd = TR - TC$$

Where :
Pd = Oil Palm Farm Income (IDR)
TR = Total Revenue (IDR)
TC = Total Cost (IDR).

To analize factors efecting oil palm farm income was analized by using multiple linear regression analysis, with matemathical model:

$$Y = a + b_1X_1 + b_2X_2 + b_3X_3 + b_4X_4 + b_5D + e$$

Where:

Y	= Oil Palm Plantation Total income (IDR/Year)
X_1	= Depreciation Cost (IDR/Year)
X_2	= Fertilizer Cost (IDR/Year /Ha)
X_3	= Herbicide Cost (IDR/Year /Ha)
X_4	= Labour Cost (IDR/Year /Ha)
D	= Farm System
D	= 1 for PIR Plasm
D	= 0 for smallholder Oil Plantation
$b_1, b_2, \ldots, b_5$	= Estimated Parameters.
e	= Error

Depreciation Cost become one factor effecting the Palm Oil Income in this study is to represent the fix cost. Some authors may be write that fix cost is related to the sunken cost that need not tobe counted in effecting annual plant farm production. Never the less [6] said that both fix cost and variable cost should be counted to be variables in analyzing farm income. Depreciation could affect the farm income through the size of the farm. The bigger is the farm size the higher will be expenditure paid to buy utilities and services.

Fertilizer, herbicide, and labor are known as production inputs. As long as the production process is in efficient phase, the higher the inputs is the higher the production will be [3], [4], [5], [6]. These three kinds of inputs then should be counted as variables in the model.

The last variable, farm system, adopted as a variable in the model for it is scientifically believed that the way of large entrepreneur runs business is differ to smallholder does. Large enterprise mostly run its business in commercial manner while small holder farm tends to be act in subsistence manner [7]. Farm system is then eligible to be counted in the model.

To know the effect of variables simultaneously is analyzed by using F test with the formula:

$$F_{value} = \frac{JK\ Regresi}{JK\ Residu}$$

Hypothesis build to be proved in this study was:

H_O : $R^2 = 0$ (no effect of hypotesized variables)

H_1 : $R^2 \neq 0$ (there is a significant effect of independent variable in to dependent variable)

The decision rule was:

if $F_{value} < F_{table}$ ($\alpha = 5\%$ db = k ; n-k-1); eccept H_o

if $F_{value} \geq F_{tabel}$ ($\alpha = 5\%$ db = k ; n-k-1); reject H_O

Furthermore to evaluate the effect of the independent variable to the dependent variable partially was approached by using $t_{student}$ test with the formula:

$$t = \frac{b^1}{Se\ (b^1)}$$

Where:

t	= t_{-value} for each independent variable
b	= regression coefficient
Se	= standard error

With the built hypothesys:
$H_o : b_i = 0$
$H_1 : b_i \neq 0$

The decision rule was:
if $t_{value} < t_{\alpha = 5\%\ db = 79}$; eccept H_o
if $t_{value} \geq t_{\alpha = 5\%\ db = 79}$; reject H_o

3. Results and discussion

Farm management is a science that studies the ways farmers combine and operate various factors of production such as land, labour, and capital as the basis for how farmers choose the type and size of farm branches in the form of crops or livestock to provide maximum and continuous revenue [3]. Further more, According [4], revenue is all income earned from farming for a period calculated from the sale or reassessment. Farming is a series of processes that aim to obtain optimal production results by using certain production factors. According to [5], low farm income is mostly influenced by several factors, namely the small size of the farm and the use of farming production elements, and the low skills of farmers in managing their farming.

Oil palm plantation has good prospects that promise good income to the farmers. Data showed that oil palm plantation is one of the mayor estated crop after rubber which is grown in Jambi Province as well as in the research area. Just like other agribusiness management, oil palm farming is inseparable from farming costs, where the costs incurred in the production process [4], [5]. Just like other farm cost schedule, oil palm farming cost can be classified into 2 (two) categories, namely fixed costs and variable costs. Fixed cost is any cost incurred in the farming production process that is not affected by the amount of production. In this study, fixed costs consist of *dodos*, *gancu*, *eggrek*, *angkong*, axes, machetes, and hoes used in the production process. The depreciation cost of the tool referred to in this study is the cost that has been reduced in value (depreciation) or the amount of money spent by farmers to buy farming equipment used in a relatively long time. Non-fixed costs (variable costs), namely the cost of which the size is affected by the product obtained. For example, costs for production facilities. If you want higher production, then labour needs to be added, fertilizer needs to be added, and so on. So that these non-fixed costs are variable depending on the size of the desired product. Non-fixed costs referred to in this study are the cost of fertilizer, herbicide costs and labour costs.

Farm income is farm revenue from all sources of farming, including the number of additional inventories, the value of sales of products, the value of home use and consumption [5]. In general, farmers expect the income earned from their farming activities will always be greater than the costs incurred on the farm. Revenue referred to in this study is the amount of FFB production per year multiplied by the price of FFB at the farmer level at the time of sale, while the production costs incurred are equipment depreciation costs, fertilizer costs, medical expenses and labour costs [4], [5]. The average partnership schemed (PIR) farmer income from oil palm farming is IDR 58,959,129.26/year. Moreover, the income of smallholder farmers from oil palm farming is IDR 92,204,588.24/year.

Further more, according to [6], farm income is the difference between revenue and all costs incurred by a certain period. The income received by farmers is affected by revenues and production costs. If the income received by large farmers and production costs incurred is small, then the income that will be received by farmers will be higher and vice versa if the income received by small farmers and production costs incurred is large then what will be received by farmers will be smaller. As it was also found by [8], [10], what greatly affects the size of the farmer's income level is revenue and production costs. If the revenue is large while the production costs are small then the income will be even greater, and vice versa if the costs are greater than the income gets smaller. Sample farmer income in the study area comes from oil palm farming. Where income is derived from the difference between receiving and total production costs. Production costs incurred are equipment depreciation costs, fertilizer costs, medical expenses and labour costs. The oil palm farming income of the sample farmers in the study area can be seen in table 1 below:

Table 1. Average Revenue, Total Production Costs and Income of Oil Palm Farming in 2017 Research Area

Description	Partnership Farmer (IDR/Ha)	Smallholder Farmer (IDR/Ha)
Revenue	58.959.129,26	92.204.588,24
Total cost	3.064.179,94	2.514.083,33
Depresiasion Cost	233.346,60	237.862,75
Fertilizer Cost	470.000,00	270.129,63
Herbicide Cost	369.722,22	360.764,71
Labour Cost	1.991.111,11	935.898,15
Benefit	55.894.949,32	89.690.504,90

At table 1 above, it can be seen that the average oil palm farming income of partnership schemed farmer is IDR. 55,894,949.32/year/ha, the largest production costs incurred were labour costs, which amounted to IDR. 1,991,111.11/year/ha. That labour cost contribute more to the smalholder oil farm cost also agree to [8], [9] found. While the average oil palm farming income of independent smallholders is IDR. 89,690,504.90/year/ha, the largest production costs incurred are labour costs, which are IDR. 935,898,15/year/ha. To find out the magnitude of the influence of FFB price factors, equipment depreciation costs, fertilizer costs, herbicide costs, labour costs and dummy (business pattern). The results of data analysis from multiple linear regression can be seen in table 2 below:

Table 2. Estimating Regression Parameters Results of Factors Affecting Oil Palm Farm Income in Jambi Provice, Year 2017

Variable	**Coefficients (b)**	**Standard Error (S_{bi})**	**t-value**
Intercept	198.771.705,2	24.056.676,56	8,263
FFB Price	75.134,281	17.075,193	4,400
Depreciation Cost	-32,849	43,816	-0,750
Fertilizer Cost	-85,582	33,046	-2,590
Herbicide Cost	-19,676	17,169	-1,146
Labour Cost	-91,690	7,433	-12,335
Farming System (Dummy),	-8.625.240,342	6.996.215,694	-1,233
R Square	0,922		
F_{value}	73,274		

Note: D=1 for Partnership Farmer,
D=0 For Smallholder Farmer

Based on the regression result as presented on table 2 above, it is known that the oil palm (FFB) independent variable has a positive effect, that any additional change in FFB prices will cause an increase in farming income as much as IDR 75.134 per Ha per year. On the other hand, the other variables which represent the cost expenditure in managing oil palm plantation effect negatively into the farming income. That price of FFB effect positively and labour cost effect negatively are also agre to what [8], [10] found. The limitation of labour available in household made them to pay commercial labour force which is made higher expenditure and reduce income gained from the oil farm.

The other variable, dummy variable, goes to the direction with input cost also shows negative coefficient which means that there is a tendency though it was not significant that the income of Partnership Schemed Farmer or Plasm PIR is smaller than the income of Smallholder Oil Farm farmer.

Based on the regression results of factors influencing oil palm farming income, on table 3 is presented the effect of dummy simulating on smallholders and partnership schemed farmer's income in Jambi Province.

Table 3. Regression Estimating Result of factors influencing oil palm farming income without Dummy Variable in Jambi Province Year 2017

Variable	Regression Coefficients
FFB Price	58.789,333
Depreciation Cost	-25,213
Fertilizer Cost	-84,479
Herbicide Cost	-19,799
Labour Cost	-94,806
Dummy	-
R^2	0,919
Fhitung	86,442

Comparing to the value of regression estimating results presented on table 2, the regression estimation results in table 3 is different each other, this explicitly exposes that the slope of independent variable also somehow different between the two farming system.

Based on the multiple linear regression analysis, it was obtained that the determination coefficient (R_2) value was 0.922, showed a high determination coefficient value. This coefficient means that 92.2 percent of the variation in oil palm farming income is influenced by the price of FFB, depreciation costs, fertilizer costs, herbicide costs, labour costs and dummy simultaneously, while the remaining 7.8 percent is caused by other variables. Based on the analysis of variance it was obtained that the F_{value} is 73.274 is greater than F_{table} at α = 0.05: 2.356, means that the R^2 determination coefficient was significantly different to zero. This value indicates that the independent variables, FFB price (X_1), depreciation cost (X_2), fertilizer cost (X_3), herbicide cost (X_4), and labour costs (X_5) as well as dummy variables (D), simultaneously affect the oil palm farming income in Jambi Province in the year 2017.

4. Conclusion

1. The research showed that the average income received by Smallholders and Partnership Farmer Farmers in Jambi Province in the year 2017 was different each other
2. Some independent variables namely: the price of FFB, fertilizer costs and labour costs were significantly effect on the oil palm farming income, while the depreciation costs, medical expenses and the farming system were not significantly affect the oil palm farming income in Jambi Province (2017).

References

[1] Syamsulbahri, 1996. Bercocok Tanam Tanaman Perkebunan Tahunan (Annual Planting Crops). Penerbit Gajah Mada University Press. Yogyakarta

[2] Fauzi, yan. 2002. Kelapa Sawit (Oil Palm). Penebar Swadaya. Jakarta

[3] Daniel, M. 2009. Pengantar Ekonomi Pertanian (Introduction to Agriculture Economics). Bumi Aksara. Jakarta

[4] Suratiyah, Ken, 2009. Ilmu Usaha Tani (Farm Management). Penerbit Penebar Swadaya. Jakarta

[5] Hernanto, Fadholi, 1996. Ilmu Usaha Tani. Cetakan pertama. Penerbit PT. Penebar Swadaya. Jakarta

[6] Soekartawi, 1995. Analisis Usahatani. Universitas Indonesia. Jakarta

[7] Scott, J. C., 1983. Moral Ekonomi Petani. Pergolakan dan Subsistensi di Asia Tenggara. Ed. Repro. Lp3ES. Jakarta

[8] Choiriyah, Siti. 2005. Analisis Faktor-Faktor Yang Mempengaruhi Produksi Kelapa Sawit. Skripsi Fakultas Pertanian (Analysis of Factors Affecting Palm Oil Production. Faculty of Agriculture Thesis). Universitas Jambi

[9] Noraniza Yusoff, 2016. Analysis on Cost and Profit in Farming Activity in Malaysia. Journal of

Modern Accounting and Auditing, April 2016, Vol. **12**, No. 4, 183-207
[10] Kunaifi. A. 2008. Analisis Faktor-Faktor Yang Mempengaruhi Pendapatan Usahatani Kelapa Sawit Peserta Pir-Trans Pt.Inti Indosawit Subur Dikecamatan Merlung (Analysis of Factors Affecting Pir-Trans Participants of Palm Farming Income of Pt.Inti Indosawit Subur in Merlung District). Skripsi Fakultas Pertanian. Universitas Jambi

Carbon stock change dynamics of oil palm plantation in Sembilang Dangku Landscape, South Sumatra

Azar Rachdian[1], Hariyadi[2], Yudi Setiawan[3] and Kresno Dwi Santoso[4]

[1]Study Program of Natural Resource and Environmental Management Science, Postgraduate School, Bogor Agricultural University, Baranangsiang IPB Campus, Bogor 16143, Indonesia
[2]Department of Agronomy and Horticulture, Faculty of Agriculture, Bogor Agricultural University, Darmaga IPB Campus, Bogor 16680, Indonesia
[3]Department of Forest Resource Conservation and Ecotourism, Faculty of Forestry, Bogor Agricultural University, Darmaga IPB Campus, Bogor 16680, Indonesia
[4]Zoological Society of London (ZSL) Indonesia Programme, Palembang 30144, Indonesia

*E-mail: azarrachdian93@gmail.com

Abstract. One of the land cover type in the Sembilang Dangku Landscape is oil palm plantation, which is developed by changing the type of previous land covers. Land cover change causes changes in carbon stock. If the carbon stock increase in an area, it means the area acts as a carbon sinker, whereas if the carbon stock decrease, inconsequently the area is a carbon emitter. The purpose of this study was to analyze the dynamics of carbon stock changes in 1997-2007 and 2007-2017. The research was carried out on six private oil palm plantations. Carbon stocks were estimated based on the type of land cover interpreted from Landsat 5 imagery with the unsupervised method by using Las Palmas QGIS 2.18.0 Software. Determination of land cover in 1997 was based on the year when oil palm plantation began. Carbon stock of oil palm plantation was estimated based on the Normalized Difference Vegetation Index (NDVI) with the equation: Y = 638.13 X - 242.65 (Y = carbon stock, X = NDVI). The value of carbon sequestration or carbon emission were based on carbon stock changes. The types of land cover in 1997 and 2007 were in the form of shrub swamp, oil palm plantation, undisturbed peat swamp forest, disturbed peat swamp forest; while in 2017 land cover was dominated by oil palm plantation. The results showed respectively that carbon stocks in 1997, 2007 and 2017 were 3,333,549.6 tonC, 1,541,825.5 tonC, and 1,626, 951.8 tonC. In 1997-2007, carbon stock encountered a decrease, resulting in carbon emission of 18.68 tonCO_2-eq /ha/year. However, in 2007-2017, carbon stock encountered an increase, leading in carbon sequestration of 0.89 tonCO_2-eq /ha/year.

1. Introduction

Carbon dioxide (CO_2) is a greenhouse gas that has the most significant volume, when compared to other greenhouse gases such as CH_4, NOX, CFC [1]. The increase of CO_2 can cause a greenhouse effect which implies global warming, which can cause life disturbance on this earth. In 2016, there was an increase in global CO_2 concentrations of 3.3 ppm, which was due to some tropical countries in 2015/16 encountered El-Nino that triggered land and forest fires [2].

Responding to the above condition, the Government of Indonesia since 2011 had issued a Decree to mitigate greenhouse gases through Presidential Decree Number 61 the year 2011 concerning the National Action Plan for Reducing Greenhouse Gas Emission [3]. The Government of Indonesia has a target of reducing greenhouse gas emission by 26% on its own business and 41% with international assistance in 2020. This Decree is then followed up with Presidential Decree Number 71 the year 2011 concerning the Implementation of Greenhouse Gas Inventories [4].

Monitoring carbon stock change is important to know the role of oil palm plantations as carbon sinker or carbon source [5]. The sink is defined as oil palm plantation as a carbon sequester which is shown by the addition of carbon stock. Conversely, the source is defined as oil palm plantation as a producer of carbon emissions due to a decrease in carbon stocks. Monitoring of carbon stock change is an action of President Decree Number 61 the year 2011 and President Decree Number 71 the year 2011.

By considering the role of oil palm in carbon sequestration and landscape-based management issue, it was necessary to conduct a study of carbon stock in several oil palm plantations in the Sembilang Dangku Landscape. The focus of the research was carbon stock above the soil surface, because this section has large biomass that is 87% [6] and or 81.9-93.4% [7] of total biomass. The purpose of this paper was to analysis carbon stock change dynamics in 1997-2007 and 2007-2017. This paper discussed changes in land cover type and carbon stock, carbon source/carbon sink, and carbon management in oil palm plantation.

2. Materials and methods

2.1. Time and location of research

This research was a follow-up of the research conducted by Rachdian *et al.* [8] that collected data in January 2018. The study included six private oil palm plantations with an area of 35 211 ha located in Sembilang Dangku Landscape; it consisted of one oil palm plantation in Banyuasin District and five oil palm plantations in Musi Banyuasin District. This area was dominated by peatland with a maturity level of 91% and a hemic maturity level of 9% [9].

2.2. Research procedure

2.2.1. Land use change analysis

Land cover type was obtained by interpreting Landsat 5 imagery with path/row 124/062 with the acquisition date on May 23, 1997, path/row 125/062 with the acquisition date on May 30, 1997, and path/row 125/061 with the acquisition date on August 18, 1997 and Landsat 5 imagery 124/62 path/row on April 1, 2007 with acquisition date and path/row 125/61 and 125/62 with acquisition date on October 10, 2007. Type of land cover in 2017 refers to the results of [8]. The land cover type classification used the unsupervised method with QGIS Las Palmas Software 2.18.0. *Carbon stock change analysis*. The unsupervised method was possibly used because this method is grouping pixels with similar spectral characteristics according to specific statistical criteria [10].

Analysis of carbon stock change used the stock-difference method. This method was carried out with a stock-based approach, namely carbon stock change at a certain time interval [11]. Analysis of changes in carbon stock was carried out by comparing carbon stock in 1997, 2007, and 2017. Carbon stock was estimated based on the type of land cover. The 10-year interval was used to estimate the change of land cover area and carbon stock accurately of the beginning, middle, and final condition of oil palm plantations. The value of carbon stock for each land cover type was approached with various existing research results. If there was oil palm, carbon stock was estimated by regression equation Y= 638.13X-242.65 (Y=carbon stock, X=NDVI) [8]. Carbon stock in each type of land cover is presented in table 1.

Table 1. Carbon stock of several land cover types

Land cover	Carbon Stock (ton/ha)	Source
Shrub swamp	10.68	[5]
Oil palm		Data Processing
Disturbed swamp forest	84.00	[12]
Undisturbed swamp forest	164.00	[12]

[a]Using the average value of the estimation results from the regression equation of NDVI relationship with carbon stock

Calculation of the percentage change in carbon stocks was carried out from 1997 to 2007 and 2007-2017 by using this following formula:

$$\Delta C = \frac{C_t - C_i}{Ci} x\ 100\ \% \quad (1)$$

Description:

ΔC = Carbon stock change (%)
C_t = Carbon stock of t-th year
C_i = Carbon stock of i-th year

Then to know the increase or the decrease of carbon stock per ha per year it was used the formula as follows:

$$\mathrm{C} = \frac{\mathrm{C_t} - \mathrm{C_i}}{\mathrm{L}/(t-i)} \quad (2)$$

Description:

C = The average of carbon stock increase/decrease per ha per year (ton/ha/year).
C_t = Carbon stock of t-th year
C_i = Carbon stock of i-th year
L = Oil palm plantation area

2.2.2. Determination of Carbon Dioxide Equivalent (CO_2-eq)

Carbon source occurred when carbon stock if there is a decrease in carbon stock in a certain period. Conversely, carbon sink will occur if there is an increase in carbon stock in a certain period. The amount of carbon lost or added can be converted to equivalent carbon dioxide (CO_2-eq) with the equation:

$$CO_2\text{-eq} = \text{Carbon stock x } 44/12 \quad (3)$$

3. Results and discussion

3.1. Land use and carbon stock change

Land use changes occur due to limited land accompanied by an increase in population and its interests. Land changes occur because there are efforts to utilize space appropriately to get optimum benefits for various interests, both agricultural and non-agricultural [13]. Growth and transformation of change in the socio-economic structure of developing communities can lead to land use change, both permanent and temporary [14]. One of the phenomena of land use change is the change of land from natural cover to oil palm; this happened in the Sembilang Dangku Landscape. The land use change began dramatically since 1990, was occurred with a decrease in the area of dryland forests, mangrove forests, swamp forests and swamp areas caused by the conversion of forest to intensive agriculture, one of which was the development of good oil palm plantations by farmers and companies that now number up to 98 oil palm companies [9].

Land use change in Sembilang Dangku Landscape into oil palm plantation has implications for changes in the value of carbon stocks. This is caused by the quantity of carbon stocks in a landscape influenced by (a) vegetation conditions, namely the type, amount and density of trees, and (b) environmental factors, which affect photosynthesis and growth processes, namely sun radiation, water content, temperature and soil fertility [15]. The increase of carbon stocks can occur if the previous area has low carbon stock such as open land and shrubs. Conversely, the decrease of carbon stock can occur if the previous area is in the form of high-density forests. Land changes from shrubs to oil palm can increase carbon stock [16], but on the contrary, the change from forest land to oil palm reduce carbon stocks [17]. Increased carbon stock can occur in areas that are purely oil palm, due to the expansion of plant area and/ or the size of the plant.

In 1997 and 2007, land cover in Sembilang Dangku Landscape was shrub swamp, oil palm, disturbed peat swamp forest and undisturbed peat swamp forest. In 2017, land cover consisted of shrub swamp and oil palm. In 1997 land cover was still dominated by undisturbed peat swamp forest covering an area of 18 366 ha or 52% of the total area, then in 2007 it was dominated by oil palm by 15 885 ha or 45% of the total area, and in 2017 oil palm was still dominated by 30 744 ha or 87% of the total area. Type of land cover in Sembilang Dangku Landscape in 1997, 2007, and 2017 as in figure 1 with details of the area and value of carbon stock for each land cover type in each location as in table 2.

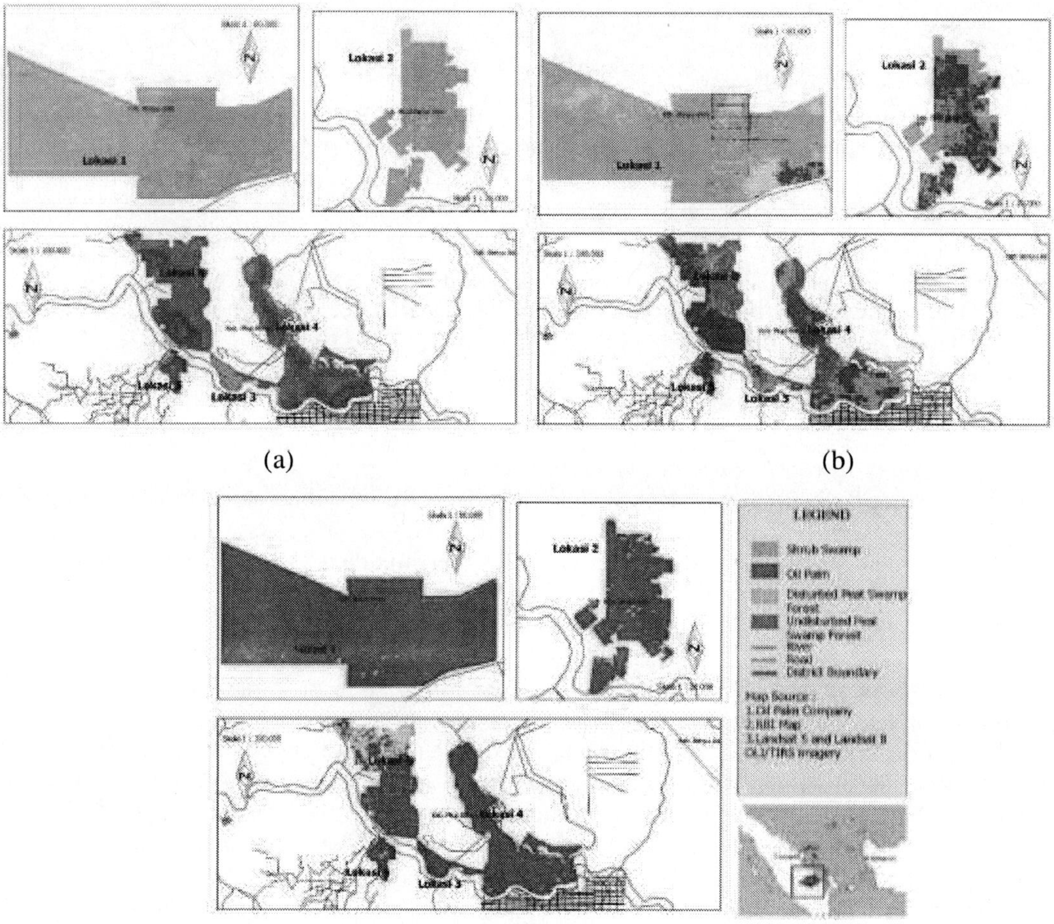

(a) (b) (c)

Figure 1. Type of land cover of oil palm plantations in Sembilang Dangku Landscape: (a) year 1997; (b) year 2007; (c) year 2017

Table 2. Land cover and carbon stock change of oil palm plantation in Sembilang Dangku Landscape

Land cover type	Year (ha)			Carbon stock (ton)		
	1997	2007	2017	1997	2007	2017
Location 1						
Shrub Swamp	692	1 571	60	7 391	16 778	641
Oil Palm	0	112	2 000	0	2 912	58 508
Disturbed Peat Swamp Forest	1 368	377	0	114 912	31 668	0
Undisturbed Peat Swamp Forest	0	0	0	0	0	0
Total	2 060	2 060	2 060	122 303	51 358	59 149
Location 2						
Shrub Swamp	6 148	3 351	475	65 661	35 789	5 073
Oil Palm	2	3 392	6 894	52	88 192	330 912
Disturbed Peat Swamp Forest	1 217	626	0	102 228	52 584	0
Undisturbed Peat Swamp Forest	2	0	0	328	0	0
Total	7 369	7 369	7 369	168 269	176 565	335 985

Land cover type	Year (ha)			Carbon stock (ton)		
	1997	2007	2017	1997	2007	2017
Location 3						
Shrub Swamp	1 164	2 158	198	12 432	23 047	2 115
Oil Palm	1 160	1 817	3 807	30 160	47 242	236 579
Disturbed Peat Swamp Forest	0	20	0	0	2 170	0
Undisturbed Peat Swamp Forest	1 681	10	0	275 684	1 640	0
Total	4 005	4 005	4 005	318 276	74 099	238 694
Location 4						
Shrub Swamp	716	3 426	0	7 647	36 590	0
Oil Palm	1 581	45 91	9 856	41 106	119 366	578 752
Disturbed Peat Swamp Forest	6	399	0	504	33 516	0
Undisturbed Peat Swamp Forest	7 553	1 440	0	1 238 692	236 160	0
Total	9 856	9 856	9 856	1 287 949	425 632	578 752
Location 5						
Shrub Swamp	180	260	146	1 922	2 777	1 559
Oil Palm	262	868	1 022	6 812	22 568	72 562
Disturbed Peat Swamp Forest	0	0	0	0	0	0
Undisturbed Peat Swamp Forest	726	40	0	119 064	6 560	0
Total	1 168	1 168	1 168	127 798	31 905	74 121
Location 6						
Shrub Swamp	277	563	3 588	2 958	6 013	38 320
Perkebunan Oil Palm	2 061	5 105	7 165	53 586	132 730	375 534
Disturbed Peat Swamp Forest	11	1 998	0	924	167 832	0
Undisturbed Peat Swamp Forest	8 404	3 087	0	1 378 256	506 268	0
Total	10 753	10 753	10 753	1 435 724	812 843	413 854
Total of Location 1-6	35 211	35 211	35 211	3 460 318	1 572 402	1 700 555

Table 2 showed the change in land cover types from 1997, 2007 and 2017 in the six oil palm plantation locations. The dynamic of land cover type and area had implications for the magnitude of change in carbon stock. The amount of carbon stock in each type of land cover was determined by the extent and carbon stock per ha. The carbon stock value per ha for undisturbed peat swamp forest, disturbed peat swamp forest, and shrub swamp land cover refers to Table 1. Oil palm carbon stock in 1997 and 2007 was thought to be based on the average carbon stock value of the regression equation, $Y = 638.13X - 242.65$ (Y = carbon reserve, X = NDVI) with the results of 26.00 tonC / ha. The dynamics of carbon stock changes in each oil palm plantation location as in figure 2.

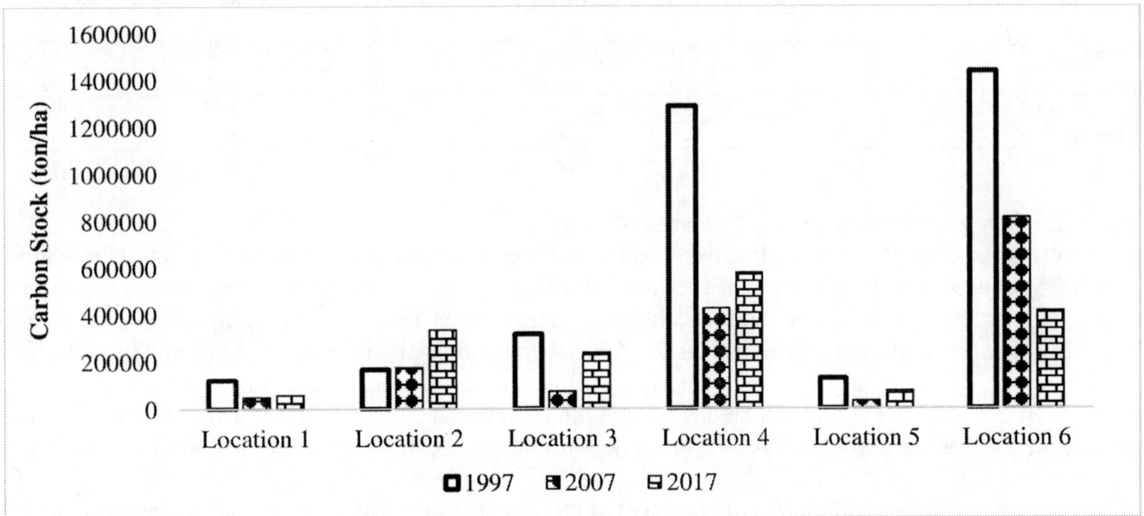

Figure 2. The carbon stock change dynamics of oil palm plantations in Sembilang Dangku Landscape

Figure 2 showed that location which encountered a continuous increase in carbon stock was Location 2 and the location that encountered a decrease continuously in carbon stock was Location 6. The continuous increase in carbon stocks in Location 2 was due to the location being initially in the form of shrub swamp which later turned into a dominated area by oil palm with a plant age of > 10 years [8]. At that age, oil palm can store a large amount of carbon stock.

Location 2 was different in conditions with Location 6. Location 6 was originally dominated by undisturbed peat swamp forest with relatively high carbon stock. Change in the type of land cover become oil palm with implication for reduced carbon stocks. There was an area that had not been planted, so the condition is still in the form of shrub swamp with relatively low carbon stocks.

3.2. *The role of oil palm plantation in CO_2 dynamics*

In 1997 - 2007 the decline in carbon stocks occurred in almost locations. The highest decrease in carbon stock was in Location 3, which was 77%. The decrease in carbon stocks was due to a decrease in the area of undisturbed peat swamp forest of 99% of the total area in 1997 or decreased by 1 671 ha. Carbon stock increase occurred in Location 2 by 5%; this was due to a decrease in the shrub swamp area of 45% or 2 797 ha and an increase in oil palm plantations that were very large, it was 390 ha from 1997 which was only 2 ha.

In 2007 - 2017 there was an increase in carbon stock in almost all locations caused by increasing age of oil palm which was able to absorb and to store carbon in more significant quantities. The biggest increase in carbon stock occurred in Location 3 of 222% caused by the aging of oil palm which grew older and the decline only occurred in Location 6 by 49% due to the burning of land into shrub swamp in 2015.

An area can act as a source or CO_2 sink depending on the difference between carbon stock before and after land use change [5]. Source means CO_2 emission occurred due to a decrease in carbon stock in a given period, while sink means CO_2 sequestration occurs due to an increase in carbon stock. Period 1997-2007 five locations act as source CO_2, but from 2007 to 2017 five locations have acted as CO_2 sinks. More fully the percentage changes in carbon stock, the role of oil palm plantation, and CO_2-equivalent values as in table 3.

Table 3. Percentage of oil palm carbon stock change in Sembilang Dangku Landscape

Location	Carbon stock change percentage (%)		Role		CO_2-equivalen (ton/ha/year)	
	1997-2007	2007-2017	1997-2007	2007-2017	1997-2007	2007-2017
Location 1	-58	15	*Source*	*Sink*	3.44	0.38
Location 2	5	90	*Sink*	*Sink*	0.11	2.16
Location 3	-77	222	*Source*	*Sink*	6.09	4.11
Location 4	-67	36	*Source*	*Sink*	8.75	1.55
Location 5	-75	132	*Source*	*Sink*	8.21	3.61
Location 6	-43	-49	*Source*	*Source*	5.79	3.71
Location (1-6)	-54.56	8.15	*Source*	*Sink*	18.68	0.89

3.3. *Carbon management in an oil palm plantation*

As a cultivation area, the primary location is oil palm area. The similar condition of the oil palm planting area with relatively wide spacing will produce lower carbon stock compared to the area in the form of trees with various compositions and vegetation structures that resemble with natural forests. Carbon stock of oil palm plantation in Malaysia is 2 – 60 tonC/ha and tropical forest 4 – 384 tonC/ha [18]. Such condition was certainly difficult for carbon stock to equal or exceed carbon stock before opening the forest land cover. However, it will be different if the initial land cover is in the form of bare land, reeds, and shrubs, the oil palm plantation carbon stock is higher than carbon stock in the condition of the land cover [16].

Referring to the description above, the level of increase in carbon stock in oil palm plantation varies, depending on the condition of the initial land cover. There is a significant decrease in carbon stock when the start of the oil palm plantation development comes from forest land cover. This condition can occur, the rate of increase in carbon stock can not equal the initial condition. This is different if the type of land cover is in the form of reeds or shrubs. At the beginning of oil palm planting, the carbon stock can equal initial carbon stock is possible. Increasing time and oil palm area and carbon stock is increasing.

Oil palm plantation can increase carbon stock by immediately planting or rehabilitating an area that still has low carbon stocks. Planting area of oil palm after land clearing can immediately be planted. Other areas that are not productive and the HCV (high conservation value) area that is not optimal can also be immediately rehabilitated with various types of fast-growing plants and have a high ability in CO_2 sequestration.

4. Conclusion

The dynamic of carbon stock in oil palm plantations has a pattern that varies depending on the initial condition of the type of land cover and the speed of planting or rehabilitation. In general, oil palm plantations in 1997-2007 had a role as a source of CO_2 emissions, but in 2007-2017 most of them had acted as CO_2 sinkers. From six examples of oil palm plantation in Sembilang Dangku Landscape showed that the amount of CO_2-equivalent emission in 1997-2007 amounted to 18.68 tonCO_2-eq /ha/year, then in 2007-2017, there was a change in role as sinker with a value of 0.89 tonCO_2- eq /ha/ year.

Acknowledgments

Thanks are conveyed to the Zoological Society of London (ZSL) Indonesia Program which had provided funding for the implementation of this research, the management of oil palm plantations that had facilitated data collection. Also, thanks are also conveyed to the KELOLA Sendang Management Project Team, namely Mr La Ode Syamsul Iman SP, MSi from P4W IPB, Mr Ir Sofyan Zaman, MS from the Department of Agronomy and Horticulture of IPB, Mr Ir. Sri Hermawan, Mr Rio Octarizza S, SP for all the help and support.

References

[1] IPCC. 2006, 2006 IPCC *Guidelines for National Greenhouse Gas Inventories*, Prepared by the National Greenhouse Gas Inventories Programme, Eggleston HS, Buendia L, Miwa K, Ngara T, Tanabe K. (eds). Published: IGES, Japan

[2] Olivier JGJ, Schure, KM, Peters JAHW. 2017. *Trends in Global CO_2 and Total Greenhouse Gas Emissions: 2017 Report*. PBL Netherlands Environmental Assessment Agency, the Hague

[3] Pemerintah Republik Indonesia. 2011. Peraturan Presiden No. 61 tahun 2011 tentang Rencana Aksi Nasional Penurunan Gas Rumah Kaca. Jakarta (ID): Sekretariat Negara RI

[4] Pemerintah Republik Indonesia. 2011. Peraturan Presiden No. 71 tahun 2011 tentang Rencana Aksi Nasional Penurunan Gas Rumah Kaca. Jakarta (ID): Sekretariat Negara RI

[5] Agus F, Gunarso P, Sahardjo BH, Harris N, Van Noordwijk M, Killen T J. 2013.Historical CO_2 Emissions from Land Use and Land Use Change from the Oil Palm industry in Indonesia, Malaysia and Papua New Guinea [Internet]. [Downloaded on 2018 Sep 17]. Available on: http://www.tropenbos.org/resources/publications/historical+co2+emissions+from+land+use+and+land+use+change+from+the+oil+palm+industry+in+indonesia,+malaysia+and+papua+new+guinea

[6] Zhang H, Wang K, Xu X, Song T, Xu Y, Zeng F. 2015. Biogeographical patterns of biomass allocation in leaves, stems and roots in China's forests. *Scientific Report.* **5**(15997): 1-12

[7] Syahrinudin. 2005. The potential of oil palm and forest plantations for carbon sequestration on degraded land in Indonesia. *Ecology and Development Series*. (28): 1-112

[8] Rachdian A, Hariyadi, Setiawan Y. 2018. Estimasi cadangan karbon kelapa sawit di Lanskap Sembilang Dangku, Sumatera Selatan. *Media Konservasi*. **23** (3). Inpress

[9] [KELOLA Sendang] *Kemitraan Pengelolaan Lanskap Sembilang Dangku*. 2017. Bogor (ID): Zoological Society of London

[10] Al-Doski J, Mansor SB, Shafri HZM. 2013. Image classification in remote sensing. Journal of Environment and Earth Science. **3**(10) : 141-147

[11] Direktorat Inventarisasi dan Pemantauan Sumberdaya Hutan, Direktorat Jenderal Planologi Kehutanan dan Tata Lingkungan, Kementerian Lingkungan Hidup dan Kehutanan. 2015. *Buku Kegiatan Serapan dan Emisi Karbon.* Jakarta (ID): Direktorat Inventarisasi dan Pemantauan Sumberdaya Hutan

[12] Syam'ani, Agustina A, Susilawati, Nugroho Y. 2012. Cadangan karbon di atas permukaan tanah pada berbagai sistem penutupan lahan. *Jurnal Hutan Tropis* **13**(2): 148-158

[13] Junaedi A. 2008. Konsistensi dan inkonsistensi pemanfaatan ruang dan implikasinya terhadap pelaksanaan rencana tata ruang wilayah Kabupaten Sumedang [tesis]. Bogor (ID): Institut Pertanian Bogor

[14] Nasoetion LI, Winoto J. 1996. Masalah Alih Fungsi Lahan Pertanian dan Dampaknya terhadap Keberlangsungan Swasembada Pangan. Prosiding Lokakarya: Persaingan Dalam Pemanfaatan Sumberdaya Lahan dan Air; Bogor, Indonesia [Waktu tidak diketahui]. Yogyakarta (ID): Hlm 113-120

[15] Sugirahayu L, Rusdiana O. 2011. Perbandingan simpanan karbon pada beberapa penutupan lahan di Kabupaten Paser Kalimantan Timur berdasarkan sifat fisik dan sifat kimia tanahnya. *Jurnal Silvikultur Tropika.* **2**(3): 149 – 155

[16] Asmani N. 2014. *Kelapa Sawit Komoditas Unggulan Sumatera Selatan yangRamah Lingkungan.* Makalah pada Seminar Pelantikan Pengurus Gabungan Pengusaha Kelapa Sawit Indonesia (GAPKI) Sumatera Selatan, Palembang 16 Januari 2014

[17] Ramdani F, Hino M. 2013.Land use changes and GHG emissions from tropical forest conversion by oil palm plantations in Riau Province, Indonesia. *PLOS ONE.* **8**(7): e70323. https://doi.org/10.1371/journal.pone.0070323

[18] Kho LP. 2015. Carbon stock of oil palm plantations and tropical forests in Malaysia: A review. Singapore journal of tropical geography. https://doi.org/10.1111/sjtg.12100

Impact of inorganic fertilizer to soil biological activity in an oil palm plantation

Resti Wahyuningsih[1], Leo Marchand[1], Pujianto, Suhardi[1], JP Caliman[1]

[1]SMART Research Institute, Jl. Tengku Umar 19, Pekanbaru, 28112, Indonesia

*E-mail: caliman@indo.net.id

Abstract. Oil palm can produce the highest quantity of vegetable oil per unit area. In order to reach this potential, as well as to maintain, and when necessary to improve the chemical fertility of soils, inorganic fertilizers are regularly applied in plantations. As it has been reported that the addition of inorganic fertilizers in agriculture, could negatively affect the biological characteristics of the soil, we have initiated a first research program to evaluate and confirm such possibility related to oil palm cultivation. This paper presents a study about the impact of inorganic fertilizer applications of soil biological activity in oil palm plantation located in the province of Riau in Sumatra, Indonesia. Observations have been conducted in a factorial NxPxK fertilizer trial, since 1995. For this study, our observations aimed to determine the short-term impact of these inorganic fertilizers, i.e. the impact during the following days after applications, as well as the long-term impact of similar fertilizers that have been applied repetitively, every six months since 1995. Bait lamina sticks were used to assess the biological soil activity (soil fauna feeding activity) from July to December 2017. Three types of fertilizer have been used for the study, Urea (as nitrogen source), Triple Super Phosphate (as phosphorus source) and Muriate of Potash (as potassium source) at various rates, and several combinations have been tested. The bait laminas were inserted in the palm circles, where fertilizers are commonly applied. The results show that in the short-term applications of inorganic fertilizers result in an increase of the soil fauna feeding activity: two days after urea applications, soil fauna feeding activity increased significantly compared the values before application. Phosphate and potassium fertilizer applications also increased the feeding activity, but this happened five days after the respective fertilizer applications. The intensity of the increase varied depending on the rate of applications. The long-term application of potassium fertilizer significantly increased the soil fauna feeding activity, while the application of phosphorus and nitrogen fertilizers did not show a significant increase.

1. Introduction

The use of inorganic fertilizers such as NPK fertilizers in oil palm plantation is a common standard practice [1]. About 50-70% of field operational costs are accounted for fertilizers [2]. The use of inorganic fertilizers as an agricultural intensification will create its own challenge, on the one hand, productivity will increase, but on the other hand, for example, the environment will face a serious threat. Some issues of soil degradation appeared in plantation such as reduction of ecosystem services, a decline in soil biological activity, and greenhouse gas emission, etc [3]. The conservation for soil quality in the plantation is a major because soil quality plays an essential role in ecosystem service [4]. In another hand, it is important to conserve the soil quality, because the soil is an investment for commercial plantation, for example, oil palm plantation [1].

The soil quality refers to the physical, chemical, and biological properties which provide the ecosystem function and give benefit to human [5]. Study about the impact of fertilizers application (long-term and short-term observation) in soil quality on chemical and physical aspect are largely explored [6], [7], [8]. However, to the best our knowledge, the information about the impact of inorganic fertilizers application to the soil biological activity in oil palm plantation was limited. In this study, we would like to observe the impact of inorganic fertilizers application to the soil biological activity focusing on soil fauna feeding activity using bait lamina method. This test was enabled to monitoring of biotic processes in the soil within a short period [9]. Feeding activity by soil fauna as one of the biological indicator of soil quality is sensitive to the changes of soil fertility and have a strong association with soil nutrient availability which affected by organic or inorganic fertilizers [7] The aim of this study is to determine the short-term impact of these inorganic fertilizers, i.e. the impact during the following days after applications, as well as the long-term impact of similar fertilizers that have been applied repetitively, every 6 months since 1995.

2. Materials and methods

2.1. Site description

The experiment was conducted on a mature oil palm plantation at Kandista estate, Siak, Riau, Indonesia. The study was carried out in July-December 2017 at mature (planted on 1993) oil palm plantation in Riau, Sumatera. This experiment was conducted in an experimental plot that previously used for NPK factorial trial that aimed to find the optimum rates and types of NPK fertilizers application to reach the optimum yield of oil palm. Fertilizer application started from January 1995 and continued until 2017. The fertilizers applied repetitively every six months in the circle for 22 years. The oil palm plantation's age was 26 years old, with palm density 136 palms per hectare, planted in the equilateral triangular arrangement at a distance of 9 m.

2.2. Fertilizers application

The rate of NPK fertilizers application followed the table 1, and general fertilizer was applied at the same rate in all palms following table 2.

Table 1. The rate of factorial NPK Fertilizer application (start from 1996 to 2017)

Fertilizers	**Amount of fertilizer (kg/palm/year)**		
	1	**2**	**3**
Urea (46 % N)	0.375	1.75	3.125
TSP (45 % P_2O_5)	0.25	1.25	2.25
MOP (60 % K_2O)	0.5	2.5	4.5

Table 2. The rate of General fertilizer application

Fertilizers	Dosage **(kg/palm/year)**	
	2000	2001 -2017
Kieserite (27 % MgO)	0.75	1.00
HGFBorat	-	0.10

2.3. Bait lamina sampling

In this study, we used bait lamina method for assessing the soil fauna feeding activity in the impact of the long and short term of NPK fertilizer application in the soil. Bait lamina was first introduced by Törne in 1990 [9] to measure the soil biological activity. Bait lamina stick is a thin PVC strips (1mm x 6 mm x 160 mm) with 16 apertures (1,5 mm diameter) bored at 3 mm intervals. The apertures were filled with standard bait, containing cellulose powder, bran flakes and activated carbon (70:27:3). The test was based on visual assessment of feeding portion of laminated bait substrate exposed (empty holes indicate some biological activities in the soil).

The sampling took place from July to December 2017. Seven treatments were chosen to evaluate the impact of fertilizer application (treatments listed on table 3).

Table 3. The combination of NPK fertilizers

No.	Treatments
1	NIP2K2
2	N2P2K2
3	N3P2K2
4	N2P1K2
5	N2P3K2
6	N2P2K1
7	N2P2K3

Three palms per treatment were chosen as replicate. Three groups of bait lamina each contain five baits were inserted in the circle of oil palm (see figure 1). The circle is an area which applied fertilizer. At each sampling time, a total of 315 sticks were inserted for six days of exposure (3 days before and harvested three days after time sequences).

The sticks were inserted until the top aperture was just bellowed to the soil/EFB surface (±1mm). Each aperture was recorded as 0 (without perforation) and 1 (partial or complete perforation). To calculate the rate of feeding activity per bait, the number of holes (partial or complete hole) are counted then divided by 16 (the total number of holes per bait). Therefore the minimum score will be 0%, and the maximum will be 100%.

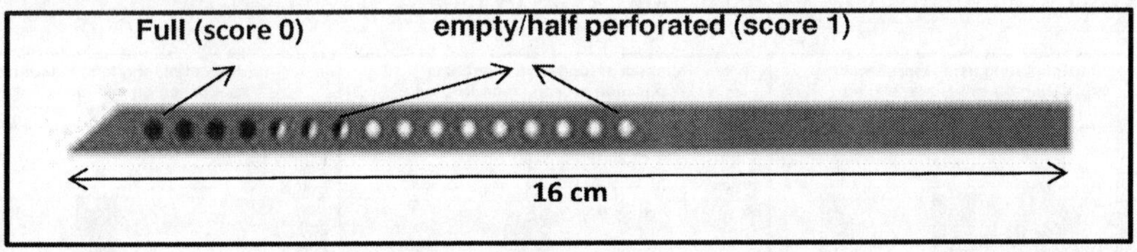

Figure 1. Scoring of bait lamina stick 0 (without perforation) and 1 (partial or complete perforation)

Additional observations such as the rainfall are recorded every time sequences during the period of the trial. Soil humidity data collected by use WET sensor, which rainfall is collected using automatic water station.

2.4. Observation

Long-term effect: Soil fauna feeding activity was observed after 22 years NPK fertilizers applied every six months since 1995.

Short-term effect: Soil fauna feeding activity was observed after NPK fertilizers application based on nine times sequence (table 4).

Table 4. Time sequences of observation

No.	Days after fertilizer's application	
1	T0	Before
2	T2	2 days
3	T5	5 days
4	T10	10 days
5	T15	15 days

No.	Days after fertilizer's application	
6	T30	30 days
7	T45	45 days
8	T60	60 days
9	T75	75 days

2.5. *Statistical analysis*
The average percentage of perforated holes of the 15 x 3 replicates bait lamina was used to represent the soil fauna feeding activity at each treatment. Repeated measure analysis was used to test the impact of fertilizer application on soil fauna feeding activity during 0-75 days. One way ANOVA was performed to test the difference in feeding activity between rates in each treatment. Significant differences in mean values were compared using the Turkey test. The Pearson correlation was used to see the correlation between soil fauna feeding activity and rainfall and other parameters that observed.

3. Results
The first trial was to study the long-term impact of NPK fertilizer to soil fauna feeding activity. Fertilizers have been applied since 22 years ago, regularly applied every semester (twice per year). The last fertilizer application was during March- June 2017. The long-term N fertilizers application in various rates (N1, N2, N3) gave an impact to soil fauna feeding activity. However, there was no statistically significant difference (P value: 0,769). The highest value showed by rate N1 (47%), and the lowest was showed by rate N2 (38%). The long-term application of P fertilizers in various rates (P1, P2, P3) tend to give a positive impact on soil fauna feeding activity. However, there was no statistically significant impact (P value: 0,245). The highest value showed by rate P2 (38%), and the lowest activity was by rate P1 (21%). The long-term K fertilizers application in various rates (K1, K2, K3) gave positive statistical significant impact to soil fauna feeding activity (P value: 0,021). The highest value of feeding activity showed by rate K3 (44%), and the lowest was by rate K1 (12%) (see figure 2).

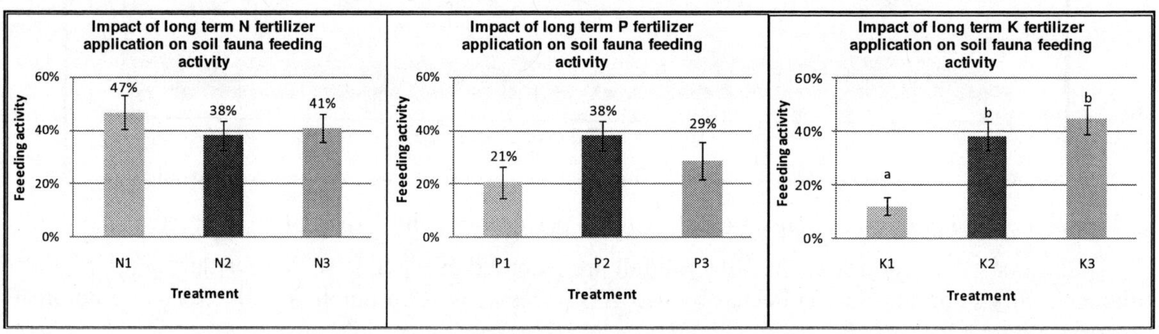

Figure 2. The soil fauna feeding activity under Effect of various amount of N fertilizer (N1:N1P2K2, N2:N2P2K2, N3:N3P2K2), P fertilizer (P1:N2P1K2, P2:N2P2K2, P3:N2P3K2), and K fertilizer (K1:N2P2K1, K2:N2P2K2, K3:N2P2K3).

The short-term impact of NPK fertilizer was observed during the following days after fertilizer application based on time sequences (table 4). After 75 days of observation, the result showed that the application of N fertilizer has a statistically significant impact on soil fauna feeding activity (P value: 0,000). Two days after N fertilizer application, the soil fauna feeding activity on rates N1 and N3 increased significantly (P value: 0.012) compared to the soil feeding activity before fertilizers application (T0). Five days after N fertilizer application rates N2 and N3 remain stable. However, rate N1 has decreased. The feeding activity in all rates N fertilizers were decreased after ten days of application and increased gradually from 15 days until 60 days after application (see figure 3).

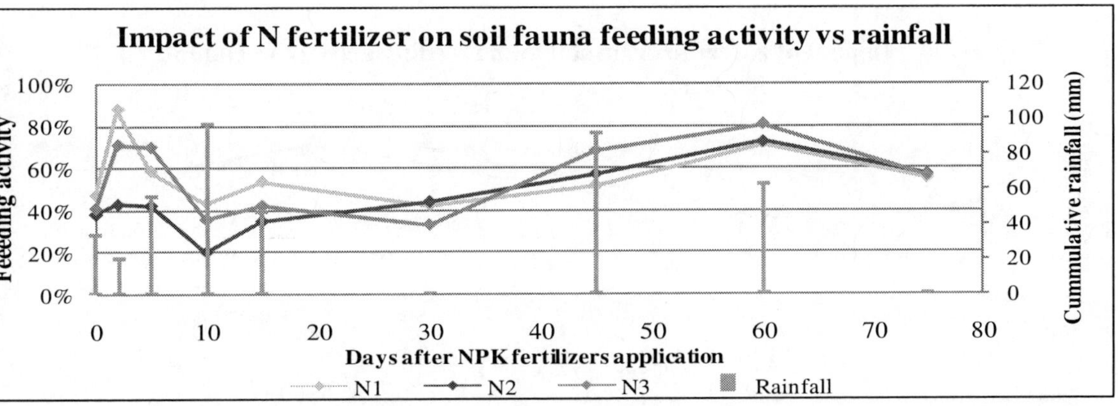

Figure 3. The soil fauna feeding activity under effect of various amount of N fertilizer (N1:N1P2K2, N2:N2P2K2, N3:N3P2K2) after 75 days of application.

The result showed that the application of P fertilizer has a statistically significant impact on soil fauna feeding activity (P value: 0,000). Five days after P fertilizer application, the feeding activity on rate P3 increased significantly compared to the soil fauna feeding activity on T0 (P value: 0,002), P1 also increased but not significant. The feeding activity in all rates P fertilizers were decreased after ten days of application and increased gradually from 15 days until 60 days after application. The feeding activity in all P rates were increased gradually; however, treatment P1 always showed the lowest feeding activity compared to the other P2 and P3 (see figure 4).

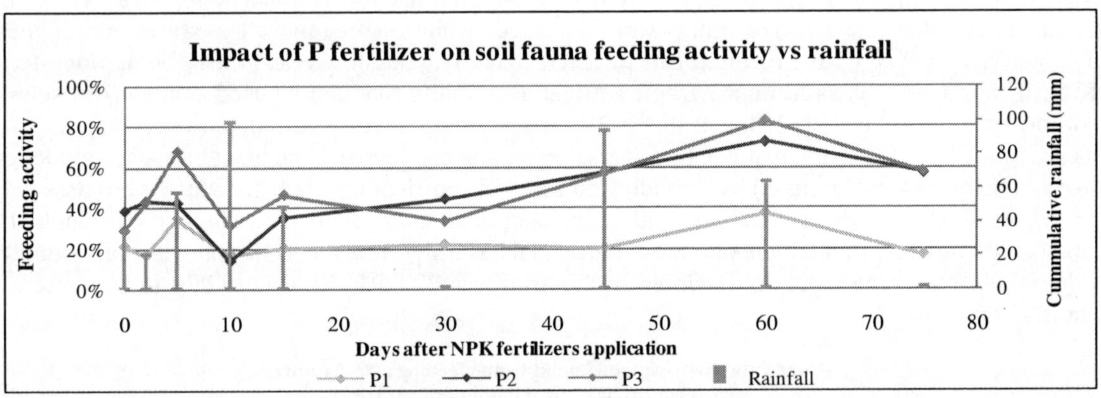

Figure 4. The soil fauna feeding activity under effect of various amount of P fertilizer (P1:N2P1K2, P2:N2P2K2, P3:N2P3K2) after 75 days of application.

The result showed that the application of K fertilizer application has a statistically significant impact on soil fauna feeding activity (P value: 0,005). Five days after K fertilizer application, the feeding activity on treatment K1 increased significantly compared to the feeding activity on T0 (P value: 0,007) while in rates K2 and K3 tend to decrease. The feeding activity in all rates K fertilizers were decreased after ten days of application and increased gradually from 15 days until 60 days after application. The feeding activity in all K rates was increased gradually, however rate K1 always showed the lowest feeding activity, and rate K3 always showed the highest between all K fertilizer rates. All treatments tend to decrease after 75 days of observation. However, the decreased was still not equal with the T0 (long-term effect) (figure 5). Rainfall during observation was varied between 0-73 mm per day. The highest cumulative rainfall during observation observed after ten days of fertilizer application (97,4 mm). Surprisingly the correlation between rainfall and soil fauna feeding activity showed no significant correlation.

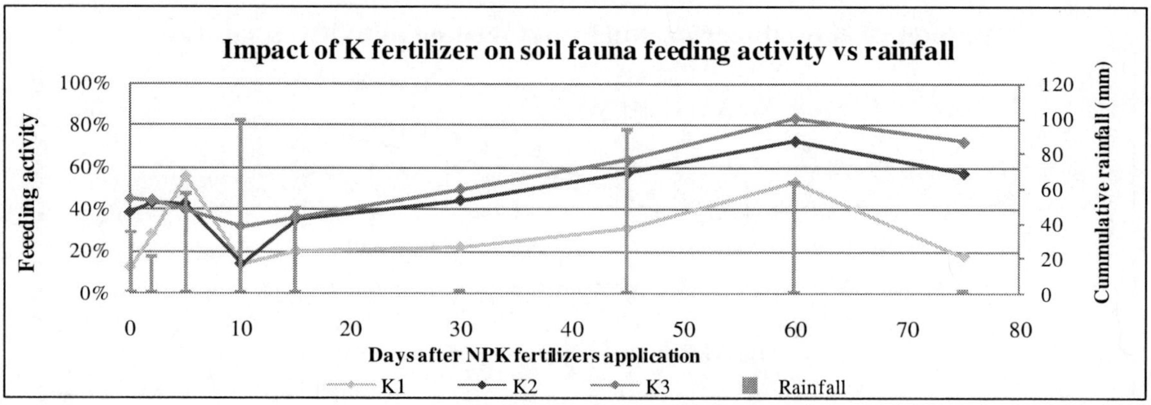

Figure 5. The soil fauna feeding activity under Effect of various amount of K fertilizer (K1:N2P2K1, K2:N2P2K2, K3:N2P2K3) after 75 days of application.

4. Discussion

4.1. Long-term fertilization effect on soil fauna feeding activity

Result showed that treatment K fertilizer gave a significant impact on soil fauna feeding activity, however treatment N and P fertilizers did not show a significant impact, this result was related with observation made by reference [11] that found if the long-term PK and N fertilizers application had no statistically significant effect on soil biological indicator (bacterial biomass or fungal biomass). Feeding activity increased when N fertilizer rates were decreased, contrary to treatment K fertilizer which the value increased when the fertilizer rates were increased, while on treatment P fertilizer the highest feeding activity showed by the medium rate of P fertilizer. The possible reason may be the long-term NPK fertilization already could improve soil nutrient availability that can be used as a nutrient source by soil organism. Reference [12] found that NPK fertilization generally stimulated microbial and soil fauna growth, while [13] found that, the fertilizers application can increase the plant biomass production (above and below ground biomass) by providing an available nutrient for plant growth, the biomass may bring higher food sources or SOC for soil faunal populations and microbial activity through the decomposition process. Bait lamina observed in this trial has a significant correlation with soil humidity (p value: 0.000), average humidity recorded in this trial was 22,84%. Ideal humidity is the preferred habitat for soil fauna.

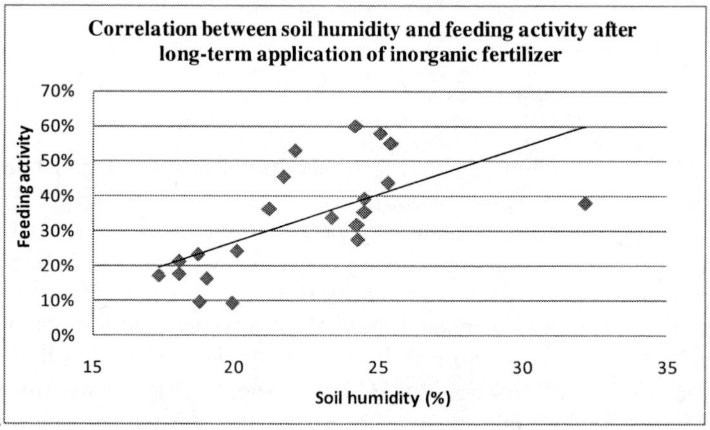

Figure 6. The correlation between soil humidity and soil fauna feeding activity after long-term application of inorganic fertilizer.

4.2. Short-term fertilization effect on soil fauna feeding activity

We found those two days after N fertilizer application; the feeding activity was increased significantly and then decreased until the minimum level on ten days after application, and then increased gradually until the end of observation. The NPK fertilizers application gave impact on soil fauna feeding activity. Treatment N fertilizer already showed a significantly different result on two days after fertilizer and then after two days until 75 days; there was no significant difference of feeding activity value. Treatment P fertilizer showed a significantly different result on five days after fertilizer application. Treatment P fertilizer showed a tendency started from 5 days until 75 days after application that the feeding activity value of rate P1 always showed the lowest value compared to P2 and P3 rates, and between P2 and P3 did not show a significant difference. This tendency also showed by treatment K, which K1 rate always showed the lowest value compared to K2 and K3. However, the significant different impact of K fertilizer application started after five days of fertilizer application. Ten days after fertilizer application the feeding activity was decreased, the possible reason may be the rainfall during the bait lamina installation was too high, the rainfall recorded on this period was about 113 mm. If the rainfall were too high, this situation would make soil pores saturated of water, thus limiting the movement of soil fauna and effected a low activity [14]. In this study, the long-term and short-term fertilizer application give impact to the soil fauna feeding activity. This might offer a chance for a scientist to find the effectiveness of inorganic fertilization rate that can give an optimal oil palm yield but still support the habitat of soil fauna (biological soil properties).

5. Conclusion

In our study, we found that after long-term inorganic fertilizers application there was no negative impact of the high rate of inorganic fertilizers application. Long-term K fertilizers gave a significant positive impact on soil fauna feeding activity. However, treatment N and P fertilizer did not show a statistically significant impact. The short-term applications of inorganic fertilizers increased the soil fauna feeding activity: two days after urea applications, soil fauna feeding activity increased significantly compared the values before application. Phosphate and potassium fertilizer applications also increased the feeding activity, but this happened five days after fertilizer applications.

References

[1] A reference Khatun, R., Reza, M. I. H., Moniruzzaman, M., & Yaakob, Z. (2017). Sustainable oil palm industry: The possibilities. *Renewable and Sustainable Energy Reviews*, *76* (December 2016), 608–619. https://doi.org/10.1016/j.rser.2017.03.077

[2] Caliman, J. P., Martha, B., & Saletes, S. (2001). Dynamics of nutrient release from empty fruit bunches in field conditions and soil characteristics changes. *Proceedings of the 2001 PIPOC International Palm Oil Congress. MPOB, Bangi*, 550–556

[3] Feller, C., Blanchart, E., Bernoux, M., Lalb, R., & Anlay, R. M. (2012). Soil fertility concepts over the past two centuries: the importance attributed to soil organic matter in developed and developing countries. *Agronomy and Soil Science*, *58*(S1), S3–S21

[4] Lavelle, P., Decaëns, T., Aubert, M., Barot, S., Blouin, M., Bureau, F., Rossi, J.P. (2006). Soil invertebrates and ecosystem services. *European Journal of Soil Biology*, *42*, S3–S15. https://doi.org/10.1016/j.ejsobi.2006.10.002

[5] Carron, M. P., Auriac, Q., Snoeck, D., Villenave, C., Blanchart, E., Ribeyre, F., Caliman, J. P. (2016). Do the impact of organic residues on soil quality extend beyond the deposition area under oil palm *European Journal of Soil Biology*, *75*, 54–61. https://doi.org/10.1016/j.ejsobi.2016.04.011

[6] NING, C. chuan, GAO, P. dong, WANG, B. qing, LIN, W. peng, JIANG, N. hao, & CAI, K. zheng. (2017). Impacts of chemical fertilizer reduction and organic amendments supplementation on soil nutrient, enzyme activity and heavy metal content. *Journal of Integrative Agriculture*, *16*(8), 1819–1831. https://doi.org/10.1016/S2095-3119(16)61476-4

[7] Šimon, T., & Czakó, A. (2014). Influence of long-term application of organic and inorganic

fertilizers on soil properties. *Plant Soil Environment*, *60*(7), 314–319. Retrieved from http://www.agriculturejournals.cz/publicFiles/125840.pdf

[8] Wang, S., Chen, H. Y. H., Tan, Y., Fan, H., & Ruan, H. (2016). Fertilizer regime impacts on abundance and diversity of soil fauna across a poplar plantation chronosequence in coastal Eastern China. *Scientific Reports*, *6*(1), 20816. https://doi.org/10.1038/srep20816

[9] Kratz, W., (1998). The bait lamina test. Soil Sciences-Research article. *Environmental, Science, and pollution*, 5 (2) 94-96

[10] Williams, A., Börjesson, G., & Hedlund, K. (2013). Soil Biology & Biochemistry The effects of 55 years of different inorganic fertiliser regimes on soil properties and microbial community composition. *Soil Biology and Biochemistry*, *67*, 41–46. https://doi.org/10.1016/j.soilbio.2013.08.008

[11] Wei, M., Hu, G., Wang, H., Bai, E., Lou, Y., Zhang, A., & Zhuge, Y. (2017). 35 years of manure and chemical fertilizer application alters soil microbial community composition in a Fluvo-aquic soil in Northern China. *European Journal of Soil Biology*, *82*, 27–34. https://doi.org/10.1016/j.ejsobi.2017.08.002

[12] Kätterer, T., Bolinder, M.A., Andrén, O., Kirchmann, H., Menichetti, L. (2011). Roots contribute more to refractory soil organic matter than above-ground crop residues, as revealed by a long-term field experiment. *Agriculture, Ecosystem, Environment*. 141, 184-192

[13] Gongalsky, K. B., Persson, T., & Pokarzhevskii, A. D. (2008). Effects of soil temperature and moisture on the feeding activity of soil animals as determined by the bait-lamina test, *39*, 84–90. https://doi.org/10.1016/j.apsoil.2007.11.007

IOP Conf. Series: Earth and Environmental Science **336** (2019) 012018 doi:10.1088/1755-1315/336/1/012018

Analysis of the feasibility of small-biomass power generation from the palm oil mill – study case: palm oil mill in Riau-Indonesia

Agus P Nuryadi[1,2], Teguh B Pratomo[2], Alfonsus A Raksodewanto[2]

[1]CAME, RWTH Aachen University

[2]Laboratory of Fuel Technology and Design Engineering, BPPT, 480 Puspiptek, South Tangerang,15314, Indonesia

*E-mail: agus.nuryadi@rwth-aachen.de, agus.prasetyo@bppt.go.id

Abstract. Solid waste from a palm oil plant in Indonesia is a tremendous potential for power generation, especially on Riau-Sumatra. The potential of electrical energy that can generate is determined by measuring or calculating the calorific value and amount of solid waste palm oil available. To be burned in the boiler continuously and more efficiently, the solid waste water content of palm oil must reduce to no more than 30%. The conservative substantial waste palm oil calorific value (HHV) is 3,500 kcal/kg. As a result, the mass and power plant balance are capable of producing a gross power of 4.6 MW using 80% efficient fluidised bed boilers and a steam turbine capacity of 5 MW with an isentropic efficiency of 85% with assumption percentages combination fibre: 30 % shell: 70 % fibre. The environmental impacts of power generation from empty fruit bunch (EFB) based on steam turbine generator technology evaluated. The results revealed that EFB utilisation for power generation is environmental friendly than fossil fuel power plants regarding global warming potential and acidification potential with GWP are 89.5 kg CO_2 eq per MWh, and 0.515 kg SO_2 eq per MWh. Also, the financial analysis of the Biomass Power Plant with a lifetime duration of 25 years, which the project is feasible to implement because IRR is 22.70%, higher than Weighted Average Cost of Capital (WACC) which amounted to 15.61%.

1. Introduction

The source energy of empty bunches is enormous potential to rise capacity electricity, especially on Sumatra Indonesia. At least about 70 palm oil mills in Sumatra managed by PT Perkebunan Nusantara (PTPN), which is a State-Owned Enterprise of Indonesia (BUMN). Based on data, the potency of energy from empty fruit bunch (EFB) is equivalent approximately 510 MW [1]. Also, the environmental impacts of power generation from EFB based on steam turbine generator technology have been evaluated. Base on study literature, the waste palm oil solid fuel is less emission than coal power plants [2].

The palm oil mill waste consists of EFB, shell and fibre bunches for utilisation as solid fuel boilers in palm oil mills to supply electricity process plant or in condition when surplus electricity they can sell electricity to other as additional profit. Empty bunches are containing high water moisture about 50% - 60% and potassium (K) 2.4% also corrosion effect will increase with Cl content, and potassium may reduce heat transfer process in the furnace [3].

For generating power generation in Indonesia there are many regulations, one is Build Operate Transfer (BOT) Scheme. BOT Scheme regulation consists in the utilisation of local property by other parties by constructing buildings or facilities and facilities. For utilised by other parties within the agreed period, then returning land and buildings and facilities and facilities after the term expiration the time of the Minister of Home Affairs Regulation Number 19 of 2016 concerning Guidelines for the Management of Regional Property Indonesia.

Thus, the study case, the location of PTPN V palm oil mill with a capacity of approximately 60 tons of fresh fruit bunches (FFB/hour will produce 12.6 tons/hour of empty palm bunches (EFB). The objective of this paper is to present an analysis of the feasibility of using the biomass resulting from residues palm oil. In addition that is combination between shell and fibre as solid fuel in a boiler in the palm oil plant where they are produced, for the generation of electric power considering economic analysis base on BOT, also, study review comparison emission value between coal and biomass base one Global warming potential and Acidification value. The main study case is calculation heat and mass balance to make sure the feasibility efficiency of biomass (shell and fibre) power generation. It is also intended for recommendation to PTPN V, to render such biomass utilisation better feasible.

2. Methodology

The methodology of determining the feasibility of small-biomass power generation from the palm oil mill is similar to coal-fired power plants method. First is location aspects, the location indicated by general conditions, geological conditions, environmental conditions and supporting facilities located at the location that influence with the transportation system. As a result, they will a significant influence for economic feasibility calculation [4] — the financial analysis of the Biomass Power Plant base one BOT scheme as Indonesia policy that will predict revenue time with feasible analysis.

The second is an analysis of fuel for boiler, from that analysis can show characteristic of biomass from palm oil, calorific value and determine percentage ratio content between shell and fibre base on heat and mass balance calculation, moreover, from this analysis it would know an efficiency and can decide best type of boiler for biomass. The environment analysis of this paper is a comparison emission value between coal and biomass from palm oil base on Global warming potential and Acidification value based on study literature.

3. Results and discussion

The capacity of palm oil mills is already operating approximate range from 30 to 90 tons of EFB/hour [5]. For this study, sources of raw materials from oil palm mills with a capacity of 60 tons of FFB/hour operate at 100% capacity. The palm oil mills with a design capacity of fewer than 60 tons of FFB/hour were modified to be increased to 60 tons of FFB/hour by PTPN V in North Sumatra. Wastes or by-products of palm oil factories that can utilise as fuel for boiler plants are shells, fibre and EFB. Table 1 shows the design of the palm oil mill PTPN V [5].

Referring to the balance sheet table 1, the number of empty bunches that used as fuel is about 12.6 tons/hour with a moisture content of about 60% and palm oil mill operations average about 20 hours per day and 285 days per year. Thus, the annual solid waste palm oil production is about 72,000 tons (60% moisture content). The potential of electrical energy that can generate is determined by measuring or calculating the calorific value and amount of solid waste palm oil. To be burned in the boiler continuously and more efficiently, the solid waste water content of palm oil must reduce to no more than 30% through the drying process.

Table 1. Mass balance palm mill plant with a capacity of 60 tons EFB / hour [6]

Input		Output	
Processed FFB	60.0	CPO	14.6
Boiling Steam	13.4	PKO	3.3
Steam Digester	2.4	EFB	12.6
Hot Water Pressing	10.8	Fibre	8.4

Input		Output	
		Shell	4.2
		Sludge (drain)	30.8
		H_2O (atmosphere)	12.7
Total Input (Ton)	86.6	**Total Output (Ton)**	86.6

The heating value of solid waste palm oil in table 2 is calculated using a formula developed [7] as a result of the average value from 4 samples of the solid waste palm oil (HHV) value of 3,927 kcal/kg. Nevertheless, for this study, the conservative solid waste palm oil (HHV) calorific value (HHV) is 3,500 kcal/kg.

Table 2. Mass balance of palm mill plant capacity 60 tons EFB / hour

Test Parameter	Source of EFB			
	Source Plant A	Source Plant B	Source Plant C	Source Plant D
Ultimate Analysis (wt.%):				
Water content	30.00	30.00	30.00	30.00
Ash (A)	4.11	3.79	3.36	4.12
Carbon (C)	37.58	40.32	36.92	34.04
Hydrogen (H)	4.61	6.32	5.04	4.07
Sulphur (S)	0.11	0.11	0.14	0.06
Nitrogen (N)	3.26	0	0	0.44
Oxygen (O)	20.32	19.46	24.55	27.27
Heating Value (HHV) (kcal/kg)				
Channiwala formula	3,901	4,643	3,876	3,289

The capacity of 60 tons of fruit palm oil/hour from palm oil of 12.6 tons of solid waste palm oil/hour or equivalent to 21% of the weight of fruit palm oil that is the factory weighing with condition waste palm oil obtains from oil palm mills with an average operating time of 20 hours/day. Alternatively, it can be compared with 285 days/year same as 5,700 hours/year, while on the other side coal-fired power plant is planned to operate for 24 hours/days and 300 days/years or a total of 7,200 hours/year. Therefore, there must be a correction factor of 5,700 / 7,200 or equal to 0.79167 so that the actual production of solid waste palm oil from the palm oil plant capacity of 60-ton fruit palm oil /hour to about 12.6 tons/hour x 0.79167 or 9.975 tons/hour.

Furthermore, boiler selection for solid waste palm oil is a comparison between stoker boiler type and the fluidised bed type with operated at 50% excess air with a final temperature of 177°C (350°F) exhaust gas. As shown in Table 3, the most significant energy loss in the boiler is heat loss through the chimney. This loss can contribute up to 30-35% of the fuel input for an old type boiler with minimal maintenance also it decreased moisture content from 35% to 0% increased efficiency by 6%.

Table 3, The efficiency of the fluidised boiler bed is higher when compared to the stoker boiler because it has lower burning losses. Stocker boilers can produce 30-40% carbon in ash and additional volatiles and CO in the flue gas, while fluidised bed boilers can usually achieve 100% combustion. The turbulence conditions in the combustion chamber combined with the inertia of the heat of the bed material resulted in complete, controllable and uniform combustion. These factors are vital in maximising thermal efficiency, minimising char and controlling emissions.

Table 3. The efficiency of Biomass Boiler Based on Combustion Parameters and Fuel Characteristics

Characteristics	Biomass Stoker		Fluidized Bed Biomass	
	Dry	As received	Dry	As received
Excess air (%)	50.00	50.00	50.00	50.00
Dry flue gas (kg/kg bb)	15.25	10.68	15.25	10.68

Characteristics	Biomass Stoker		Fluidized Bed Biomass	
	Dry	**As received**	**Dry**	**As received**
Final tem. Exhaust (°C)	177.00	177.00	177.00	177.00
HHV (kJ/kg)	19.77	13.84	19.77	13.84
Water content bb(%)	0.00	30.00	0.00	30.00
Hydrogen content bb(%)	4.59	3.21	4.59	3.21
Loss of fuel efficiency				
Dry flue gas losses (%)	11.63	11.63	11.63	11.63
Water content bb(%)	0.00	5.90	0.00	5.90
Latent heat (%)	5.69	5.69	5.69	5.69
Unburned fuel (%)[1]	3.50	3.50	0.25	0.25
Radiation, etc (%)[2]	2.03	2.03	2.03	2.03
Total burning loss (%)	**22.85**	**28.74**	**19.60**	**25.49**
HHV based boiler efficiency (%)	**77.15**	**71.26**	**80.40**	**74.51**

Notes: (1) Estimation
(2) Include radiation, steam in the air. [6]

In figure 1, the design of the mass and power plant balance is capable of producing a gross power of 4.6 MW using 80% efficient fluidised bed boilers and a steam turbine capacity of 5 MW with an isentropic efficiency of 85%. The total generator efficiency alone is 19.71%, by the estimates in which assumes 20% total plant efficiency. The net heating rate that achieves is 4,363 kcal / kWh; this value is relatively proportional to small-scale coal-fired power plants using low-quality stoker and coal boilers.

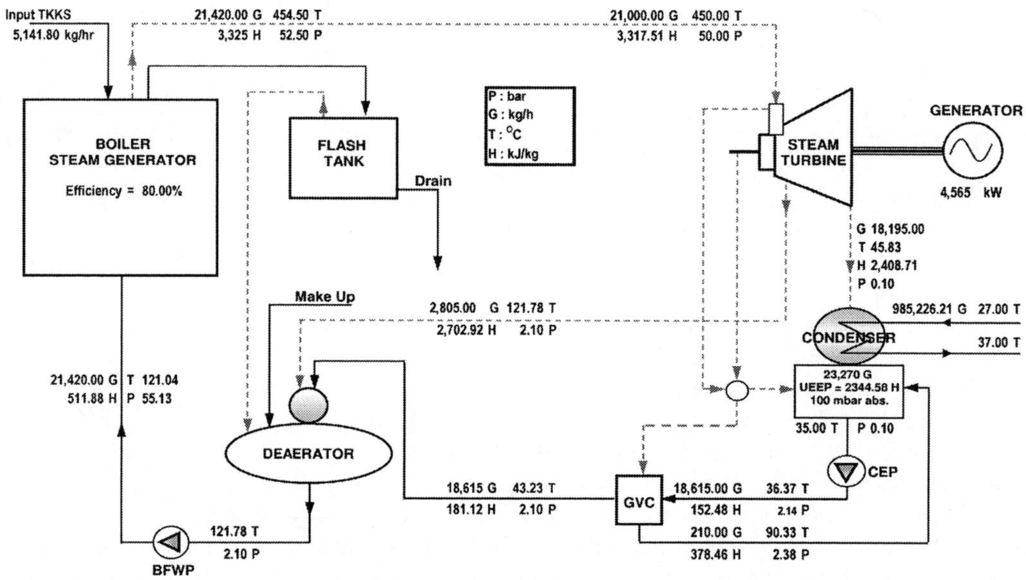

Figure 1. Mass and heat balance design solid waste palm oil-power generation 4.5 MW

Furthermore, based on study literature the environmental impact categories selected in this study consists of global warming potential (GWP100), and acidification potential (AP). The criteria for evaluating the environmental impacts selected from environmental problems that significantly in this case study. The emission factor is used from this case study from IPCC-2006 [8] and EMEP/EEA-2013 [9]. GWP is related to emissions of greenhouse gases to the air. Global warming potential with 100 years (GWP100) [10].

Table 4. Global warming potential and Acidification impact assessment of all stages from 1MWh of EFB utilisation for power generation

Impact Assessment	Global Warming (GWP100) kg CO_2 eq	Acidification kg SO_2 eq
Oil Palm Plantation	35.38	0.312
CPO Extraction	14.89	0.093
Power Generation	89.51	0.515
Transportation	9.92	0.066

By considering the whole environment analysis of the EFB utilisation for power plants, there is a significant potential resulting from power generation, oil palm plantation, CPO extraction, and transportation table 5. This category concerns the effects of toxic substances on the human environment. Health risks of exposure in the working environment are not included. For each toxic substance, HTP's are expressed as 1, 4-dichlorobenzene equivalents. Table 5 shows AP is expressed as kg SO_2 equivalents [11]. AP is contributed by NOx and SO_2 emissions. For the EFB utilisation for power generation process.

For a coal-fired power plant, when coal is burned, carbon dioxide, sulphur dioxide, nitrogen oxides, and mercury compounds are released. Especially sulphur dioxide is a high volume from this stage, and it can cause acid rain [12]. For that reason, the emission rate is high more than biomass, With AP 5.15E-01 kg SO_2 eq with EFB and 1.50E+01 kg SO_2 eq with coal, also GWP 100 3.54E+01 kg CO_2 eq for EFB and 1.32E+03 kg CO_2 eq for coal.

Finally, The financial analysis of the Biomass Power Plant with a lifetime duration of 25 years is shown in Graph 6. The power plant base on solid waste palm oil is feasible to be implemented because IRR is 22.70%, higher than the Weighted Average Cost of Capital (WACC) which amounted to 15.61% and this is understandable because the owner of coal-fired power plant biomass has a relatively long revenue-taking time (25 years) in operating the plant. Also, loan payments are distributed within 15 years so that the annual instalment to be paid becomes small. The magnitude of this instalment effect on Generating Cost which for this simulation is in the range of IDR 805.37/kWh. If a comparison is made to the electricity selling rate of IDR 1.150/kWh, then there is sufficient margin for the loan repayment. Under these conditions, the total balance sheet is positive in the fifth year in which the payback period has been fulfilled in 5 years and 2 months [13].

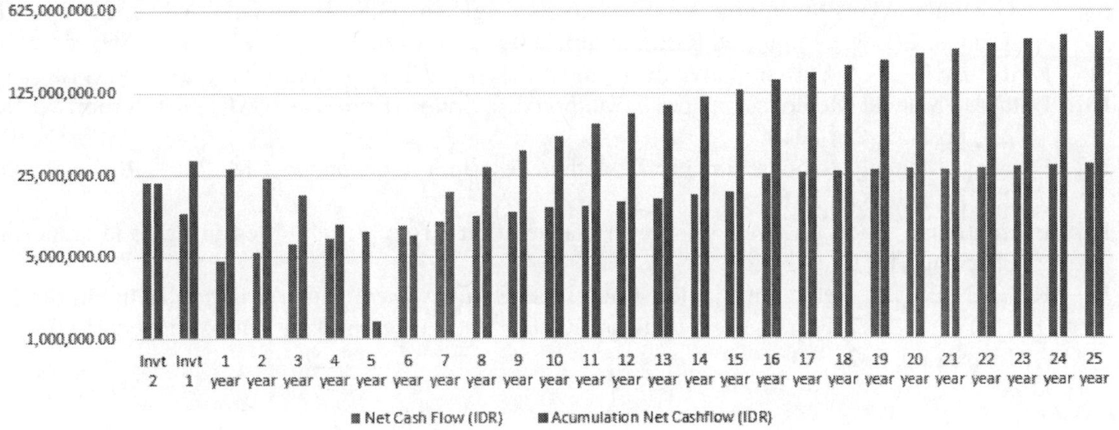

Figure 2. Financial analysis

The increase of investment cost to 20% turns out the project is still feasible to be implemented if based on IRR value (20,32%) which is still higher than WACC (15,61%). This increase in investment costs may occur if there is a strengthening of the exchange rate of foreign currency (US $) against rupiah because most of the mechanical, electrical and instrumentation equipment at the plant is imported. The underlying assumption of an escalation of electricity tariff increase of 3% per year, if this assumption is reduced to 2% is still feasible because the project IRR value (19.29%) is still higher than WACC (15.61%).

4. Conclusion

Solid waste palm oil considered renewable and less emission which is a substantial potential possibility in Indonesia especially at Riau, with source rang about 30 – 90-ton EFB. This study approved the feasibility of power generation as solid fuel fibre and shell with BOT 25 years and IRR 22.70%. Base on study literature has shown that biomass power generation is less emission than coal-fired power generation with acidification potential is 5.15E-01 kg SO_2 eq for empty fruit bunch (EFB) and 1.50E+01 kg SO_2 eq for coal, and Global warming potential (GWP-100) is 3.54E+01 kg CO_2 eq for empty fruit bunch (EFB) and 1.32E+03 kg CO_2 eq for coal.

Acknowledgements
The research described in this paper was financially supported by Laboratory Fuel Technology and Engineering Design-(BTBRD) BPPT, and PTPN V (PERSERO), Indonesia.

References

[1] BPP Teknologi. 2011. PTPN 2 Biomass Feasibility Study. Presentation material. Jakarta.
[2] Sarunya C. Life cycle assessment of palm empty fruit bunch utilisation for power plants in Thailand. International Conference on Biological, Environment and Food Engineering (BEFE-2015) May 15-16, 2015 Singapore
[3] Kantor Menko Perekonomian. 2004. Kajian pembiayaan investasi transportasi perkotaan dengan skema debt-for-natute swap. Jakarta
[4] Black & Veath. 1996. *Power Plant Engineering*. New York (US): Chapman & Hall
[5] PTPN V. 2013. Data Tandan Buah Segar Olah di Pabrik Kelapa Sawit PT Perkebunan Nusantara V. Riau
[6] BTBRD. 2017. Studi kelayakan pembangkit listrik tenaga biomasa tandan kosong kelapa sawit.
[7] Parikh J., Channiwal S.A. and Ghosal G.K., 2005. A correlation for calculating HHV from proximate analysis of solid fuels. *Fuel Vol.84*. Pages. 487-494
[8] IPCC, IPCC Guidelines for National Greenhouse Gas Inventories, Prepared by the National Greenhouse Gas Inventories Programme. In: Eggleston,H.S., Buendia,L., Miwa,K., Ngara,T., Tanabe,K.(Eds.), Published IGES, no.pp. 2009
[9] EMEP/EEA. 2013. Emission inventory guidebook. *1.A.1 Energy industries*, no. pp. 1-118
[10] Database Manual Methods, simapro 7, Method description, European, CML2001, Report version 2.4, pp.1-57, Oct 2010
[11] Database Manual Methods, simapro 7, Method description, European, CML2001, Report version 2.4, pp.1-57, Oct 2010
[12] International Energy Agency. Power Generation from Coal. Measuring and reporting efficiency.1-114
[13] KEMEN ESDM. Peraturan Menteri Energi dan Sumber Daya Mineral Republik Indonesia No. 12 Tahun 2017 Tentang Pemnafaatan Sumber Energi Terbarukan Untuk Penyediaan Tenaga Listrik. pp. 2–22, 2017

The role of oil palm biomass recycling on soil health in oil palm plantations

R.S. Tarigan[1], Suhardi[1], R. Wahyuningsih[1], Pujianto[1], P.A. Wahyuningsih[1], R. Marichal[2], J.P. Caliman[1]

[1]SMART Research Institute, Jl. Tengku Umar 19, Pekanbaru, 28112, Indonesia
[2]CIRAD, UPR Systèmes de pérennes, 34398 Montpellier, France

*E-mail: caliman@indo.net.id

Abstract. Oil palm cultivation produces a massive amount of biomass (e.g., empty fruit bunches (EFB), frond and trunk) which could be recycled directly in the plantation to maintain/improve the quality of soils. The impact of this biomass recycling, i.e., EFB and frond, - in term of chemical, physical and biological characteristics, will be analyzed. The fresh application of the EFB could alternate most of the soil parameters in line with the kinetics of the decomposition of the EFB and the release of nutrients. For example, the pH of the soil after the EFB application was higher during more than 18 months compared to the initial situation. Soil exchangeable potassium increased dramatically within a few days after EFB spreading. Frond recycling, at harvesting and pruning, modifies the chemical and physical characteristics of soils whether they were applied on heaps between palms in dedicated inter-rows, or spread on harvesting paths where they could reduce run-off and erosion. The systematic distribution of fronds in inter-rows reduced rainwater run-off by more than 30%, even on 5% gentle slopes. Subsequently, soil loss through erosion was reduced by 65% depending on the slope intensity of the terrain. Consequently, the loss of nutrients was considerably reduced, resulting in higher fertilizer efficiency. Similar measurements have been done when applying fresh EFB to the soil. The impact on soil physical fertility, i.e., rainwater infiltration rate and subsequently soil humidity are improved. We have also recorded a reduction of the soil resistance to penetration, as well as an improvement of the soil aggregate stability. Several representative parameters of soil biological activity, i.e., soil fauna feeding activity using bait-lamina test system and earthworm' population showed the positive impact of EFB applications as well as frond spreading in inter-rows.

1. Introduction

Thousands of years ago, before the development of agriculture, the soil was considered merely physical support for crops. With the development of modern agriculture, a wider role for soil has emerged. A greater interest in soil has been associated with the discovery of the role nutrients and water play in growth and development, and subsequently their intensive use as part of the development of intensive agriculture in the middle of the 20th century, required to meet exponential population growth. This transformation, later called 'the Green Revolution' in agriculture after the Second World War, was based on extensive use of ploughing of soil, an-organic fertilizers and irrigation, combined with the high-yielding cultivars. Concerns to reduce a perceived over-reliance on fertilizers lead to a new interest in soil health. This new understanding of the role and importance of soil has led to the development of new soil management systems required to achieve sustainable agriculture. It is not surprising that the Food and Agriculture Organization of the United Nations (FAO) declared 2016 the year of soils, and 2015-24 the decade of soils [1].

Soils contribute to disease and pest control, as well as carbon sequestration and storage. They are home to microorganisms that contribute to plant health and regulate nutrient availability and to the decomposition and mineralization of organic residues and waste. Soil fertility was recognized as a key factor for plant growth, while soil degradation, such as erosion, undermines ecosystem services and biomes [2], [3].

The oil palm is a heavy feeder and requires a quite large quantity of fertilizers to achieve its potential yield [4]. Every tonne of crude palm oil produced generates by way of by-product, one tonne of empty fruit bunches (EFB). In the last decade, an increasing number of planters have been considering the utilization of EFB as a means of reducing mineral fertilizer inputs while improving soil fertility [5]. The decomposition of EFB in the soil is facilitated by soil faunas and microorganisms [6].

An important way of reducing costs in the industry is to improve the efficiency of use of the fertilizer applied. Part of the loss cannot be regained if it is in run-off, erosion, leaching and volatilisation, and this must be minimized [7]. Land cover crops play an important role in influencing erosion. Cover crops provide protection against the destruction of soil aggregates by rain and run-off. The soil conservation techniques in oil palm cultivation can reduce the rate of surface run-off, soil erosion and nutrient loss [8].

The objectives of this study were to study and quantify the impact of biomass recycling (e.g., EFB and frond) on soil health (chemical, physical and biological characteristics) under oil palm plantation.

2. Materials and methods

The experiment was conducted in a mature oil palm plantation planted in 1989, located in Riau (Sumatra-Indonesia). The soil type was a typic dystropept, with a sandy loam texture. There are several methods to observe the impact of biomass recycling to maintain the quality of soils:

2.1. Physical characteristics

In this study, we used a penetrologger to measure the soil compaction (presented in penetration resistance value) until ± 80 cm of soil depth. The distribution of water content (humidity) in the soil were measured every 10 cm soil depth until 20 cm soil depth using Gravimetri method. These observations carried out in with and without EFB application. To quantify the volume of water run-off and soil erosion, we were installed erosion plot. Each plot covered 18 m x 18 m of the area and included four palms. Each plot was isolated from outside water run-off intrusion by a 20-cm high wall built with bricks. The amounts of run-off water and soil eroded were collected in a series of 200-liter plastic drums after each rainfall event.

2.2. Chemical characteristics

In this study, we collected water and soil eroded samples from erosion plot to quantify the impact of biomass recycling on nutrient losses (e.g., Nitrogen, Phosphorus, Potassium, Total organic carbon and soil C-organic). The impact of EFB application to maintain the soil quality was observed with soil sampling until ± 60 cm soil depth for soil pH (using H_2O method), soil K-exchangeable and soil Al-exchangeable. While, for soil C-organic and soil base saturation, the soil sampling until ± 160 cm soil depth.

2.3. Biological characteristics

In this study, we used bait-lamina method and soil macrofauna samplings (earthworm and soil insect) for assessing the soil fauna feeding activity and macrofauna abundance in the soil as the impact of various slopes and agriculture practices to the biological soil properties. Bait-lamina stick is thin PVC strips (1 mm x 6 mm x 160 mm) with 16 apertures (1.5 mm diameter) bored at 3 mm intervals. The apertures filled with standard bait, containing cellulose powder, bran flakes and activated carbon (70:27:3). The test was based on visual assessment of the feeding portion of laminated bait substrate

IOP Conf. Series: Earth and Environmental Science **336** (2019) 012019 doi:10.1088/1755-1315/336/1/012019

exposed (empty holes indicate some biological activities in the soil). A matrix of six baits lamina sticks inserted vertically and randomly into the soil on three management zones (circle, harvesting path, and windrow) each sticks about 20 cm apart. The sticks were inserted until the top aperture was just below the soil surface (± 1 mm) and baits exposed for 6 days. Each aperture was recorded as zero (without perforation) and one (partial or complete perforation). To calculate the rate of feeding activity per bait, the number of holes (partial or complete hole) counted then divided by 16 (the total number of holes per bait). Therefore the minimum score will be 0%, and the maximum will be 100%. A Tropical Soil Biology and Fertility (TSBF) method used to assess the macrofauna abundance in the soil. A square of 25 cm x 25 cm x 10 cm randomly placed in the soil, then dug the soil in 10 cm depth and put the soil into closed bag. The macro-invertebrates defined as the fauna larger than 2 mm in body length, were collected by hand sorting. The major taxa of macrofauna such as Formicidae, earthworms, Diptera, Dermaptera, Coleoptera, Isopoda, Myriapoda, Chilopoda, Diplopoda, Blattaria, Aranea extracted from the soil using grips and stored in tubes containing 70% alcohol.

Statistical analysis: The analysis of variance was used to compare the effect of each treatment; whereas treatment means were compared, using Tukey test at $P \leq 0.05$ with SPSS software.

3. Results and discussion

The results on the impact of this biomass recycling, mainly the EFB, applied fresh from the mill or after composting, but also of the fronds, on several components of soil health, i.e. the chemical, physical and biological characteristics.

EFB fresh applications modify most of the soil parameters with specific patterns in line with the kinetics of the decomposition of the EFB and the release of nutrients. For example, the pH of the soil after the EFB application was higher during more than 18 months compared to the initial situation. Soil exchangeable potassium increased dramatically within a few days after EFB spreading. **Figure 1** indicates that the application of EFB tends to increase pH mainly in the upper layer of soil. The impact of EFB on soil pH is observed until about 30 cm of soil depth. The application of EFB increases soil pH from 5.0 (acid) to 7.0 (neutral) after 21 days of application in 0-5 cm soil upper layer and becomes 7.2 after 90 days application. The rise soil pH is observed until 120 days after application and tends to decrease after this period. The application of EFB also increases the soil K-exchangeable significantly, mainly in the upper layer of soil (0-6 cm) and reduces the soil Al-exchangeable during decomposition.

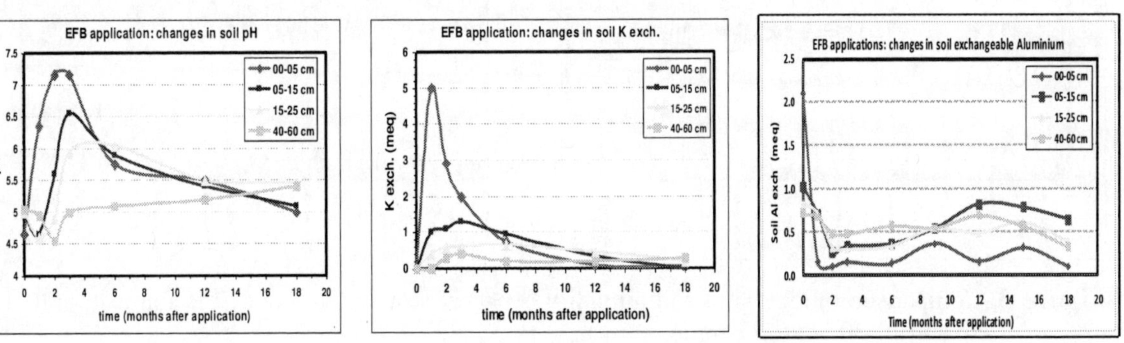

Figure 1. Dynamic of soil pH (H_2O method), K_{exch} and Al_{exch} during EFB decomposition according to soil depth under EFB

The application of EFB gives slightly higher on soil C-organic content mainly in the upper soil layer. The observation also indicates that the concentration of C-organic in upper layer soil is higher than in the lower layer, which may relate with the relatively high organic source in this area (root, leaf, etc.). The application of EFB increases the average of soil base saturation significantly mainly in the soil upper layer until about 75 cm depth. The increase of this value may relate with the release of basic cations such as K, Mg, and Ca from EFB during its decomposition (figure 2).

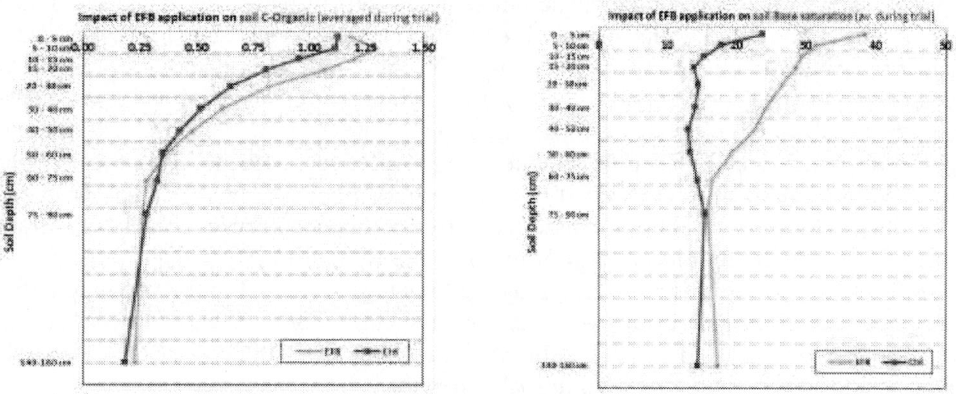

Figure 2. Soil c-organic and soil base saturation with EFB application according to soil depth

The impact on soil physical fertility, for example, rainwater infiltration rate and subsequently soil humidity are improved. We have also recorded a reduction of the soil resistance to penetration, as well as an improvement of the soil aggregate stability. The observation indicates that the application of EFB affects the distribution of water content (humidity) mainly in the upper soil layer. In an area with EFB application, the distribution of soil water content is relatively continuous along soil depth; while in soil without EFB application (control site), the distribution of water content is discontinuous. In an area with EFB application, the humidity of soil in the upper layer tends to decrease smoothly with soil depth. During wet condition (after rainfall), the difference of humidity between upper and lower soil layer increases; and this gap of difference decreases during dry condition, but the humidity of soil in the upper layer is still higher than the lower layer (figure 3).

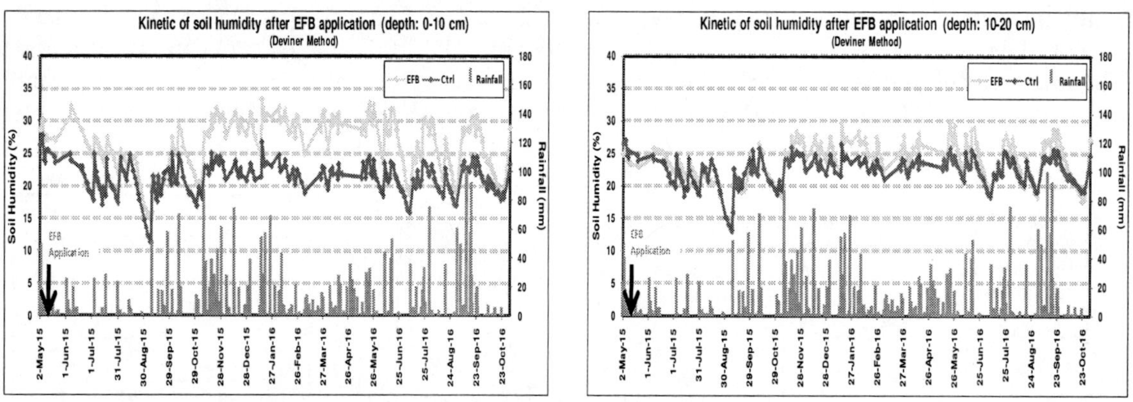

Figure 3. Comparison of kinetic of soil humidity between with and without EFB application (0-10 cm and 10-20 cm soil depths)

The soil compaction (presented in penetration resistance value) is relatively equal in area with and without EFB application during the initial condition (before EFB application). The application of EFB tends to reduce the value of soil compaction. This decrease is more intense in the upper soil layer. Our observation indicates that the ratio value between soil compaction with EFB application compared to control after 1-month application decreases on 0-10 cm soil depth. After 3 months of application, the decrease of soil compaction increases until 40 cm soil depth. The decrease of penetration resistance observed in the area with EFB application may mainly relate to the higher soil humidity in the soil as mentioned earlier. This decrease is more intense in the upper layer since the humidity is higher. In fact, the value of soil compaction (penetration resistance) is a physical characteristic of soil relating to other physical soil characteristics such as moisture content, bulk density, soil structure, soil texture and organic matter content. The application of EFB may increase soil moisture content, reduces bulk density, improves soil structure and increases organic matter content. Hence, the application of this organic matter may undoubtedly reduce the soil compaction and facilitates the development of root mainly in the upper soil layer (figure 4).

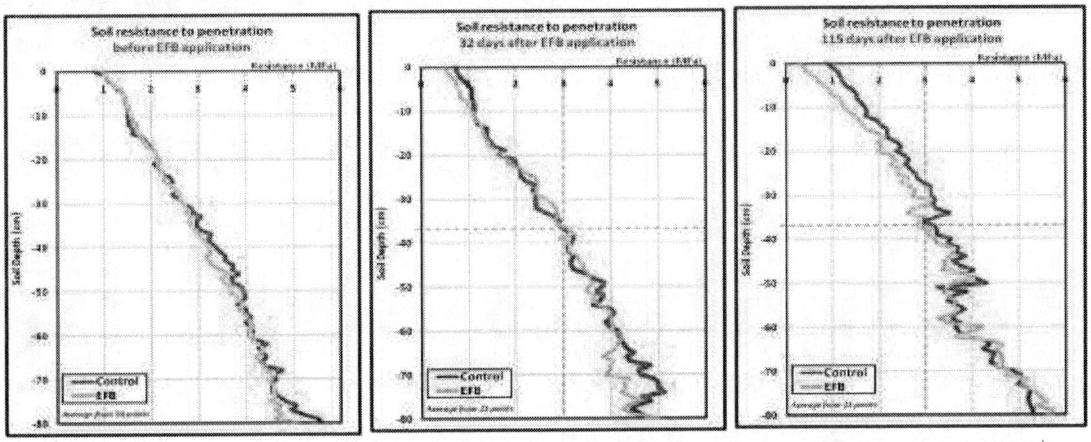

A B C

Figure 4. Soil penetration resistance in initial condition (A), after 32 days (B) and after 115 days of EFB application (C)

Frond recycling, at harvesting and pruning, modifies the chemical and physical characteristics of soils whether they were applied on heaps between palms in dedicated inter-rows, or whether they were spread on harvesting paths where they could reduce run-off and erosion. These impacts have also been quantified in our field experiment: the systematic distribution of fronds in inter-rows reduced rainwater run-off by more than 30%, even on 5% gentle slopes. Subsequently, soil loss through erosion was reduced by 65% depending on the slope intensity of the terrain (figure 5). This shows the importance of ground cover because it can furnish run-off and erosion protection in several ways such as: protection of soil surface against disruptive action of falling raindrops, offer resistance to moving water and slows down its rate of run-off, the plant roots also help to hold the soil in place. The plant roots and crop residues help to improve soil structure which makes it more porous and better able to absorb rainfall [9]. Land cover crops play an important role in influencing erosion. Cover crops provide protection against the destruction of soil aggregates by rain and run-off.

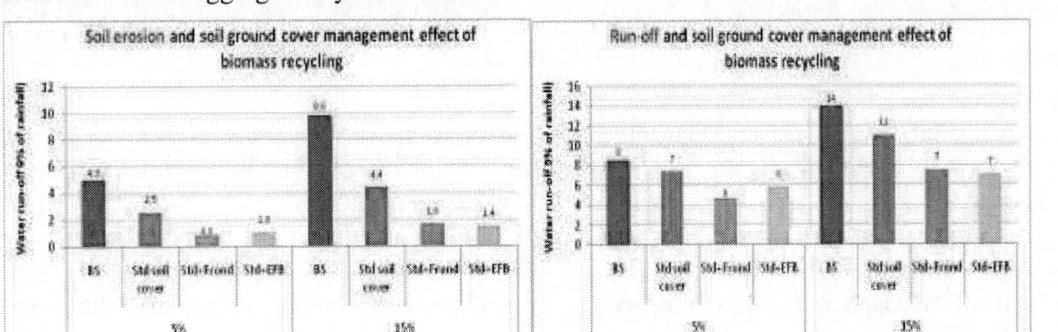

Figure 5. Soil Ground Cover Management Effect of Biomass Recycling on Water Run-off and Soil Erosion

Nutrient losses can be expressed as concentration in the soil solution that lost as dissolved forms in water run-off, or the sediment carried in run-off water, and these losses are affected by soil type, cover and fertility; slope of the terrain; amount, method and time of fertilizers application; intensity, amount and moment of occurrence of rainfall [10]; as well as by soil management practices [11]. Consequently, the loss of nutrients was considerably reduced, resulting in higher fertilizer efficiency. The effect of biomass recycling such as fronds can reduce the nutrient losses for nitrogen about 72% and 70%, for phosphorus about 54% and 48%, for potassium about 58% and 70% for 5% and 15% respectively compared to the standard soil cover. Similar measurements have been done when applying fresh EFB to the soil (figure 6). The result also indicated that the average of total net nutrient losses increased with the slope and the impact of the slope is more intense in bare soil (BS) than the others treatment.

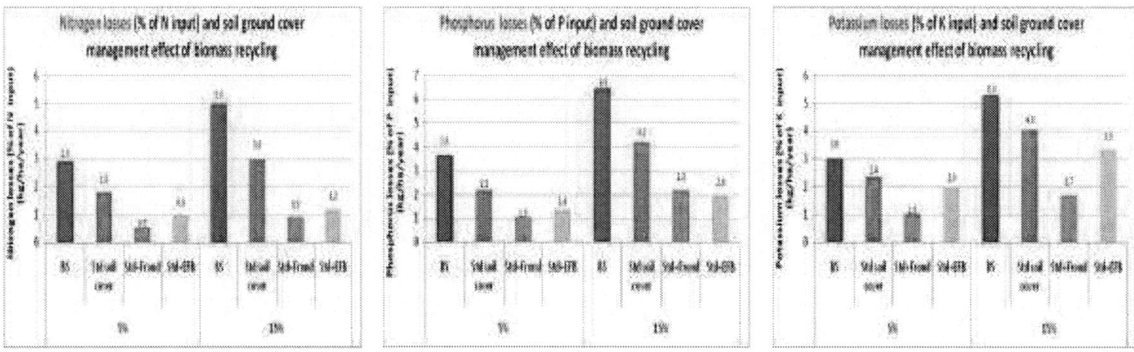

Figure 6. Soil Ground Cover Management Effect of Biomass Recycling on Nitrogen, Phosphorus and Potassium Losses

Besides nitrogen, phosphorus and potassium; fronds and EFB can reduce the total organic carbon from water run-off and soil eroded about 57% and 24% for 5% and 15% respectively compared to the standard soil cover as shown in figure 7.

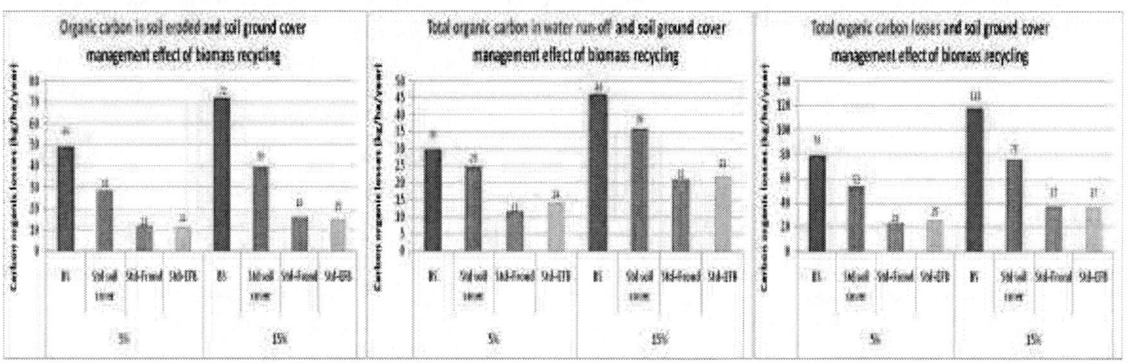

Figure 7. Soil Ground Cover Management Effect of Biomass Recycling on Total organic carbon losses, Organic carbon in soil eroded and Total organic carbon in water run-off

Several representative parameters of soil biological activity have been recorded. Soil fauna feeding activity using bait-lamina test system, insect and earthworm' population showed the positive impact of EFB applications as well as frond spreading in inter-rows (figure 8). The feeding activity, soil insect and earthworm abundance in harvesting path, in the soil under applied organic material, showed a higher result compared to the bare soil and standard soil cover. In this study, we found that the addition of organic material has an impact on soil biological property (soil fauna feeding activity as one of the biological soil indicator). The soil fauna was dominantly found in the upper soil layer; this might affect the depth distribution of soil fauna feeding activity [12].

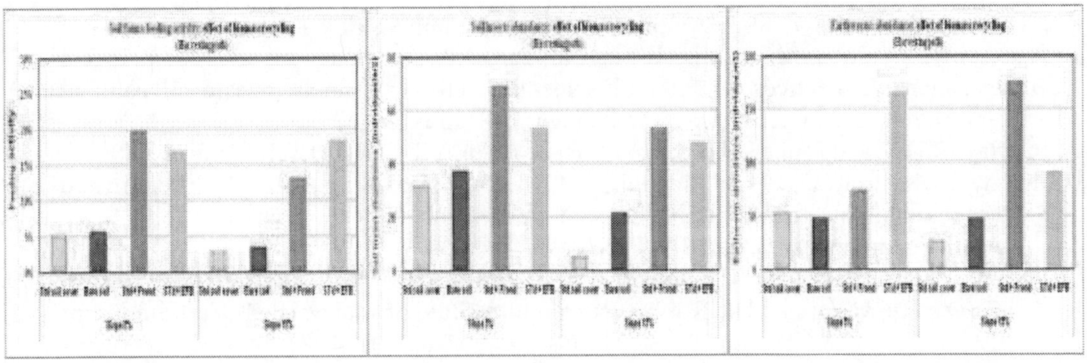

Figure 8. Effect of Biomass Recycling on Harvesting Path to Soil Fauna Feeding Activity, Soil Insect and Earthworm Abundance

This result showed that there might be a relationship between the organic material and soil macrofauna abundance, this result expected that the organic material provides food for the soil macrofauna, for example, earthworm. The earthworm is one of the examples of soil macrofauna, which affects the dynamics of changes of organic matter in the soil, as they decomposed the bigger fragments of organic material and turned in to cast (smaller fragments). Soil fauna feeding activity has a positive correlation with the earthworm and soil insect abundance. This result was linked with the observations made by reference [12] that bait perforation was not found when only mesofauna (such as collembolans) was present but seemed to be dependent on medium and large fauna like earthworms.

4. Conclusion

In this study, we found that the application of EFB gives a significant effect on chemical soil characteristic such as soil pH mainly in 0-30 cm layer, C-organic, K-exchangeable, Al-exchangeable and base saturation. The increase of base saturation was observed until 75 cm soil depth. The existence of EFB can maintain soil moisture content and reduce soil compaction mainly in the upper layer. The effect of slope on water run-off, soil erosion and nutrient losses could be minimized using addition application EFB or Frond on the surface of the soil. The addition organic material such as EFB and frond increased the value of soil fauna feeding activity, the earthworm, and soil insect abundance.

Acknowledgments

The authors would like to express their appreciation to SMART-Research Institute for the support especially all the workers who helped in the field. We would also like to thank Mr Tauvik R, Mr Dias K, Mr Edy N, and Mr Egyfaldy B and all the technicians in the laboratory who helped to analyze the samples also Mr Divo D. Silalahi and Ms Dwivahyani Sentosa who helped to analysis statistic.

References

[1] Caliman, J.P, Suhardi, Pujianto. 2018. Achieving sustainable cultivation of oil palm, Volume 1: Introduction, breeding and cultivation techniques, Chapter 11: Improving soil health and crop nutrition in oil palm cultivation, 229-249. Burleigh Dodds Series in Agriculture Science.

[2] Roose, E. J., Lal, R., Feller, C., Barthes, B. and Stewart, D. A. 2006. Soil erosion and carbon dynamics. In: *Advances in Soil Science*. CRC Press, Boca Raton, FL, 352pp

[3] Roose,E. J. 2017. *Restauration de la productivité des sols tropicaux et méditerranéens*. IRD Editions. Collection Synthèses. ISBN: 978-2-7099-2277-7

[4] Comte, I., Colin, F., Whalen, J. K., Grünberger, O. And Caliman, J.P. 2012. Agricultural practices in oil palm plantation and their impact on hydrological changes, nutrient fluxes and water quality in Indonesia. A review. *Advances in Agronomy*, 116, 71-124

[5] Caliman, J.P, Budi, M. And Saletes, S. 2001. Dynamic of nutrient release from empty fruit bunches in field conditions and soil characteristics change. In: *MPOB International Palm Oil*

Congress (PIPOC), 20-23 August 2001, Kuala Lumpur, Malaysia

[6] Tao, H.H., Slade E. M., Willis, K. J., Caliman, J.P. and Snaddon, J.L. 2015. Effects of soil management practices on soil fauna feeding activity in an Indonesia oil palm plantation. *Agriculture, Ecosystems and Environment*, 218, 133-140

[7] Corley, R.H.V and Tinker, P.B. 2016. *The Oil Palm*. Wiley Blackwell, Oxford, 639p

[8] Faudy, Z., Sastriawan, H. and Mayani, N. 2014. Runoff, erosion and nutrient sediment due to the vegetation-soil conservation on oil palm plantation. Journal Soil Science and Agroclimatology, 11(2), 95-103

[9] Soong, N.K, G. Haridas, C.S. Yeoh and P.H. Tan.1980. *Soil Erosion and Conservation in Peninsular Malaysia*. The Rubber Research Institute of Malaysia, Kuala Lumpur. p. i-viii, 1-69

[10] Gascho, G.J, R.D. Wauchope, J.G. Davis, C.C. Truman, C.C. Dowler, J.E. Hook, H.R. Sumner, A.W. Johnson. 1998. Nitrate-Nitrogen, soluble, and bioavailable phosphorus runoff from simulated rainfall after fertilizer application. *Soil Science Society of America Journal*, vol 62. pp: 1.711-1.718

[11] Mello, E.L. 2002. Erosão hídrica em diferentes sistemas de manejo do solo sob chuva simulada. Lages: UDESC/CCA. 88p. (Dissertação – Mestrado)

[12] Gongalsky, K. B., Persson, T., & Pokarzhevskii, A. D. 2008. Effects of soil temperature and moisture on the feeding activity of soil animals as determined by the bait-lamina test, 39, 84–90. https://doi.org/10.1016/j.apsoil.2007.11.007

Fresh fruit bunch production of oil palm plantation in the lowland area of Sembilang Dangku Landscape

Rio Octarizza Segara[1], Hariyadi[2], Sukarman[3] and Kresno Dwi Santoso[4]

[1]Study Program of Natural Resource and Environmental Management Science, Postgraduate School, Bogor Agricultural University, Baranangsiang IPB Campus, Bogor 16143, Indonesia
[2]Department of Agronomy and Horticulture, Faculty of Agriculture, Bogor Agricultural University, Dramaga IPB Campus, Bogor 16680, Indonesia
[3]Indonesian Agency for Agriculture Research and Development of Land Resources, Ministry of Agriculture Republic Indonesia, Cimanggu, Bogor 16680, Indonesia
[4]Zoological Society of London (ZSL) Indonesia Programme, Palembang 16128, Indonesia

*E-mail: riooctsegara@gmail.com

Abstract. The results of the study showed that Sembilang Dangku Landscape had known two types of soil namely peat soil and mineral soil. Detail information for peat soil in this area namely *Typic Haplosaprists* and *Typic Haplohemists* meanwhile, for mineral soil, there are two types of soil that have been identified at these locations namely Inceptisols (*Typic Endoaquepts* and *Typic Humaquepts*) and Entisols (*Typic Sulfaquents*). This study aimed to analyze the production of oil palm plantations in lowland areas on marginal land. This research was conducted at six locations of oil palm plantations in Musi Banyuasin Regency, South Sumatra province. The methods used were descriptive quantitative and qualitative calculation of FFB (fresh fruit bunches) production that compared the standard potential production of the Palm Oil Research Center (PPKS referred *in Indonesia*). The yield of FFB oil palm productivity (ton/ha) on marginal land showed that *Typic Haplosaprists, Typic Haplohemists and Typic Endoaquepts, Typic Humaquepts* did not have a significant effect because yield from plantation management produced FFB production oil palm was quite good. The effect that has quite a different production is Entisols (*Typic Sulfaquents*) or L6 locations. Based on this research, the type of soil is a factor that greatly affects the productivity of oil palm.

1. Introduction

Oil palm plant is one of the seed plants in the plantation industry in Indonesia. Oil palm is one of the main sources of income for foreign exchange from the non-oil and gas sector [1]. The business of oil palm plantations in Indonesia is quite open because of the increasing demand for domestic and foreign palm oil. At present, there are around 10 million hectares of oil palm plantations in the country [2]. However, the amount is still considered to be increased to meet demand. The problem faced today is the lack of mineral land resources that have optimum productivity characteristics for the growth and development of oil palm plants. One alternative that is used to overcome this problem is the use of abandoned marginal land. The current area of Indonesian peatlands is 14.85 million ha [3] whereas, the area of unproductive degraded peatlands in Indonesia is reported to be around 4.2 million ha [4]. Currently, oil palm plantations in Indonesia are generally located in degraded peatland locations, because 90% of Indonesia's peatlands are degraded [5]. This research was conducted in the lowland area of the Sembilang Dangku Landscape which was dominated by tidal areas. The locations research of oil palm plantation are outside of the wildlife reserve area (Suaka Margasatwa Dangku). This area has four areas designated as Business Use Rights (HGU) for industrial activities, including areas of oil palm plantations, Industrial Plantation Forests (HTI), and oil and coal companies [6]. In this study, there were three marginal soil types known as Histosol *(Typic Haplosaprists and Typic Haplohemists*), Inceptisols (*Typic Endoaquepts and Typic Humaquepts*), and Entisols (*Typic Sulfaquents*). This study aimed to look at the effect of marginal soil types in lowland areas on FFB productivity with potential production standards from Palm Oil Research Center (Pusat Penelitian Kelapa Sawit (PPKS) referred in Indonesia) [7].

2. Materials and methods

2.1 Place and time of research

The research was carried out in the Sembilang Dangku Landscape area in Musi Banyuasin Regency, South Sumatra Province. Sampling research was taken at six different plantation locations. A sampling of the research just was carried out on oil palm plantation blocks which were in lowland areas with a height of 5 msl (meter sea level). The ground checking method was performed in some companies. The term used in the study is Location 1 = L1, Location 2 = L2, Location 3 = L3, Location 4 = L4, Location 5 = L5 and Location 6 = L6. This research was conducted from January to March 2018.

2.2 Tools and materials

The tools and materials used in the study were GPS (Global Positioning System), company plantation maps, meters, cameras, notebooks, and stationery.

2.3 Methods

Secondary data is obtained by results of the soil map from the Indonesian Agency for Agriculture Research and Development of Land Resources by matching the company plantation map and ground check at the sampling location. The other secondary data research is obtained by soil types and FFB production data for the last three years at the sampling location of the Sembilang Dangku Landscape. The primary data was done by ground check at sampling areas. The data is processed in descriptive qualitative and descriptive quantitative depending on the results obtained and then the result of the data will be compared with standard potential production of palm oil (PPKS referred in Indonesia).

Table 1. Estimation of Oil Palm Production in the S-3 Land (marginal suitable land) based on Pusat Penelitian Kelapa Sawit (PPKS 2006)

Age (Year)	FFB (ton/ha)	ABW (kg/unit)
3	6,2	3
4	12	5,3
5	14,5	6,7

Age (Year)	FFB (ton/ha)	ABW (kg/unit)
6	17	8,5
7	22	10
8	24,5	12,7
9	26	15,5
10	26	16
11	26	17,4
12	26	18,5
13	26	19,5
14	25	20
15	24,5	20,6
16	23,5	21,8
17	22	23
18	21	24,2
19	20	25,5
20	19	26,6
21	18	27,4
22	17	28,4
23	16	29,4
24	15	30,4
25	14	31,2
Total	461,2	441,6
Average	20,1	19,2

Note : FFB = Fresh fruit Bunch (ton/ha), ABW = Average Bunch Weight (kg/unit)

3. Results and discussion

The location sampling of the Sembilang Dangku Landscape is on the flat topography area. The shape of the research location is generally flat with a slope of 0-3%. The dominance of vegetation in the research location is lower vegetation category, including the types of grass and ferns. Mineral soil types that had been identified at the site are Inceptisols (*Typic Endoaquepts* and *Typic Humaquepts*) and Entisols (*Typic Sulfaquents*) whereas for peat soil types are *Typic Haplosaprists* and *Typic Haplohemists*.

The results showed that in general peat soil (Histosols) was not significantly different compared to Inceptisols, but its production was still better than Entisols. The results of oil palm FFB productivity showed that the productivity of L1 (*Typic Haplosaprists* and *Typic Haplohemists*) sampling locations were around 26 tons/ha (85% of the PPKS standard), L2 (*Typic Endoaquepts*) productivity was around

26 tons/ha (95% of PPKS standard), L3 (*Typic Endoaquepts* and *Typic Humaquepts*) productivities were around 9 tons/ha (45% of PPKS standard), L4 (*Typic Haplosaprist* and *Typic Humaquepts*) productivities were around 10 tons/ha (50% of PPKS standard), L5 (*Typic Haplosaprist, Typic Endoaquepts* and *Typic Humaquepts*) productivity were around 12 tons/ha (55% of the PPKS standard), L6 (*Typic Endoaquepts* and *Typic Sulfaquents*) productivities were around 3 tons/ha (25% of the PPKS standard). The location of L1 and L2 productivity meets the standard PPKS on S-3 land (according to standard of marginal land). These data showed in the production of palm oil production in Sembilang Dangku Landscape which is presented in table 2.

Table 2. Production of oil palm in Sembilang Dangku Landscape compared to the potential production of FFB (PPKS 2006) [7]

Location	Type of soils	Planting year	FFB Production (tons/ha)			Standard Potential production of FFB (PPKS) based on Planting Year (tons/ha)			Production trends (%)
			2015	2016	2017	2015	2016	2017	
L1	*T-Hs*	2007	28.4	21.2	20.9	24.5	26.0	26.0	-13.0
L1	*T-Hs*	2008	27.7	22.3	21.2	22.0	24.5	26.0	-11.7
L1	*T-Hs*	2011	17.3	15.9	16.6	12.0	14.5	17.0	-2.0
L1	*T-Hh*	2007	26.9	21.8	22.9	24.5	26.0	26.0	-7.4
L1	*T-Hh*	2008	22.8	19.8	20.8	22.0	24.5	26.0	-4.5
L1	*T-Hh*	2009	18.1	17.3	18.4	17.0	22.0	24.5	1.2
L2	*T-Ea*	1997	31.3	23.9	19.9	21.0	20.0	19.0	-18.2
L2	*T-Ea*	1998	29.9	25.7	21.5	22.0	21.0	20.0	-14.0
L2	*T-Ea*	2006	25.1	22.4	20.2	26.0	26.0	26.0	-9.8
L2	*T-Ea*	2008	19.4	16.7	12.6	22.0	24.5	26.0	-17.5
L2	*T-Ea*	2009	25.0	20.2	18.2	18.3	22.6	24.9	-13.8
L3	*T-Ea*	2007	17.4	8.3	13.0	24.5	26.0	26.0	-12.8
L3	*T-Ea*	2008	12.9	9.5	10.7	22.0	24.5	26.0	-6.9
L3	*T-Ea*	2009	7.1	5.9	10.1	17.0	22.0	24.5	21.1
L3	*T-Ea*	2010	7.2	5.7	6.8	14.5	17.0	22.0	-3.1
L3	*T-Hq*	2009	8.1	6.4	8.0	17.0	22.0	24.5	-0.3
L4	*T-Hs*	2008	10.9	7.8	10.1	22.0	24.5	26.0	-1.4
L4	*T-Hq*	2007	15.5	10.0	12.4	24.5	26.0	26.0	-10.3
L4	*T-Hq*	2008	15.5	10.0	12.2	22.0	24.5	26.0	-10.6
L4	*T-Hq*	2009	8.9	5.2	5.6	17.7	22.4	24.7	-18.6

Location	Type of soils	Planting year	FFB Production (tons/ha)			Standard Potential production of FFB (PPKS) based on Planting Year (tons/ha)			Production trends (%)
			2015	2016	2017	2015	2016	2017	
L5	*T-Hs*	2008	17.7	14.3	18.5	22.0	24.5	26.0	3.6
L5	*T-Hs*	2011	3.0	2.1	4.1	12.0	14.5	17.0	19.4
L5	*T-Ea*	2007	18.1	12.9	14.9	24.5	26.0	26.0	-8.7
L5	*T-Ea*	2008	11.8	11.4	16.0	22.0	24.5	26.0	22.2
L5	*T-Hq*	2010	10.0	7.2	8.4	14.5	17.0	22.0	-8.0
L6	*T-Sq*	2007	15.2	11.1	8.6	24.5	26.0	26.0	-21.3
L6	*T-Sq*	2012	2.5	3.3	2.0	6.2	12.0	14.5	-10.3
L6	*T-Ea*	2012	0.6	1.2	0.8	6.2	12.0	14.5	-8.6

Note : L = location, *T-Hs = Typic Haplosaprists, T-Hh = Typic Haplohemists, T-Ea = Typic Endoaquepts, T-Hq = Typic Humaquepts, T-Sq = Typic Sulfaquents*

Based on the palm oil production table in L1 locations, *Typic Haplosaprists* and *Typic Haplohemists* had good FFB productivity. The highest productivity at the L1 location was *Typic Haplosaprists* which was able to achieve productivity of 28.4 tons FFB in the planting year (2007) and then continued to decline until planting year (2017) with a production of 20.9 tons. The lowest production of oil palm FFB at L1 location in 2017 which was type soil namely *Typic Haplosaprist* the planting year of 2011 (production is about 16 tons). The yield of FFB productivity in L1 locations was quite stable, as evidenced by the trend of decreasing productivity which was not too high at 2 to 13%.

The location of L2 in the form of Inceptisols (*Typic Endoaquepts*) had the highest FFB productivity. This is because the management applied by the company is very good. This can be caused by the use of suitable fertilizer doses and the right time in fertilizing. Because based on the type of land, generally this soil is not too good for plantation, but because of good management so that the product produced is also good. Based on its nature, the land has properties that are newly developed with obstructed drainage, the pH was quite acidic, base saturation was <50% [4].

Soil type at L3 location is Inceptisol (*Typic Endoaquepts* and *Typic Humaquepts*). The highest productivity of the L3 location was owned by the planting year 2007 (*Typic Endoaquepts*) in 2015 which reached 17.4 tons. This is considered to be too low, the production potential that should have been achieved according to [7] in the planting year was around 24.5 tons/year. L3 location (*Inceptisol soil type*) is should get higher productivity by *Typic Humaquepts* soil when compared to *Typic Endoaquepts*. It caused *Typic Humaquepts* to consist by the enrichment of organic material from peat that has been decayed and mixed with the mineral soil which is able to improve the physical and chemical properties of the soil resulting in better productivity of FFB. It described that better soil utilization for oil palm plantations has a wide potential productivity potential depending on the management of the plantation [8].

Locations of L4 and L5 that have type of peat soil (*Typic Haplosaprists*) and Inceptisols (*Typic Endoaquepts* and *Typic Humaquepts*) with many different of planting years. At L4 location, FFB productivity in *Typic Humaquepts* soil types was higher, namely ± 12.4 tons/ha compared to *Typic Haplosaprists* (±10.1 tons/ha) which has type of mature peat soil. This indicates that *Typic Humaquepts* is better than *Typic Haplosaprists*, so that means mineral soils mixed with organic matter (peat-mineral) produce higher values than soils which only contain organic matter. Location L5 has the highest FFB productivity in Inceptisols (*Typic Endoaquepts* and *Typic Humaquepts*), around ± 10-18 tons/ha in a year. This is due to the density of oil palm plants on peatlands which tend to be higher than marginal mineral soils (Inceptisols), so as to produce more plant populations in each hectare, which implements to the higher productivity of FFB. Compaction of plant paths (plant density) using mechanical tools and making a hole in hole planting methods with puncher is an effort that is often done by large plantations [9].

L6 locations generally consist of Entisols soil. Entisols are in the category of young soils or undeveloped soil that still similar to the main material. The results of the study on L6 locations namely *Typic Sulfaquents* and *Typic Endoaquepts* have low productivity in both. It showed that in L6 locations especially *Typic Sulfaquents* still decreasing productivity even at older planting year. The results of *Typic Endoaquepts* productivity at L6 location only ranged from 0.6 to 1.2 tons (ha-1 year) which should be at the age of the plant for the appropriate land category (S-3 field) can produce ± 6-14 tons/ha.

4. Conclusion

Oil palm plantations in the Dangku Landscape are located in the flat topographic area in the Musi Banyuasin Regency of South Sumatra Province. The object of this research is only carried out on oil palm plantation blocks which have marginal land areas at an altitude of 5 MSL. The yield of FFB oil palm productivity (ton/ha) on marginal land showed that *Typic Haplosaprists*, *Typic Haplohemists*, *Typic Endoaquepts,* and *Typic Humaquepts* had the good enough productivity although not too many the evidence. The effects that are quite different are indicated by *Typic Sulfaquents*. Soil type of Entisols that is *Typic Sulfaquents* has the lowest FFB productivity in this study when seen generally with the same planting year conditions on other marginal soils (peat soil and Inceptisols). The results of oil palm FFB productivity showed that the productivity of L1 *(Typic Haplosaprists and Typic Haplohemists)* sampling locations was around 26 tons/ha (85% of the PPKS standard), L2 (*Typic Endoaquepts*) productivity was around 26 tons/ha (95% of PPKS standard), L3 (*Typic Endoaquepts and Typic Humaquepts*) productivity was around 9 tons/ha (45% of PPKS standard), L4 (*Typic Haplosaprist and Typic Humaquepts*) productivity was around 10 tons/ha (50% of PPKS standard), Its L5 (*Typic Haplosaprist, Typic Endoaquepts,* and *Typic Humaquepts*) productivity was around 12 tons/ha (55% of the PPKS standard), L6 (*Typic Endoaquepts and Typic Sulfaquents*) productivity around 3 tons/ha (25% of the PPKS standard). The location of L1 and L2 productivity meets the standard PPKS on S-3 land (according to marginal). Based on this research can be described that type of soil is a main factor that greatly affects the productivity of oil palm.

Acknowledgment

The authors would like to express their gratitude to the supervisor Dr Ir Hariyadi MS and Dr Ir Sukarman MS for their guidance in helping to provide ideas and input on this research. Also, thanks to the ZSL (Zoological Society of London) and KELOLA SENDANG with one of the representatives Mr Kresno Dwi Santoso who has given the mandate to work on this project.

References

[1] Suryana A, Hutabarat B, Susilowati SH. 1998. Penawaran dan permintaan serta peluang pasar komoditas tanaman industri dan perkebunan. Prosiding. Pertemuan Komisi Penelitian Pertanian Bidang Perkebunan. Peremajaan Rehabilitasi dan Perluasan Tanaman Perkebunan: Kelapa, Kelapa sawit, Karet, Kopi, Kakao, The, Lada, Pala dan Jambu Mente. Pusat Penelitian dan Pengembangan Tanaman Industri, Badan Penelitian dan Pengembangan Kehutanan dan

Perkebunan

[2] Ditjenbun. Direktorat Jenderal Perkebunan. 2013. *Statistik Perkebunan Indonesia 2012-2014 : Kelapa sawit.* Jakarta (ID) : Kementerian Pertanian

[3] Wahyunto W, Nugroho K, Ritung S, Sulaeman Y. (2014). Peta lahan gambut Indonesia: Metode pembuatan, tingkat keyakinan dan penggunaan. In A. Wihardjaka, E. Maftu'ah, Salwati, Husnain, & F. Agus (Eds.), Prosiding Seminar Nasional dan Network Meeting Pengelolaan Berkelanjutan Lahan Gambut Terdegradasi untuk Mitigasi Emisi Karbon dan Peningkatan Nilai Ekonomi. (pp. 81–96). Balai Besar Litbang Sumberdaya Lahan Pertanian. Kementerian Pertanian

[4] [BBSDLP]. Balai Besar Penelitian dan Pengembangan Sumberdaya Lahan Pertanian. 2013. Peta dan Rekomendasi Pengelolaan Lahan Gambut Terdegradasi di Sumatera. Bogor (ID): Kementerian Pertanian

[5] Joosten H. 2009. The Global Peatland CO_2 Picture: Peat Land Status and Emission in all Countries of The World. Wet Land International

[6] Kelola Sendang. Kemitraan Pengelolaan Lanskap Sembilang Dangku. 2017. *Menjajal pengelolaan lanskap berkelanjutan di sub-lanskap Dangku. Factsheet.* Palembang (ID) : Hal 1-8

[7] [PPKS]. Pusat Penelitian Kelapa sawit. 2006. Petunjuk Kultur Teknis Tanaman Kelapa sawit. Medan (ID) : Pusat Penelitian Kelapa sawit

[8] Pahan I. 2006. Panduan Lengkap Kelapa Sawit: Manajemen Agribisnis dari Hulu Hingga Hilir. Penebar Swadaya. Jakarta

[9] Sutarta, Purba ESP, Darmosarkoro W. 2003. *Budidaya Kelapa Sawit*. Medan (ID) : Pusat Penelitian Kelapa Sawit Sumatera Utara

IOP Conf. Series: Earth and Environmental Science **336** (2019) 012021 doi:10.1088/1755-1315/336/1/012021

Soil physicochemical properties in oil palm plantations impacted to peatland fire

Etik Puji Handayani[1] and Maswar[2]

[1]Sekolah Tinggi Ilmu Pertanian (STIPER) Dharma Wacana Metro, Jl. Kenanga No 3 Mulyojati 16C Metro, Lampung, Indonesia
[2]Indonesian Soil Research Institute

*E-mail: etik_ph@yahoo.com; etikpuji68@gmail.com

Abstract. Currently, Indonesia has 1.7 million hectares of oil palm plantations on peatland (12% of the total area of 13.5 million ha Indonesia's peatland). As a consequence, fires on peatland become trending topics. Field observation research that aims to determine the physical and chemical characteristics of peat in oil palm plantations has been carried out on several peat locations in West Aceh District: Suakraya village, Johan Pahlawan district (15 years old), Cot Gajah Mati Village, Arongan Lambalek district (1 year old), and Suakpuntong village, Nagan Raya district (10 years old). Substantial changes in the physicochemical properties of soils from oil palm plantations were observed, then evaluated the impact these characteristics to fires on peatland. The analysis soil showed that peatlands around in West Aceh District have physical characteristic: bulk density (BD) from 0, 12 - 0.16 g cm-3; groundwater level of 6 cm (rainy seasons) and 86 cm (dry seasons); moisture or water content in the surface layer (0-20 cm) during the dry season was 505 - 985% The thickness of peat was 457 cm on deep peat in Suakraya, 157 cm on shallow peat in Suakraya, 243 cm in Cot Gajah Mati, 136 cm in Suakpuntong. The chemical characteristic of peat soil has pH H_2O 2.9 – 3.9, pH KCl 2.23 - 3.07. The level of maturity of peat in the surface layer in all observation locations is the same, namely hemic. Water content per unit dry weight depends on peatland maturity level (fibric: 540 - 1187%, hemic: 268 - 480% and sapric: 106 - 242%). Ash content per unit dry weight varied from 1.8 – 5.9%, and C-organic content 53.4 – 57.6%. Peat total acidity is 4.2 – 6.4 me g-1, COOH content 0.02 - 0.16 me g-1 and phenolic-OH content 4.2 – 6.2 me g-1. The important result of this study showed that the causes of peatland fires are not caused by the characteristics of peat (occur naturally), but there are triggers of human activities.

1. Introduction

The total area of peatland in Indonesia is about 1.7 million ha with an average yield of oil palm in peatland that can reach 23 tons of Fresh Fruit Bunches (FFB) per year, peatlands have considerable potential for the development on oil palm agribusiness. On the other side, human activity along with the impact of climate change threatens the stability of this large pool of carbon that has been increasing rapidly over the last few decades owing to deforestation, drainage, and fire. Converting primary forest to oil palm plantation decreases the stock of soil organic carbon (SOC) 23% ± 21% [1] and would change the physical, chemical and biological characteristics of the peat. These changes are needed to support the growth and development of oil palm plants, but the newly different characteristics of peatlands would easily cause peat fires.

Peat fires are influenced by many factors among them are rainfall, peat water content, level of peat decomposition and groundwater level [2], the value of bulk density in some depth of peat [3]. Field observations aimed to determine the physical and chemical characteristics of peat on oil palm plantations have been carried out in several peat locations in West Aceh. Hence, the data was used to analyze the possibility of peat fires. Based on the type of fuel and the nature of its combustion, peat fires are grouped into 3 types, namely, ground fire, surface fire, and crown fire.

As a consequence of fires on peatland produces CO_2 emissions. Based on the calculation of the fire emission amount of 0-50 cm layer is 19.850.484 t CO_2 and at layer 50-100 cm is 18.284.371 t CO_2. So total emission of the depth of peat layer 0-100 cm with fire area 6000 hectares is estimated 38,134,855 t CO_2 [3]. Thus, peat management in order to avoid fire is very important, because these ecosystems contribute to greenhouse gas emissions and global warming.

2. Materials and methods

Field observations were conducted in West Aceh and Nagan Raya Districts in which the studied sites located in the oil palm plantations having different plant ages at 1) Suakraya Village of Johan Pahlawan District (15 years old); 2) Cot Gajah Mati Village of Arongan Lambalek District (1 year old); 3) Suakpuntong Village of Nagan Raya District (10 years old), and 4) peatlands forest in Cot Gajah Mati Village (forest without drainage); and, 5) Simpang Village of Kaway XVI Sub-district (peatland forest drainage).

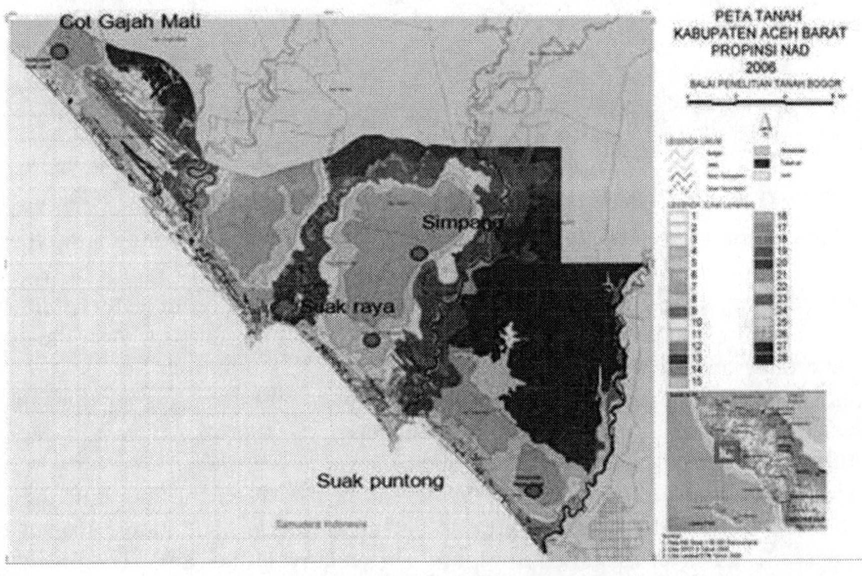

Figure 1. Observation Fields

Soil samples were collected by a peat auger to measure bulk density, water content, and soil chemical property. The water table inside dipwell was measured every 5 points from the samples site. The peat sampler of Eijkelkamp model can take semi undisturbed samples and its full length of the sample tube has a volume capacity of 500 cm^3. The samples were transferred into plastic bags, labeled and sealed to be transported to the laboratory in Bogor City of West Java Province. The samples were dried at 100 °C for 48 hours or until a constant weight has been reached. The dry weight was measured then bulk density and water content by weight was calculated.

The physicochemical analysis was carried out at the Soil Fertility Laboratory, Faculty of Agriculture, Bogor Agricultural University. Peat soil samples were taken from oil palm plantations over a specified transect using a peat drill. The composite soil samples in each transect are grouped based on the level of maturity of the peat determined by the fast method in the field to be fibric, hemic, and sapric. To determine the effect of peat maturity level on the physical and chemical characteristics of peat, analysis of variance was conducted then proceeded with Duncan's test at a significant level of 5%.

3. Results and discussion

The average thickness of peat at the forest in Simpang was about 1000 cm, at the oil palm in Suak Raya 457 cm, at the oil palm in Cot Gajah Mati 243 cm, at the forest in Cot Gajah Mati 227 cm, at the oil palm in Suak Raya 157 cm, and at the oil palm in Suak Puntong 136 cm. Bulk density in West Aceh ranges from 0.05–0.26 g cm^{-3} such as shown in the following figure.

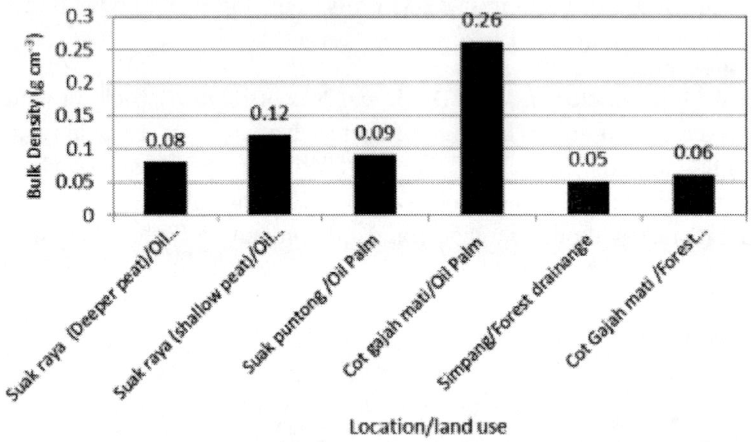

Figure 2. The bulk density of the peat at depth of 0–50 cm in 6 locations

The conversion of forests to oil palm plantations increased bulk density indicating there have been significant compressions of peat pores that could cause peat soil more compact and increase the capacity of peat capillary. Therefore, the water movement from the groundwater table to the surface layer increases that can maintain soil moisture on the surface layer. Furthermore, the compaction of peat soils could reduce oblique plant, soil nutrient leaching, CO_2 emissions and peat risk to fires. [3] in the fire area shows that the bulk density value on the peat surface is higher than the inside part. The differences in bulk density values are due to differences in peat maturity in each soil profile because of both naturally and also human activities.

Table 1 presents the results of the groundwater table measurement showing the groundwater levels on peatlands varied widely both in forests and in oil palm plantations. However, the water content in the peat surface layer as shown in Figure 3 are relatively the same. Therefore, groundwater level is not the best indicator for sustainable peat management especially in the cultivated peatlands, but soil moisture is important to maintain both for the optimum plant growth and fire prevention.

Table 1. Average groundwater levels at the study site and land use.

Village/Land use	Water table			
	Average (cm)	Minimum (cm)	Maximum (cm)	Difference (cm)
Suak Raya (Deep peat)/Oil Palm	40	24	58	34
Suak Raya (Shallow peat)/Oil Palm	45	33	53	20
Suak Puntong/Oil Palm	73	63	86	24
Cot Gajah Mati/Oil Palm	35	6	65	60
Simpang/Forest drainage	53	32	70	38
Cot Gajah Mati /Forest without drainage	13	8	17	9

Soil moisture can be a rational criterion if it is associated with the risk of fires on peatlands. According to [4] and [5], more than fifty percent of peat fires are on the soil moisture content of 100-125%. For this reason, efforts to maintain peat soil moisture are needed so that fire incidents can be avoided.

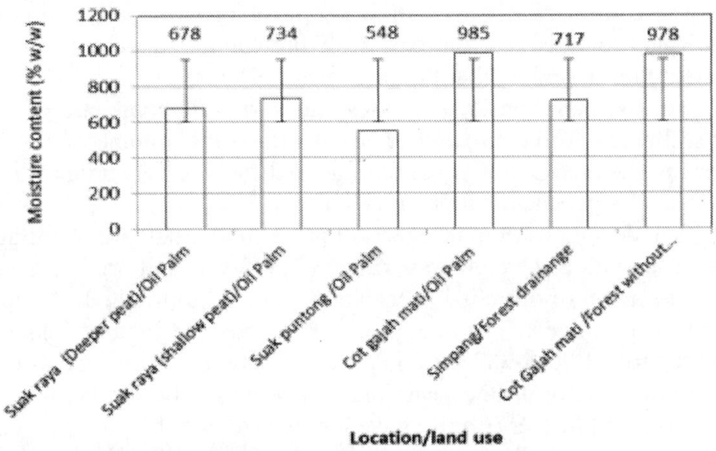

Figure 3. Soil moisture in surface soil (0-50 cm) in August 2009

Figure 3 shows the average soil moisture on the surface of the peat is relatively high above 500%. The risk of a peat fire in this condition is very small. Rainfall pattern has 2 peaks in the wet and dry seasons in a year, and the rainfall in the dry season is relatively high (Figure 4).

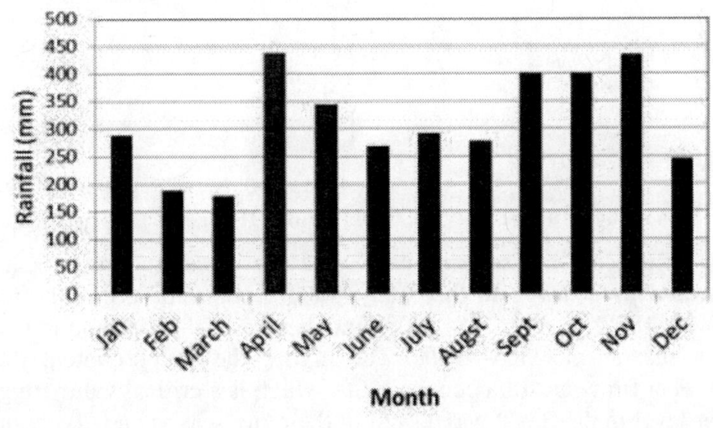

Figure 4. Rainfall patterns in the studied site (Source: *Badan Pusat Statistik* (BPS) West Aceh, 2007)

Peatlands in Meulaboh have pH H_2O 2.9–3.9 and pH KCl 2.23–3.07. Based on the level of peat maturity, the average of pH H_2O was 3.52 for peat fibric, 3.45 for hemic peat, and 3.08 for peat sapric. While the average of pH KCl was 2.73 for peat fibric, 2.62 for hemic peat, and 2.55 for peat sapric. The results of the variance analysis showed that pH H_2O was significantly different on the level of peat maturity, while the pH KCl was not significantly different. With Duncan's test, the pH H_2O on fibric is the same as hemic, but sapric is the lowest.

In this study, the water content of peat was depending on the level of maturity of the peat. Fibric peat has the highest water content, followed by hemic and sapric. The range of water content fibric: 540–1187%, hemic: 268–480% and sapric: 106–242%. Peat in Cot Gajah Mati area which is a peat forest that has just been opened and cultivated for oil palm cultivation has an average water content higher than peat in Suak Puntong and Suak Raya villages, so Cot Gajah Mati including in the fibric peat, it has not decomposed. Decomposition process tends slowly at low pH. This indicates that subsidence has occurred in peat which has long been used for oil palm cultivation such as in Suak Puntong and Suak Raya villages.

Peat subsidence was caused by changes in water level due to the position of compaction and oxidation. When the water level is low, the peat structure cannot support the above material and the pores of the structure are destroyed so that the peat material on the surface of the peatland decreases. While subsidence occurring, bulk density increases and hydraulic conductivity decreases, due to the compaction of peat soil pores. The consequences of changing the volume on the hydraulic properties will directly influence to the changes of water storage and flow of penetrates water to the subsoil [6] and lead to changes in the biogeochemical processes of peat [7].

The accelerated peat decomposition in Suak Puntong and Suak Raya villages have led to an increase in ash content per unit dry weight varied from 1.8–5.9%, and organic C content 53.4–57.6%. [8] showed that the ash content in fibric is lower than hemic and saprik and [9] reported that poor peat has lower ash content than rich peat. The increase in the content of organic matter from 0.1 to 57.4% caused BD to decrease from 1.86 to 0.33 g cm^{-3} and the total porosity increase from 38 to 90%. In addition, the pore size distribution in the peat soil is determined by the degree of decomposition of peat, the peat decomposes will further (sapric) have a pore size distribution more uniform [10].

Total acidity of peat represents the total of carboxylic acid (-COOH) and a phenolic acid (-OH) which shows the magnitude of the cation exchange capacity of peat soil. Dissociation of a carboxyl group (COOH) and phenolic -OH may result in the formation of a negative charge, thereby increasing the adsorption of water and cations. The results of measurements of total acidity of peat, carboxylic acid (-COOH) and a phenolic acid (-OH) of peat were presented in Figure 5.

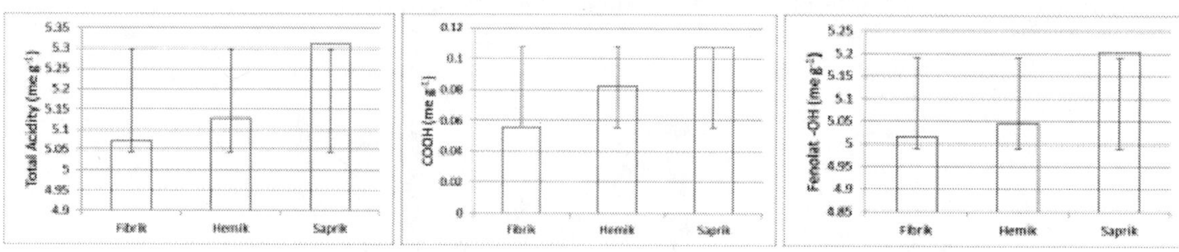

Figure 5. Average total acidity, -COOH, and -OH base on the peat maturity level

Peat fires are influenced by several peat characteristics such as peat water level, peat decomposition rate, and water level. In this study, water content based on the maturity level of peat ranges for fibric: 540–1187%, hemic: 268–480% and sapric: 106–242% and in the surface layer (0–20 cm) during the dry season was 505–985%. The higher the water content the lower the rate of combustion of peat. Peat fires can still occur at 119% which is a critical value triggering peat fire [2].

The groundwater level in the forest was 11 cm in the rainy season and 70 cm in dry season while in the oil palm plantations was 6 cm in the rainy season and 86 cm in the dry season. The water content of peat on the surface layer (0–20 cm) during the dry season was 717–978% in the forest, and 505–985% in the oil palm plantations. Thus, the conversion of the forests to the oil palm plantation has no significant impact on peat fires.

The maturity level of peat on the surface layer at all locations is hemic. The level of peat decomposition also affects peat fires, the more mature the peat (sapric type) the more difficult to burn compared to the immature peat type such as fibric and hemic types. The average bulk density in the forest (0.04 g.cm^{-3}–0.06 g.cm^{-3}) was lower than in the oil palm plantation (0.12 g.cm^{-3}–0.16 g.cm^{-3}).

Therefore, the incidence of peat fires could potentially occur in all conditions of land cover (forest and plantation), or in other words, the cause of the fire peatlands is not caused by the characteristic peat (naturally occurring), but there must be a trigger that human activity (whether intentional or accidental). Peat Fires usually begin with the ignition of fire above ground level. The fire will move in all directions, below the surface, above the surface, left, right, front and back. Fire will move up to the surface affected by wind speed and direction as the surface fire and when it reaches the tree canopy will be as crown fire.

4. Conclusion

Result in this study showed that the causes of peat fires were not caused by the characteristics of peat (occur naturally), but there are triggers of human activities. Maintaining soil moisture on peat soil is more important to avoid peat fires and must be used as a major variable over groundwater levels in peat management.

Acknowledgments

The award was given to World Agroforestry Centre (ICRAF), which has provided funds through the Rebuilding Green Infrastructure project with Trees People Want (ReGrIn) which is a partnership between the Soil Research Institute, ICRAF, the Indonesian Plantation Research Institute (LRPI) and the University of Hohenheim (Germany) until the implementation of this study.

References

[1] Shanmugam S, Dalal RC, Joosten H, Raison RJ, and Goh Kah Joo. 2018. SOC Stock Changes and Greenhouse Gas Emissions Following Tropical Land Use Conversions to Plantation Crops on Mineral Soils, with a Special Focus on Oil Palm and Rubber Plantations. Journal Agriculture 8, 133: 1-17

[2] Syaufina L. 2014. Trend analyses of forest and land fires towards climate change in Indonesia. SEAD International Workshop. 13 August 2014. Bogor

[3] Ratnaningsih, A. T., & Prasytaningsih, S. R. 2017. IOP Conference Series: Earth and Environmental Science The Characteristics of Peats and CO_2 Emission Due to Fire in Industrial Plant Forests. IOP Publishing IOP Conf. Series: Earth and Environmental Science, 97, 12029. https://doi.org/10.1088/1755-1315/97/1/012029

[4] Frandsen WH (1997). Ignition Probability of Organic Soils. Canadian Journal of Forest Research 27, 1971-1477

[5] Rein G, Cleaver N, Ashton C, Pironi P, Torero JL (2008) The Severity of Smoldering Peat Fires and Damage to the Forest Soil. Catena 74, 304-309

[6] Price JS. 2003. Role and character of seasonal peat soil deformation on the hydrology of undisturbed and cutover peatlands. Water Resources Research. 39: 1241

[7] Strack M, Waddington JM, Tuittila ES. 2004. Effect of water table drawdown on northern peatland methane dynamics: Implications for climate change. Global Biogeochemical

[8] Klemedtsson AK. 1997. Greenhouse gas emissions from farmed organic soils: a review. Soil Use and Management. 13: 245-250

[9] Berglund. 1995. Properties of cultivated gyttja soils. International pcat 3umd. 6: 5-23

[10] Walezak BW, Bieganowski A, Rovdan E. 2002. Water-air properties in peat, sand and their mixtures. Int. Agrophysics. 16: 313–318

CO_2 balance on oil palm agrosystem in Sumatra, Indonesia

Bayu Septiwibowo[1], Bram Hadiwijaya[1], Jean-Pierre Caliman[1]

[1]SMART Research Institute. PT SMART, Jalan Teuku Umar 19 Pekanbaru 28112 Riau Indonesia

*E-mail: caliman@indo.net.id

Abstract. The high demand and advantage of palm oil as vegetable oil for food and cosmetic products, and more recently for biofuels have led to the development of area dedicated to this crop, especially in South East Asia, and to some extent in South America and in Africa. In South Asia, where the highest production is recorded in Indonesia, the development of oil palm plantations has often been done at the expense of forests, causing changes in the atmospheric carbon dioxide (CO_2) balance due to the land use changes and agricultural management practices. Several studies have emphasized the importance of quantifying the oil palm contribution to the global carbon budget. A Long-term spatial and temporal measurement during its 25-30 years of life-cycle is needed to provide an accurate CO_2 balance assessment. One of the advanced methods is to directly measure CO_2 fluxes across a relatively large area using a micrometeorological approach based on eddy covariance principles. This method can estimate time series of Net Ecosystem CO_2 Exchange (NEE), Gross Primary Production (GPP) and ecosystem respiration (Reco) which are useful to assess carbon balance over an oil palm plantation. For that purpose, a flux tower of 25 m height, fully equipped with micrometeorological instruments, has been installed in 2011 in the center of a 90-ha plot of mature palms planted in Riau Province (Sumatra, Indonesia). The average oil palm canopies height was 18 m with an annual average air temperature of 26°C, 80% air humidity and a 2,500 mm average annual precipitation. The atmospheric conditions, seasonal variations and air pollutants such as haze, heavily influenced the CO_2 flux measurement. During this 6 year of study, the mature oil palm agro-system acted as a carbon sink that assimilates CO_2 from the atmosphere between 36 to 40 t CO_2 ha^{-1} y^{-1} as NEE. This result was the balance between emission assimilation of 209 t CO_2 ha^{-1} y^{-1} (GPP) and release back to the atmosphere of between 138 and 173 t CO_2 ha^{-1} y^{-1} through respiration (Reco). These variations on the CO_2 fluxes were mainly caused by seasonal patterns with a significant contribution of wet and dry days.

1. Introduction

Southeast Asia is most expanding agriculture and practices in the last decade, the palm oil plantation is the most common plantation found from the main island of Papua New Guinea to Thailand [1]. The expansion of the palm oil plantation and as other crops have been directly and indirectly affecting biodiversity [2], the conversion of the crops and under-story vegetations leading to a global question of further impact to greenhouse gas emission to the environment [3]. Indonesia's largest mainland Sumatra and Kalimantan involved developing palm oil plantation, the large-scale and intensive agro-system playing a significant role in the global carbon balance. However, the palm oil as the perennial crop with 25 to 30 years lifespan producing huge amount of biomass, some parts are recycled on site as the cultivation practices. The assessment of the net sequestration of the biomass due to palm oil stem growth is estimable from the balance of CO_2 uptake from the atmosphere and CO_2 released to the atmosphere, therefore estimating carbon sequestration of oil palm is a critical step to improve management practices.

The eddy covariance techniques along with micrometeorological measurement taking advantage of the improved technology of sonic anemometer and open path gas analyzer capable rapid measurement over 10 times per second enable detecting air parcel with low disturbance [4], monitoring environmental variation that governing vegetative growth and decomposition process over oil palm plantation. Flux measurements above the oil palm canopy using a micrometeorological method allow estimating the CO_2 exchange in the oil palm agro-system continuously through large spatial and temporal scales [5]. The daytime CO_2 exchange mostly is influenced by a photosynthetic variable such as concentration CO_2 in the air and light energy availability [6].

This study aims to obtain the value of the annual CO_2 uptake and emission rate trough long-term continuous CO_2 flux measurement, on a representative mature oil palm plantation using the micrometeorological method and the response to the climatic condition.

2. Materials and methods

2.1. Measurement site and parameters

This study located at a homogenous oil palm plantation in Riau-Indonesia, the 25-m tower was constructed in the center of the site, 90 ha of almost flat terrain at 71 asl, planted at 1993 with deli x lame 136 palms per ha. The mean palm height was 18.4 m with mean diameter 55.3 cm at breast high. Under-storey vegetation dominated with *Asystasia gangetica S.sp Micramtha*, *Neprolepis bissearata*, and *Asplenium longissium*. The soil classified as Typic Dystrudepts with the very deep-water table.

A set of CSAT3 3D Sonic Anemometer (Campbell Sci., USA) and LI-7500 Infra-red CO_2/H_2O gas analyzer (Li-Cor Biosciences, USA) alighted to 315 degrees from the North, heading the direction of the main wind. The measurement logged with CR3000 (Campbell Sci., USA) data logger with 10 Hz frequency. The gas analyzer was calibrated monthly period with High Purity N_2 gas, 402.8 ppm CO_2 tube, and dew point generator. The seamless measurement was started from 8th September 2011.

The meteorological measurement also installed at the same height, global and reflected radiation long and shortwave was measured with CNR-4 net radiometer (Kipp & Zonen, The Netherland). The air temperature (Ta), relative humidity (RH) sensed using HMP45 temperature humidity probe (Vaisala, Finland) was installed at 25-m height respectively, the vapor pressure deficit calculated using Ta and RH. Total Rainfall (Rt) was measured using tipping bucket TE525 (Texas Instrument, USA), Photosynthetic Photon Density (PPFD) was measured using SQ110 quantum sensor (Apogee Instrument, USA) also installed at a 25-m height.

Soil heat flux plate HFP01, soil moisture CS616, and soil temperature TCAV type E thermocouple averaging soil temperature sensor (Campbell Sci., USA) were installed at 5 cm and 10 cm under the soil surface. The setup was feed by 2x125 Watt hour off-grid solar cell and 3x100 Ampere hour valve regulated acid with the 12V parallel configuration as power storages.

2.2. Flux Calculation and post data processing

Eddy Pro version 5.1 software (Li-Cor Biosciences, USA) was used to calculate fluxes from 30-minute averaging data, the process through spike removal, plausible limit test, and skewness and kurtosis test [7]. Air density correction [8], planar fit rotation [9], frequency respond correction [10], bandpass correction factor transfer function calculate after [11]. The CO_2 flux was calculated by the covariance of gas fluxes density (ρ_d), vertical wind (w) and dry mole fraction of CO_2 as fluctuated from the mean value. The fluxes were tested QA/QC steady-state test after Foken [12] and on developed turbulence condition after Foken [13] which give the flag for each calculation result.

The NEE filtering was checked with overlap 4 days windowed 2.5-time moving average for spike removal, the plausible values were checked with assuming night-time fluxes were positive and daytime fluxes were negative. The NEE gap less than 4 continuous missing data were filled using linear interpolation. The larger dataset gap, mean diurnal variation were applied with differentiating daytime (PPFD > 10 μmol m^{-2} s^{-1}) and night-time data, look-up table also apply if gap is still emerging, with multiple input environmental data such as air and soil temperature, humidity, and PPFD to ensure look-up data is equal in environmental condition.

The ecosystem photosynthetic was estimated using light-respond curve with vapor pressure deficit (VPD) limitations after Lasslop [14] Eq. 1 and 2. The night-time respiration temperature response Lloyd and Taylor model fitted to daytime temperature to calculate daytime ecosystem respiration (R_{eco}) with 4 days windows using Eq. 3 [15].

$$NEE = -\frac{\alpha\beta\ PPFD}{\alpha\ PPFD + \beta} + \gamma \tag{1}$$

$$\beta = \begin{cases} \beta.\exp^{(-k\ VPD-VPD0)} & ,VPD>VPD0 \\ \beta=\beta0 & ,VPD<VPD0 \end{cases} \tag{2}$$

Where α is the initial slope of the light response curve (μmol CO_2 J^{-1}), β is the maximum CO_2 uptake rate of the canopy light saturation (μmol CO_2 $m^{-1}s^{-1}$), and γ is ecosystem respiration (μmol CO_2 $m^{-1}s^{-1}$).

$$R_{eco} = R_{eco\ reff} \cdot Exp^{E_0 \frac{1}{Tref-T_0} - \frac{1}{T-T_0}} \tag{3}$$

Where $R_{eco\ reff}$ is nighttime respiration base data (μmol CO_2 m^{-2} s^{-1}), T_{reff} is reference temperature (°C), E_0 is the temperature sensitivity parameter, T_0 equal to -46.02°C as in the original model.

The daytime gross primary production estimated as a substitute of R_{eco} to NEE base on Desai [16] Eq. 4, the nighttime GPP assumed to zero.

$$Gpp = R_{eco} - NEE \tag{4}$$

3. Results and discussion

3.1. Environmental variable

The measured NEE values influenced by multiple droughts of the dry period with monthly precipitation lower than100 mm [17] and wet season with high rain and cloudy days decrease the amount of PPFD reaching crop canopies. El Niño and La Niña variations caused drier and wetter than normal weather, the strong El Niño events at 2014 and 2015 and heavy peat fire in the neighboring region sent thick haze to the site influencing environment condition that drives NEE. The haze series reduce the amount of measured PPFD by 11% might be lowering plant physiological process and reduce CO_2 uptake potential [19]. The air temperature oscillating between 23 °C to 34 °C, the dry season air temperature both maximum (Tmax) and minimum (Tmin) increased by 1.2 °C, the highest air temperature recorded in April 2016 reaching 35 °C and the lowest daily average air temperature recorded at 20 °C, the annual mean air temperature, however, it remains very close to constant. The wettest month recorded in October with 550 mm to November 2012 with 690 mm, follow the series of rain in the second semester of 2013, this phenomenon caused an increasing water content on the mineral soil reaching saturated maximum 50% (figure 1).

3.2. Seasonal variation

Regular wet and dry seasons show different in the daily sum of NEE, the wet season NEE tend to be more negative then shifted to less negative in the dry period. The dry season at the end of June to early of July 2013 follows with a thick haze on February 2014, in 2015 haze event occur from August to September. However, the haze was difficult to be determined using radiation base sensor due to cloud cover determination, haze event was visually observed. During this period the PPFD were declining followed by the less fluctuation daytime air temperature this condition suppressing both R_{eco} and NEE.

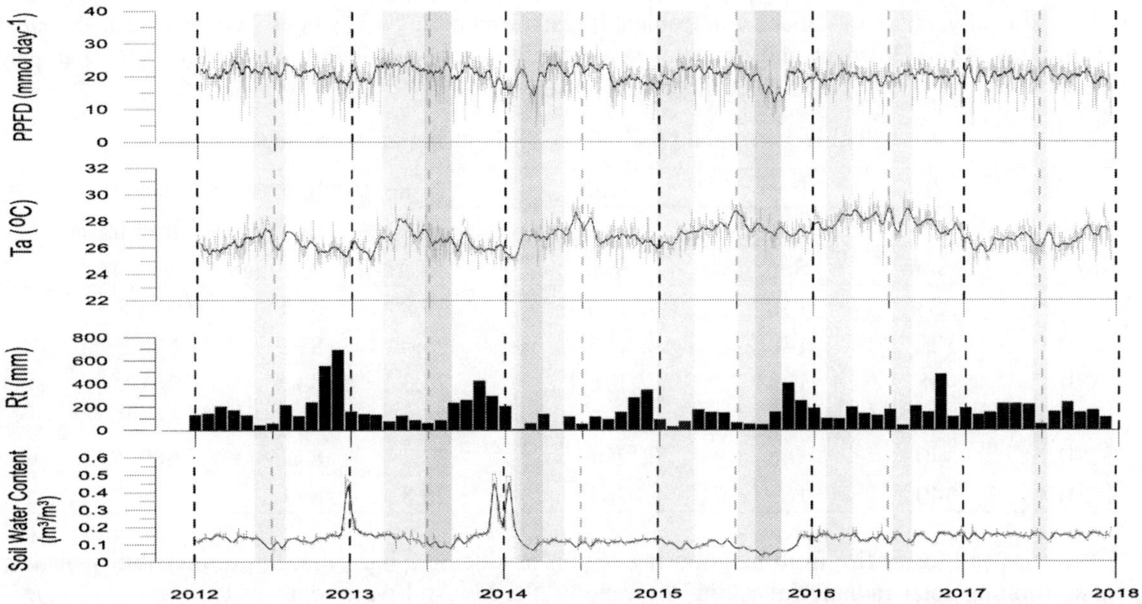

Figure 1. Seasonal variations of daily photosynthetic photon flux density (PPFD), daily mean air temperature (T_a), monthly rain total (Rt), soil water content (SWC) from January 2012 to December 2017. The black line is 15 days moving averages, bold grey shade indicating haze and light grey shades were indicating a dry period.

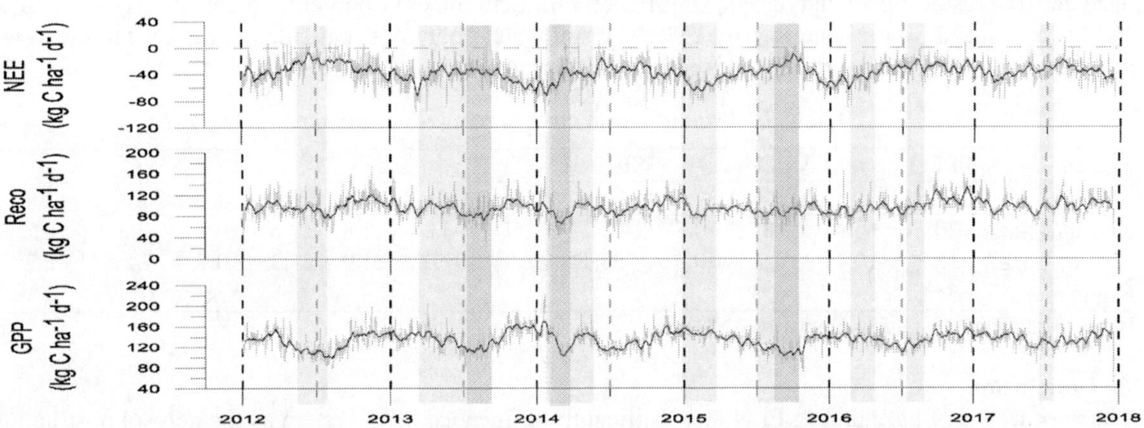

Figure 2. Seasonal variations of daily sum of NEE, R_{eco}, GPP. The black line is 15 days moving averages, bold grey shade indicating haze and light grey shades were indicating a dry period.

3.3. *Annual CO_2 Flux*

The palm oil in mineral soil showed annual NEE is negative ranging from lowest 38 to 40 t CO_2 ha^{-1} y^{-1} this result is almost consistent, highest CO_2 uptake was in 2017 which is 40 t CO_2 ha^{-1} y^{-1} and the lowest was on 2014 due to increasing R_{eco} following the temperature increase. Both R_{eco} and GPP is increasing during the measurement with different rate (table 1), The net carbon uptake of oil palm agro-system in Riau is higher than net carbon uptake by coconut (*Cocos nucifera* L.) in Vanuatu 17–30 t CO_2 ha^{-1} yr^{-1}[20].

The mean daily value of NEE was also similar range compared to other measurements on tall canopy vegetation located at the tropical continent. A study from Pasoh Forest Reserve, Malaysia has shown net ecosystem CO_2 uptake by the tropical rain was around 8.95 g CO_2 m^{-2} day^{-1} [21]. This rate is much higher than daily values of Amazonian tropical forests, which the NEE values were around 1.98–3.96 g CO_2 m^{-2} day^{-1} [22]. However, those results were obtained from calculation of short-term measurement data that may ignore the temporal variability of NEE.

Table 1. Annual NEE, Reco, GPP during measurement

Year	NEE	Reco	GPP	Temp	RH	PPFD
		tCO_2 ha^{-1}yr^{-1}		oC	%	µmolm-1s-1
	Sum	Sum	Sum	Average	Average	Average
2012	-38	138	176	25.8	81.8	622.8
2013	-39	146	185	26.0	81.9	623.5
2014	-38	173	196	26.3	83.5	661.8
2015	-38	142	180	26.8	87.6	669.7
2016	-40	164	204	27.4	84.6	689.8
2017	-40	163	203	26.8	84.0	599.5

Peat fire and forest fire from neighboring region producing a haze that limiting visibility distance and so limiting solar radiation reaching the canopy, the low soil water content tend to correlate with haze event when soil biomass supports the low-temperature combustion. The peat fire also produces numerous GHG pollutant such as carbon dioxide, carbon monoxide, nitric oxide and nitrogen dioxide [23] and aerosol that less than 10 microns in diameter.

The haze series occur in dry season 2013, El Niño 2014, and strong El Niño 2015 with different thickness and duration, we tested haze series with simple one-way ANOVA before haze versus during haze data the result shows that haze is significantly influent the environmental parameter (P>0.01). The NEE was reduced by 19% and R_{eco} by 13%. The soil water content is coincidently differed to the lower value due to haze apparent mostly caused by long drought.

Table 2. Haze effect on environmental parameter.

Season	PPFD		VPD		NEE		R_{eco}		T_a		SWC	
	mmol d^{-1}		hPa		g C ha^{-1} d^{-1}		g C ha^{-1} d^{-1}		°C		m^3 m^{-3}	
Dry w/o haze	20.7	±0.4	8.8	±0.3	-44.7	±1.7	94.0	±1.8	27.6	±0.1	13.3	±1.5
Dry w/ haze	18.3	±0.4	10.9	±0.3	-35.9	±1.7	81.9	±1.9	27.2	±0.1	7.2	±1.5
Sig test	***		***		***		***		***		***	

Note: Significant test using One-way ANOVA

4. Conclusion

The appearance of haze during El Niño significantly influenced agro-system as the aerosol obstructing PPFD to reach the palm canopy. It also affected the temperature and VPD that control the rate of photosynthesis. The prolonged dry season might contribute to low-temperature biomass combustion that causing haze.

The mature oil palm agro-system could be part or importance role of carbon sequester about 40 t CO_2 ha^{-1} y^{-1} or equal to 10.5 t C ha^{-1}yr^{-1} as NEE, at the same time the release back to the ecosystem about 160 t CO_2 ha^{-1} y^{-1} equal to 42 t C ha^{-1} yr^{-1} as respiration. Total CO_2 assimilate on the oil agro-system were about 200 t CO_2 ha^{-1} y^{-1} equal to 50 t C ha^{-1} yr^{-1}. However, this assumption and measurement result using the EC method should be complemented by field measurement to provide substantial supportive data [18].

Acknowledgements

This project is fully supported by PT. SMART Tbk. Partial funding was also given by CIRAD, France. We would like to thanks Winthrop Prof. Jason Beringer, University of Western Australia, and Ian

McHugh, Monash University, Australia, for instruments and installation set up as well as valuable advice. Moreover, we also would like to thanks Prof. Takashi Hirano, Dr. Ryuichi Hirata and Dr. Masahito Ueyama from AsiaFlux for sharing the processing and gap-filling strategy. Lastly, we thanks to Dr. Hiroki Ikawa, University of Alaska, USA, for discussion about co-spectra and Dr. Gerardo Fratini, Li-Cor Biosciences, USA, for data processing script discussion.

References

[1] Colchester, M., and Chao S. 2011. Oil Palm Expansion in South East Asia: trends and implications for local communities and indigenous peoples

[2] Emily B. Fitzherbert, Matthew J. Struebig, Morel. A. Danielsen F, Carsten A. Bruhl, Paul F. Donald and Phalan B. How will oil palm expansion affect biodiversity? Trends in Ecology and Evolution Vol.23 No.10 2008 -539

[3] Page, S.E., Morrison, R., Malins, C., Hooijer, A., Rieley, J.O., and Jauhiainen, J., 2011. Review of peat surface greenhouse gas emissions from oil palm plantations in South East Asia. ICCT White Paper no. 15.t

[4] Baldocchi, D.D., 2003b. Assessing the eddy covariance technique for evaluating carbon dioxide exchange rates of ecosystems: past, present, and future. Global Change Biology. 479: 492.

[5] Henson, I.E. and Harun, M.H., 2005. The influence of climatic conditions on gas and energy exchanges above a young oil palm stand in North Kedah, Malaysia. Journal of Oil Palm Research Vol. 17, 73:91.

[6] Chapin, F.S., Matson, P.A., and Mooney, H.A., 2002. Principles of terrestrial ecosystem ecology. Principles of Terrestrial Ecosystem Ecology. 2002 Springer-Verlag New York, Inc

[7] Vickers, D. and Mahrt, L. (1996) Quality Control and Flux Sampling Problems for Tower and Aircraft Data. Journal of Atmospheric and Oceanic Technology. 512:526

[8] Webb, E.K., Pearman, G.I., and Leuning, R., 1980. Correction of the flux measurements for density effects due to heat and water vapor transfer. Quarterly Journal of the Royal Meteorological Society, 106. 85:100

[9] Wilczak, J.M., Oncley, P., and Stage, S.A., 2011. Sonic anemometer tilt correction algorithms. Boundary-Layer Meteorology, 99. 127:150

[10] W.J. Massman .2000. A simple method for estimating frequency response corrections for eddy covariance systems. Agricultural and Forest Meteorology 104. 185:198

[11] Moncrieff, J., Clement, R., Finnigan, J., Meyers, T. 2004. Handbook of micrometeorology -2004.

[12] Thomas Foken. Tools for quality assessment of surface-based flux measurements. Agricultural and Forest Meteorology 1996. Volume 78, Issues 1–2, 83:105

[13] Thomas Foken, Micrometeorology. 2008

[14] Gitta Lasslop, Markus Reichstein, Dario Papale, Andrew d. Richardson, Almut Rneth, Alan Barr, Paul s Toy and Georg Wohlfahrt . 2010. Separation of net ecosystem exchange into assimilation and respiration using a light response curve approach: critical issues and global evaluation. Global Change Biology.16. 187:208

[15] Reichstein M., Falge E., Baldocchi D., Papale, D., Aubinet, M., Berbigier, P., Bernhofer, C., Buchmann, N., Gilmanov, T., Granier, A., Grünwald, T., Havránková, K., Ilvesniemi, H., Janous, D., Knohl, A., Laurila, T., Lohila, A., Loustau, D., Matteucci, G., Meyers, T., Miglietta, F., Ourcival, J.-M. Pumpanen, J., Rambal, S., Rotenberg, E., Sanz, M., Tenhunen, J., Seufert, G., Vaccari, F., Vesala, T., Yakir, D., and Valentini, R., 2005. On the separation of net ecosystem exchange into assimilation and ecosystem respiration: review and improved algorithm. Global Change Biology, 11. 1424:1439

[16] Ankur R. Desai, Andrew D. Richardson, Antje M. Moffat, Jens Kattge, David Y. Hollinger, Alan Barr, Eva Falge, Asko Noormets, Dario Papale, Markus Reichstein, Vanessa J. Stauch. 2007. Cross-site evaluation of eddy covariance GPP and RE decomposition techniques. Agricultural and Forest Meteorology, 148. 821:838

[17] Malhi, Y., Pegoraro, E., Nobre, A.D., Pereira, M.G.P., Grace, J., Culf, A.D., Clement, R., 2002

Energy and water dynamics of a central Amazonian rain forest. J. Geophys. Res. Atmos. 107, 8061. http://dx.doi.org/10.1029/2001c5JD000623.

[18] Malhi, Y. and Grace, J., 2000. Tropical forests and atmospheric carbon dioxide. TREE, 15.333:337.

[19] Graham, E. A, Mulkey, S. S., Kitajima, K., Phillips N. G., and Wright S. J. 2002. Cloud cover limits net CO_2 uptake and growth of a rainforest tree during tropical rainy seasons. www.pnas.org/cgi/doi/10.1073/pnas.0133045100. 572:576.

[20] Roupsard, O., Lamanda, N., Jourdan, C., Navarro, M.N.V., Mialet-Serra, I., Dauzat, J., and Sileye, T., 2008. Coconut carbon sequestration, Part I / Highlights on the carbon cycle in coconut plantations. Cord, 24, 1:15.

[21] Yasuda, Y., Ohtani, Watanabe, T., Okano, M., Yokota, T., Liang, N., Tang, Y., Nik, A.R., Tani, M., and Okuda, T., 2002. Measurement of CO_2 flux above a tropical rainforest at Pasoh in Peninsular Malaysia. Agricultural and Forest Meteorology, 114, 23:244.

[22] Grace, J., Malhi, Y., Lloyd, J., McIntyre, J., Miranda, A.C., Meir, P., Miranda, H.S., 1996. The use of eddy covariance to infer the net carbon dioxide uptake of the Brazilian rainforest. Global Change Biology, 2, 209-218.

[23] Levine. J. S., Biomass Burning and the Production of Greenhouse Gases. Climate Biosphere Interaction: Biogenic Emissions and Environmental Effects of Climate. https://www.researchgate.net/publication/24161310

Regional heat capacity changes on oil palm plantation development in 1994-2010 based on Landsat-5 TM satellite data

Idung Risdiyanto[1], Winda Ariyani[1], Odjat Sujatnika[2]

[1]Department Geophysics and Meteorology, Bogor Agricultural University. Jalan Meranti Kampus IPB Darmaga, Bogor, Indonesia, 16680
[2]Aksenta Socio-Enviro Consultant, Jakarta, Indonesia

*E-mail: idungris@ipb.ac.id

Abstract. Regional heat capacity change is calculated from the ratio between the addition or subtraction of heat (ΔQ) with the increase or decrease in temperature (ΔT) region. The purpose of this study is to calculate the regional heat capacity change due to the changes of land cover composition with forest, shrubs, oil palm plantation and bare soil using Landsat-5 TM satellite data on 1994, 2000 and 2010. The total area used in this study is 12971 ha. In 1994-2000, 4 % of forest area and 2% shrubs were increased, followed by addition biomass forest 4.01 tons/ha and 2.83 tons/ha for shrubs. The increased of forest area and biomass (tons/ha) caused by forest and shrubs growth processing towards a climax that added the canopy volume. So that, the regional heat capacity in 1994 amounted 19384 MJ $^{o}C^{-1}$ increased to 19929 MJ $^{o}C^{-1}$ in 2000. Data observation for 2000-2010 showed that forest area decreased by 66% due to forest's clearing into oil palm plantations (47%), shrubs (8%) and bare soil (11%). However, plant's biomass continues to increased, i.e. 1.48 ton/ha for the forest, 2.73 tons/ha for shrubs and 4.63 tons/ha for bare soil. Before 2000, there was no land cover by oil palm plantations, so the increasing rate from this land was the biggest than the three other lands, amounting to 122.29 tons/ha. Decreasing in the percentage of forest area does not cause a decrease in the heat capacity of the region. Intensive maintenance on oil palm plantation such as water management, fertilizer, and planting space made it biomass productivity, and ability to save the heat is greater than the forest. As a result, in 2010 regional heat capacity increased to 22508 MJ $^{o}C^{-1}$.

1. Introduction

The utilization of forests for settlements, mines, and plantations in Indonesia began intensively since 1970. Deforestation during the period 1985-1997 has covered about 20 million ha. Sumatra and Kalimantan are the areas with the largest deforestation of 6.7 million ha and 8.5 million hectares [1]. The forest area has been reduced from 103.33 million hectares to 88.17 million hectares in the period of 2000-2009, in Kalimantan about 4.95 million hectares and Sumatra 3.33 million hectares [2]. In 1984-1998 there were approximately 2.4 million hectares of forest converted to oil palm plantations, mainly in Kalimantan and Sumatra. The area of oil palm plantations increased from 4.16 million hectares in 2000 to 8.25 million hectares in 2009 [3].

Many studies demonstrated land-use changes is one of the causal factors to climate changes [4], [5], [6], [7], [8], [9], [10], [11]. The impact of land-use change on climate can be described in biogeophysical and biogeochemical effects [4]. Land cover changes have been impacted to surface albedo values so that the ability of the earth's surface to absorb, reflect and forward the incoming solar radiation also changes. Landcover patterns changes influence the altered mass and energy flux [6]. The albedo value is influenced by the type and nature of surface radiation, atmospheric conditions, as well as the physical properties of surfaces such as color, water content and surface roughness [12]. This results in changes in surface energy balance and heat capacity, thus affecting the spatial distribution of microclimate elements, such as surface temperatures.

Multispectral remote sensing data such as Landsat TM can be used to estimate surface temperature changes due to urban area expansion and land cover changes. In [13] showed that urban growth increased the surface temperature by 13.01K and positively correlated with decreasing plant biomass marked by changes in vegetation index. In [14] also estimated the surface temperature of vegetation and urban areas by the vegetation index approach and found that there was a negative correlation between surface temperature and vegetation index of -0.73 for forest and -0.67 for urban. In [15] has also demonstrated that vegetation can weaken the urban heat island effect and describe it as a negative correlation between land surface temperature (LST) and Normal difference vegetation Index (NDVI) o -0.41.

The small vegetation index shows low vegetation and biomass densities. [16], [17], [18] demonstrated a positive correlation between the vegetation index and vegetation density, and [19] used it to estimate the biomass that also correlated positively. Changes in the distribution of vegetation density and biomass in a region will have an impact on land surface temperature distribution [20]. The region is experiencing changes in heat capacity, namely the ability of the region to store energy derived from solar radiation.

The purpose of this study was to calculate the change in heat capacity due to changes in the composition of the cover area as a result of land conversion to oil palm plantations in the period 1994-2010. The approach used is the surface energy balance method and temperature change using Landsat-5 TM remote sensing satellite data. This data is used because Landsat-5 TM provides the longest data series based on satellite observation so it can be used as a resource to monitor the earth on a global scale and see the effect of land-use change on the value of the region's heat capacity.

2. Materials and Methods

2.1. Satellite imagery and interpretation

The study area is 112°24'29"- 112°34'1"E and 2°30'54"- 2°37'37"S which is administratively located in Seruyan district and Kotawaringin Timur, Central Kalimantan. The topography of the study area ranges from 0-150 mdpl, with most of the area is lowland, swamp and slightly hilly area in the North. This study uses Landsat 5 TM Path / Row 119/62 data, the date of acquisition of 1 September 1994, 9 March 2000 and 16 January 2010 [21]. The data management and analysis process uses some software such as Minitab 15, ER Mapper 7.1 and Arc Map 10.0 (No. IPB License: EFL588104064).

The data analysis procedure consists of preparing satellite image data, calculating energy balance component, biomass calculation, and heat capacity calculation area. Preparation of image data processing starts from geometric correction, taking of study area and classification of land cover. Geometric correction is done to reduce the geometry error so that the image projection matches the map [22]. The classification of land cover is done by unclassified classification method, which is the process of grouping the digital number (DN) image into several classes using cluster analysis [23].

Classification of the land cover class consists of forests, shrubs, oil palm plantations, and open land. Using a composite band 542 Landsat 5 TM for vegetation [21]. This composite image can detect and distinguish visual image capture by placing it on three primary colors (red green blue, RGB) in sequence. Band 5 captures mid-infrared waves (1.567-1.784 μm) that are sensitive to reflections of rock and soil moisture. Band 4 captures the near infrared wave (0.776-0.904 μm) that is sensitive to the reflection of the leaf's internal structure so it can be used to determine the biomass content, vegetation type and area with high vegetation reflections. Band 2 captures green waves (0.528-0.609μm) sensitive to fertility and reflection of green vegetation values, sediment concentration estimators and bathymetry mapping [24].

2.2. The surface energy balance components

All Landsat 5 TM satellite image data is DN in 8 bits format which must be converted to spectral radiance (Wm-2sr-1) with equations from [21]. The channels used include (i) the visible spectral channel (band 1,2,3) used in the energy balance calculations to obtain shortwave EM radiation and albedo, (ii) Red and NIR channels (bands 3 and 4) to calculate NDVI and (iii) thermal channel (band 6) for surface temperature (Ts, ^{0}C).

Albedo (α) is the ratio of reflected solar radiation (Rs out) to the incoming radiation (Rs in) on an object on the surface [25] and has a range of values 0-1. The value α = 0 indicates that the object absorbs all radiation shortwave that comes, while α = 1 indicates the object reflects entirely. Albedo is a function of wavelength, so its value gives physical information that occurs at the wavelength.

The surface temperature is the outermost temperature of an object. The surface temperature was extracted from Landsat-5 TM band 6-thermal (wavelength range are 10.45-12.42 μm). Band 6 can detect natural phenomena related to heat so that it can be used for mapping and thermal geological information [24]. The temperature of the image data extraction result is the brightness temperature that is converted from the spectral radiance value with the assumption that the surface of the earth is black (emissivity = 1) [26], so it is necessary to do the emissivity correction with NDVI approach. According to [27], Ts is a function of Tb (brightness temperature) and emissivity (ε) objects ((vegetation 0.96, open land 0.95) [28];[29], and wavelength (Λ) emission radiation (11.5 μm) with the following equation (1):

$$Ts = \frac{T_B}{\left(1+\frac{\lambda T_B}{\partial}\ln \varepsilon\right)} \quad (1)$$

An energy balance approach was conducted to obtain a net radiation value (RN, watt.m^{-2}). It was the difference between the in and out coming (longwave and shortwave) radiation of the earth's surface system by the following equation:

$$RN = Rs\ in - Rs\ out - RL\ out \quad (2)$$
$$RN = G+H+LE \quad (3)$$

The value of incoming shortwave radiation from the equation $\alpha=\frac{Rs\ out}{Rs\ in}$ (eq.4), albedo (α) is calculated by the equation, and the RL out value is derived from the Stefan-Boltzmann equation [21].

The calculated soil heat flux (G, watt.m-2), from RN, Ts, α and NDVI, as [30]:

$$\frac{G}{Rn}=\frac{Ts}{\alpha}(0.0038\alpha+0.0074\alpha^2)(1\text{-}0.98NDVI^4) \quad (5)$$

where,

$$NDVI = \frac{(NIR - Red)}{(NIR + Red)} \quad (6)$$

Furthermore, H is calculated by eq.3 and Bowen ratio $\beta=\frac{H}{LE}$ (eq.7), so the equation is obtained $H=\frac{\beta(R_n\text{-}G)}{1+\beta}$ (eq.8); β of tropical forest = 0.47 and β of palm plantations = 0.29 [31];[32], β of shrubs = 0.93 [33], and β of open forestland = 0.4 [34]. The LE was calculated as the remaining energy of Rn (3). After all the energy balance components are known, the air temperature (Ta, ^{0}C) can be calculated by modifying the [35], as:

$$Ta = Ts - \left(\frac{Hr_{aH})}{\rho_{air}C_p}\right) \quad (9)$$

Humid air density (ρ_{air}) is 1.27 kg.m^{-3}, specific heat of air at constant pressure (c_p) 1004 J.Kg^{-1}K^{-1}. The aerodynamic resistance (r_{aH}) is calculated by the formula r_{aH} = 31.9.u$^{-0.96}$ (u = normal wind speed at altitude 1-2 m, u vegetation = 1.41 m.s^{-1} and u non-vegetation = 1.79 m.s^{-1}) [36].

2.3. Regional heat capacity

The regional heat capacity is the ability of an area to store the heat received from solar radiation. The common approach of heat capacity is obtained from the product of the specific heat to the mass of the object, since information on the heat of the type of land cover does not exist, the regional heat capacity in this study is calculated by the ratio of heat changes (ΔQ) to the temperature changes (ΔT),

$$C = \frac{\Delta Q}{\Delta T} \tag{10}$$

3. Results and Discussion

3.1. Land cover changes and Albedo (α)

Forest area increased by 4% from 1994 to 2000; it came from bush and shrubs succession to the forest. Open land in 1994 also has occurred a succession to the bush and shrubs. As a result, in the year 1994-2000, the extent of bush cover increased 2%, and open land area decreased 6%. In 2010 the forest area was reduced by 66%, which changed to oil palm plantations (47%), shrubs (8%), and open land (11%), leaving only 4% of the total area covered. Part of the open land in 2000 was the result of forest and bush conversion that was suspected as preparation for oil palm plantations (figure 1).

Land cover changes since 1994, 2000 and 2010 led to a change in the solar radiation balance received by the surface. The value of Rs in for all types of land cover tends to be the same, which is 765-766 W.m^{-2} for 1994, 799-801 W.m^{-2} for 2000, and 804-806 W.m^{-2} for 2010. Rs out each type of cover is different, oil palm (52.9 W.m^{-2}), forests (53-66 Wm^{-2}), bush and shrubs (62-72 W.m^{-2}) and open land (70-79 W.m^{-2}). The Rs out value of each land cover in different years cannot be compared because of the difference in solar radiation reception time corresponding to the date of satellite image acquisition.

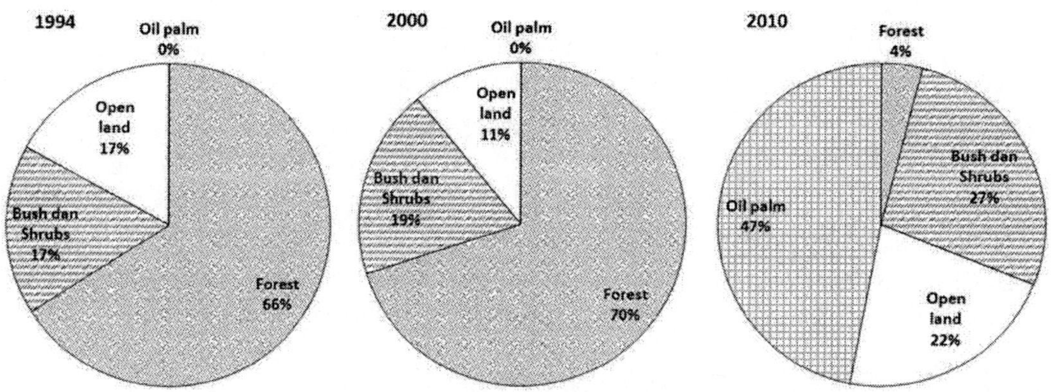

Figure 1. Area percentage of each type of land cover from Landsat-5 TM image interpretation

Radiation parameters on the surface that can be compared to land cover are α values. Oil palm plantation cover is the smallest compared to others, ie 0.07, forest-ie 0.07-0.08, bush and shrubs ie 0.08-0.09, and open land has the largest albedo with a range of 0.09-0.10 (Table 1). Changes in land cover types from 1994 to 2010 were not significant because the conversion of forest land is still used for vegetated land cover, so the change in albedo value from year to year is not large. This indicates that the land cover of palm oil plantations that absorb more homogeneous solar radiation compared to others.

The range of albedo extraction of image data in this study is still within the range of albedo previous research results so that the results obtained can describe albedo in the study area. The distribution of albedo values from 1994, 2000 and 2010 shows lower (figure 2). The mean albedo values at the regional level are 0.090, 0.089 and 0.086 respectively. This condition means that in 2010, the proportion of the amount of solar radiation reflected in 2010 is smaller than in previous years. It also shows that in 2010 more surface absorbs incoming solar radiation.

The albedo values of each land cover type vary due to differences in surface physical properties such as vegetation fraction, water content, surface color and roughness [37];[38]. The soil albedo is determined by particle size, mineral composition, moisture, organic matter content and surface roughness, in [12] adds that the surface of a subtle object is capable of reflecting greater radiation than a rough surface. Small albedo values in vegetated land cover are also caused by multilevel canopies that can absorb solar radiation at multiple levels of the canopy [39]. That is when the incoming solar radiation is reflected by the leaves/stems of plants in diffuse, canopy levels and canopy plant at the bottom of the canopy can absorb back the radiation.

Evaluation of albedo value according to the physical properties of the surface that influence it, indicating this value can be used as one of the indicators to know the change of regional heat capacity. Smaller albedo values are caused by changes in the land cover composition area, where oil palm plantations and forests absorb more incoming solar radiation than shrubs and bare land.

Table 1. Albedo values extracted Landsat-5 TM satellite image data and references

Albedo from satellite data				Albedo reference			
land cover type	1994	2000	2010	vegetation cover		non-vegetation cover	
forest	0.08	0.08	0.07	Bush and Shrubs [43]	0.06 - 0.08	Road [37]	0.05 - 0.20
Bush and shrubs	0.09	0.09	0.08	Grassland [10]	0.10 - 0.25	Roof [37]	0.08 - 0.18
Open land	0.1	0.1	0.09	Mix forest [10]	0.05 - 0.20	Soil [37]	0.05 - 0.40
Oil palm			0.07	Oil palm [43]	0.05 - 0.07	Bare soil [42]	0.07 - 0.11

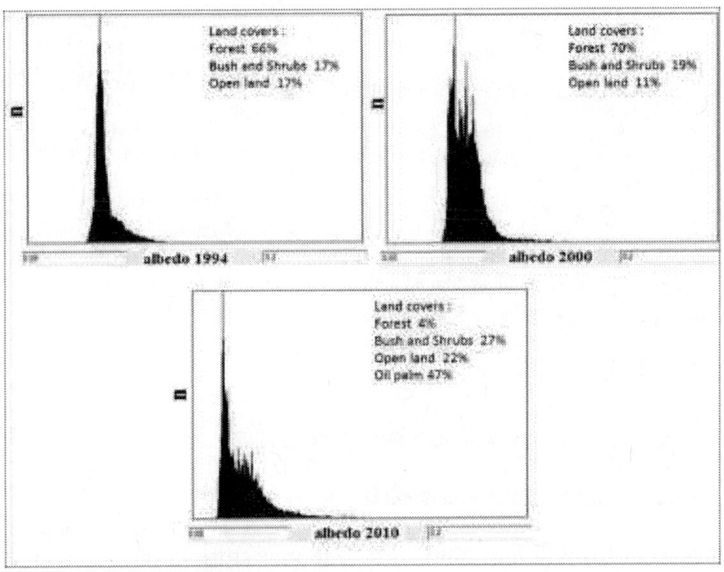

Figure 2. Distribution of albedo values in 1994, 2000 and 2010

3.2. Spatial distribution of surface temperature

The surface temperature distribution of Landsat-5 TM data extraction results in 1994, 2000 and 2010 is presented in figure 3 and figure 4. The average surface temperature shows a decrease, as well as its deviation. The change in surface temperature in this area is the impact of changes in the area and composition of land cover in the study area. Each land cover has different emissivity, conductivity and heat capacities. The higher density level of vegetation cover, it will cause the lower surface temperature. This is in accordance with the results of research by [15], [14] using NDVI to denote vegetation density and stated that the surface temperature is negatively correlated with NDVI.

. The difference in acquisition date of satellite data causes the surface temperature value to be incommensurable. Data for 1994 and 2000 were taken on September 1 and March 9 when the relative position of the Sun was around the equator, while January 16, 2010, was in the vicinity of the subtropics. In [13] also explained that the difference in solar radiation illumination in image data could affect the value of surface temperature. However, differences in the land cover composition will affect the distribution of surface temperatures [40]. Therefore, in this study comparable absolute value of the deviation between the mean surface temperature of the region with the surface temperature of each pixel

($30x30\ m^2$). In addition, relating to surface temperature can also be compared is the ratio between H with Rn and with LE. The surface temperature correlates with heat flux whose value is affected by the object or the type of surface.

Surface temperature estimation results in this study cannot be used to suggest land-use for oil palm has a lower surface temperature compared to secondary forests. Increased vegetation density from 1994-2010 is one of factor can reduce surface temperature and its deviation. The mean daily maximum of air temperature measurements was 2.5°C cooler than the secondary forest and 6.5°C cooler than the oil palm, regardless of the age [41]. It is in contrast to [31], [32] which states that oil palm with 12-meter canopy height and forest with 35-meter canopy height the same average H value and same leaf area index. The absolute value of the difference between the surface temperature of the region and the mean temperature shows that in 2010 showed the lowest deviation value compared to 1994 and 2010. Ts deviation for the entire study area was 1.3°C, 1.4°C, and 0.7°C for 1994, 2000 and 2010. Ts value of each land cover indicates that secondary forest and oil palm plantations have a lower value than another land cover. This shows that climate conditions in 2010 are more stable and cool. In [42] the daily maximum and minimum temperature differences to determine the cooling ratio.

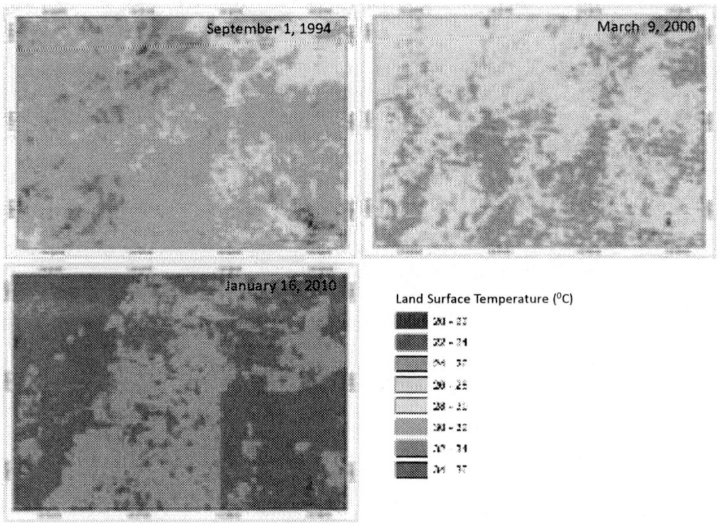

Figure 3. Distribution of land surface temperature in the study area with Landsat-5 TM satellite data on 1994, 2000 and 2010.

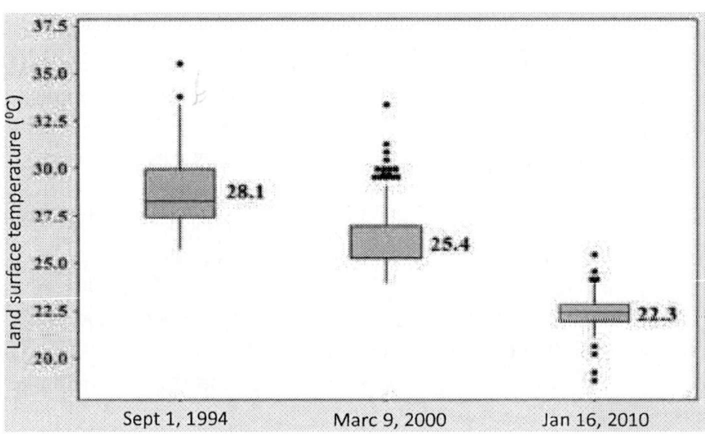

Figure 4. The surface temperature range in the study area with Landsat-5 TM satellite data

3.3. Surface energy balance

The common pattern of net radiation is inversely proportional to albedo and surface temperature (Ts). The higher value of the albedo and surface temperature, the smaller the radiation will be. Net Radiation (RN) ranging from the largest is oil palm cover, forests, shrubs, and open land. The calculation results show that the value of H is inversely proportional to LE (table 2). The H ranging value from the largest is in open land, shrubs, forests and oil palm plantations.

In contrast, the LE value of the largest is found in oil palm plantations, forests, shrubs, and open fields. In [32] concluded a difference in canopy structure between rainforest and palm oil generate the same mean of daily maximum H value 100 Wm-2, but different for LE value, ie 300 Wm-2 for rainforest and 400 for oil palm. This indicates the surface temperature between both land-use will tend to be the same, but the air humidity is different.

H value is the energy used for surface heating. The higher H-to-R ratio can show more heat flux near the surface (figure 5). The distribution of ratios between H and Rn for the whole region shows that in 2010 had a lower average value compared to the year 1994 and 2000. The addition of forest and forest area from 1994 to 2000 increased the ratio to 43.1% and 42.6%, while conversion to palm oil decreased the ratio to 36.2%.

The land cover change to vegetated land causes more incoming solar radiation to be absorbed by the surface [43], so net radiation also increases. The energy of the net radiation reaching the earth's surface is then used for the process of soil heat flux (G), sensible heat flux (H), latent heat flux (LE) and residual energy used for metabolic processes of organisms (S). According to [44], the value of residuals for terrestrial areas can be neglected as it is very small. The value of G is derived from the equation by [30], where G is directly proportional to Ts, so the energy for warming the soil on open land will be greater than the vegetated land.

Open land has a small LE value because of its high conductivity and shortwave reflection. Net radiation in open land is mostly used for heating the air so that the energy stored for the evapotranspiration process is low. By contrast, vegetated land has lower shortwave reflections and thermal conductivity [45];[46], so that the LE is larger. Much of the net energy on the surface is used for evapotranspiration and less for heating air. Oil palm plantations have the highest LE of all vegetation cover as there is a water treatment system that can retain moisture around the plantation. Vegetation land cover with high LE values has low surface temperature values. This is due to the proportion of the sensible heat flux that may cause a smaller temperature rise [47]. The increase in H value is an indication of the ongoing heating process and vice versa for cooling. This proves that the condition of 2010 experiencing surface cooling.

Table 2. Spatial distribution of the energy balance of each land cover with Landsat-5 TM satellite image data

Land cover	RN ($W.m^{-2}$)			G ($W.m^{-2}$)			H ($W.m^{-2}$)			LE($W.m^{-2}$)		
	1994	2000	2010	1994	2000	2010	1994	2000	2010	1994	2000	2010
Forest	257	303	339	28.8	30.8	27.2	73.1	87.0	99.9	155.5	185.1	212.5
Bush and shrub	251	295	327	29.1	31.1	28.6	106.8	127.2	144.0	114.8	136.8	154.9
Open land	236	282	322	31.4	34.3	31.0	163.9	197.8	233.2	41.0	49.5	58.3
Oil palm			340			25.8			70.6			243.3

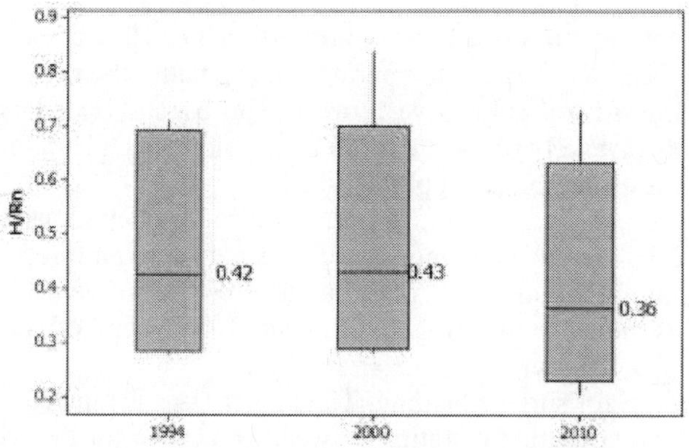

Figure 5. Changes in the ratio of H to Rn in the study area

3.4. Regional heat capacity

The resulting heat capacity value illustrates the amount of instantaneous energy contained in the study area on changes in air temperature and surface temperature. The heat capacity of each type of land cover varies (figure 6). When viewed from the type of land cover, the largest heat capacity is found in oil palm plantations because only with a percentage of 47% of the total area can store the heat of 14991 MJ $^{o}C^{-1}$ close to 1994 forest heat capacity value of 14929 MJ $^{o}C^{-1}$ whose percentage is 66 % Of the study area. The heat capacity of shrubs and open land is almost the same because although the land cover is classified as open land, its NDVI ranges from 0.35-0.44, meaning that on open ground vegetation is still encountered in the form of grass that is sufficient to withstand heat.

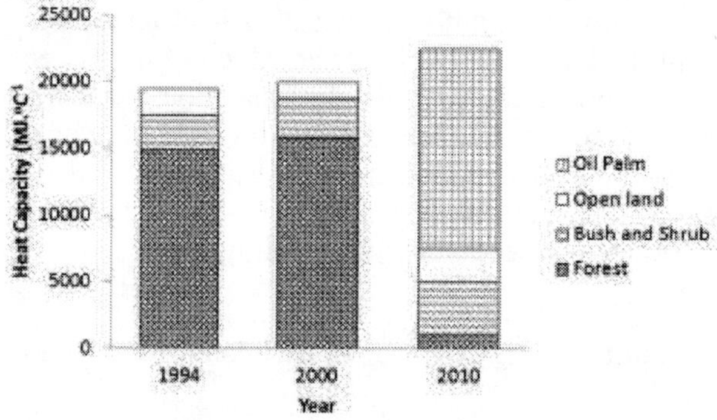

Figure 6. The heat capacity of each type of land cover with Landsat-5 TM

During the period of 1994-2010, there was an increase in regional heat capacity. In 1994, the regional heat capacity was 1494 kJ $^{o}C^{-1}$ Ha^{-1}, 2000 was 1536 kJ $^{o}C^{-1}$ ha^{-1} and in 2010 was 1738 kJ $^{o}C^{-1}$ ha^{-1} (figure 7). The increase between 1994-2000 was caused by the addition of forest area by 4% and 2% shrubs. While in 2000-2010 caused by the conversion of forest land into palm oil plantation by 47%. Oil palm plantations in 2010 have better heat storage capabilities than shrubs, forests and open land.

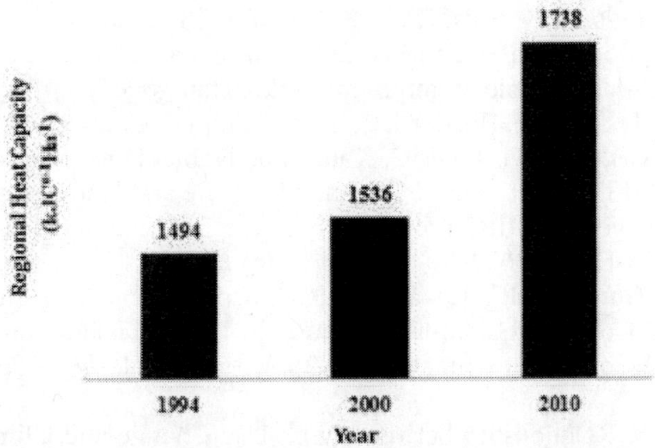

Figure 7. Changes in regional heat capacity of the study area per hectare with Landsat-5 TM

One of the factors affecting the area heat capacity in this research is potential biomass, ie the mass that comes from converting solar energy into organic material through the photosynthesis process. Biomass is also affected by net radiation (Rn 2010> Rn 2000> Rn 1994), where the greater the solar energy that can be used for the photosynthesis process, the potential biomass to be produced is also greater.

The calculation of the heat capacity value will give better results if using the function of the object mass and the heat of the object type. However, this study has not been conducted because there is no information on the types of forest cover, bushes, oil palm, and open land. Also, the calculation of biomass in any land cover is also not done. Therefore, the heat capacity calculation is based solely on the ratio of the addition or reduction of heat (ΔQ) to the increase or decrease in temperature (ΔT). The weakness of heat value calculation by using Landsat-5 TM image data is the time of repeat image coverage (the time required for satellite image to photograph the same coordinate and path/row object) is 16 days. In addition, geographically the territory of Indonesia is located around the path of ITCZ (Inter Tropical Convergence Zone), which is the area with an intensive airlifting process that causes high levels of resistance that can scatter the incoming solar radiation.

4. Conclusion

Changes in the land cover composition in the study areas in 1994, 2000 and 2010 led to an increase in regional heat capacity. Increased forest area by 4% and 2% shrub in the 1994-2000 year, increasing the region's heat capacity increased by 3%. Oil palm plantations covering 47% of the study area in 2010 could increase the region's heat capacity by 11.5% compared to 2000. Increased regional heat capacity from 1994, 2000 and 2010 led to a smaller range of surface temperature values in the study area in the addition of heat or energy which comes from solar radiation.

Increased heat capacity in a region can be used as an indication of cooling in the context of climate change on a regional scale. The absorption of solar radiation is increasing with increasing heat capacity. One factor that is thought to increase heat capacity is to increase the rate of growth that can increase the amount of biomass on the surface.

This study is a remote sensing assessment. The results obtained were the output of the equation models and parameters from other studies. Therefore, the results of this study still require verification with field observation data. In addition, it can also be improved by adding remote sensing observation data with more frequency data and different study locations.

References

[1] Holmes DA. 2002. Where Have All the Forest Gone?. Environmental and Social Development East Asia and Pasific Region Discussion Paper. The World Bank

[2] Sumargo W, Nanggara SG, Nainggolan FA, Apriani I. 2011. *Potret Keadaan Hutan Indonesia*

Periode Tahun 2000-2009. Edisi Ke-1. Forest Watch Indonesia

[3] The Ministry of Agriculture of Indonesia. 2009. Basis data statistic pertanian [internet]. [diacu 2016 Desember 6]. Available at: http://aplikasi.deptan.go.id/bdsp /newlok.asp

[4] Brovkin V, Boysen L, Arora V, Boisier J, Cadule P, Chini L, Claussen M, Friedlingstein P, Gayler V, van den Hurk B, Hurtt G, Jones C, Kato E, de Noblet Ducoudré N, Pacifico F, Pongratz J, and Weiss M. 2013. Effect of anthropogenic land-use and land cover changes on climate and land carbon storage in CMIP5 projections for the 21st century, J. Climate, **26**, 6859–6881, doi:10.1175/JCLI-D-12-00623.1

[5] Comarazamy DE, González JE, Luvall JC, Rickman DL, and Bornstein RD. 2013. Climate Impacts of Land-Cover and Land-Use Changes in Tropical Islands under Conditions of Global Climate Change. Journal of Climate 2013 **26**:5, 1535-1550, *https://doi.org/10.1175/JCLI-D-12-00087.1*

[6] Dale VH. 1997. The Relationship between Land-Use Change and Climate Change. *Ecological Applications,* Vol. **7**, No. 3 (Aug. 1997), pp. 753-769

[7] Dirmeyer PA, Niyogi D, Ducoudre DN, Dickinson RE, and Snyder PK. 2010. Impacts of land use change on climate (*editorial*). Int. J. Climatol. **30**: 1905–1907

[8] Fearnside PM. 2000. Global warming and tropical land-use change: greenhouse gas emissions from biomass burning, decomposition, and soils in forest conversion, shifting cultivation and secondary vegetation. Climatic Change **46**: 115–158

[9] Kalnay M and Cai M. 2003. Impact of urbanization and land-use change on climate. *Nature* **423**, 528–531 (29 May 2003), http://dx.doi.org/10.1038/nature01675

[10] Matthews H, Weaver A, Meissner K, Gillett N, and Eby M. 2004. Natural and anthropogenic climate change: incorporating historical land cover change, vegetation dynamics and the global carbon cycle, Clim. Dynam., 22, 461–479

[11] Oliver TH and Morecroft MD. 2014. Interactions between climate change and land use change on biodiversity: attribution problems, risks, and opportunities. WIREs Clim Change 2014, **5**:317–335. doi: 10.1002/wcc.271

[12] Dobos E. 2003. Albedo. *Encyclopedia of Soil Science*. doi:10.1081/E-ESS 120014334

[13] Weng Q. 2001. Remote sensing – GIS evaluation of urban expansion and its impact on surface temperature in the Zhujiang Delta, China. Int. *J. Remote Sens.* **22**(10): 1999-2014.

[14] Weng Q, Lu D, Schubring J. 2004. Estimation of land surface temperature-vegetation abundance relationship for urban heat island studies. *J Remote Sens. Environ.* **89** (2004): 467-383. doi:10.1016/j.rse.2003.11.005

[15] Liu L and Zhang Y. 2011. Urban Heat Island Analysis Using the Landsat TM Data and ASTER Data: A Case Study in Hong Kong. Remote Sens, **3**, 1535-1552; doi:10.3390/rs3071535

[16] Barati S, Rayegani B, Saati M, Sharifi A, Nasri M. 2011. Comparison of the accuracies of different spectral indices for estimation of vegetation cover fraction in sparsely vegetated areas. The Egyptian Journal of Remote Sensing and Space Science, Volume **14**, Issue 1, 2011, Pages 49-56, ISSN 1110-9823, https://doi.org/10.1016/j.ejrs.2011.06.001

[17] Gitelson AA. 2003. Wide Dynamic Range Vegetation Index for Remote Quantification of Biophysical Characteristics of Vegetation. Journal of Plant Physiology, Volume **161**, Issue 2, 2004, Pages 165-173, ISSN 0176-1617, https://doi.org/10.1078/0176-1617-01176

[18] Matsushita B, Yang, W, Chen J, Onda Y, and Qiu, G. 2007. Sensitivity of the Enhanced Vegetation Index (EVI) and Normalized Difference Vegetation Index (NDVI) to Topographic Effects: A Case Study in High-Density Cypress Forest. *Sensors (Basel, Switzerland)*, *7*(11), 2636–2651

[19] Hansen PM, and Schjoerring JK. 2003. Reflectance measurement of canopy biomass and nitrogen status in wheat crops using normalized difference vegetation indices and partial least squares regression. Remote Sensing of Environment, Volume **86**, Issue 4, 2003, Pages 542-553, ISSN 0034-4257, https://doi.org/10.1016/S0034-4257(03)00131-7

[20] Jim CY. 2012. Effect of vegetation biomass structure on the thermal performance of tropical

green roof. Landscape and Ecological Engineering, July 2012, Volume **8**, Issue 2, pp 173–187

[21] [USGS] United State Geological Survey. 2013. Landsat 7 science data users handbook [Internet]. [diacu 2013 February 20]. Tersedia dari: http://landsathandbook.gsfc.nasa.gov/pdfs/Landsat7_Handbook.pdf

[22] Jensen JR. 2000. Remote Sensing of the Environment: An Earth Resource Perspective. New Jersey: Prentice Hall

[23] Indarto & Faisol A. 2009. Identifikasi dan klasifikasi peruntukan lahan menggunakan Citra ASTER. *J Media Teknik Sipil.* **9**(1):1–8. ISSN: 1412-0976

[24] EROS Data Center. 1995. *Landsat-7 Technical Working Group*. Sioux Falls, USA South Dakota. October 31-November 2, 1995

[25] Stull RB. 2000. *Meteorology for Scientists and Engineers*. 2nd ed. USA. Brooks/Cole, Cengage Learning. Brooks/Cole, Cengage Learning, Belmont, CA, USA, 528 pp

[26] Chander G, Markham BL, Helder DL. 2009. Summary of current radiometric calibration coefficients for Landsat MSS, TM, ETM+ and EO-1 ALI sensors. *J Remote Sens. Environ.* **113**(2009):893–903. doi:10.1016/j.rse.2009.01. 007

[27] Artis DA & Carnahan WH. 1982. Survey of emissivity variability in thermography of urban areas. *J Remote Sens. Environ.* **12**: 313-329

[28] Jin M and Liang S. 2006. An improved land surface emissivity parameter for land surface models using global remote sensing observations. *J Climate*. **19**:2867-2881

[29] Van De Griend AA and Owe M. 1993. On the relationship between thermal emissivity and the normalized difference vegetation index for natural surfaces. *Int. J. Remote Sens.* **14:**1119–1131

[30] Allen RG, Morse A, Tasumi M, Bastiaansen W, Kramber W, Anderson H. 2001. Evapotranspiration from Landsat (SEBAL) for water right management and compliance with multi-state water compact. University of Idaho Kimberly, ID 83341

[31] Fowler D, Nemitz E, Misztal P, Marco CD, Skiba U, Ryder J, Helfter C, Cape JN, Owen S, Dorsey J *et al*. 2011a. Effects of land use on surface-atmosphere exchanges of trace gases and energy in Borneo: comparing fluxes over oil palm plantations and a rainforest. *J Environ. Science*. **366**:3196–3209. doi:10.1098/rstb.2011.0055

[32] Fowler D, Nemitz E, Misztal P, et al. 2011b Effects of land use on surface-atmosphere exchanges of trace gases and energy in Borneo: comparing fluxes over oil palm plantations and a rainforest. *Philosophical Transactions of the Royal Society B: Biological Sciences*. 2011;**366**(1582):3196-3209. doi:10.1098/rstb.2011.0055

[33] Fisch G, Tota J, Machado LAT, Dias MAFS, Lyra RF da F, NobreCA, Dolman AJ, Culf AD, Halverson J, and Fuentes JD. 2001. The convective boundary layer over pasture and forest in Amazonia. Aerospace Technical Center, Institute of Aeronautics and Space, University Sao Jose dos. Brazil

[34] Oliver JE. 1973. Climate and Man's Environment: An Introduction to Applied Climatology. New York: John Wiley & Sons, Inc

[35] Monteith JL and Unsworth MH. 1990. *Principles of Environmental Physics*. 2nd ed. London: Edward Arnold

[36] Rosenberg NJ. 1974. *Microclimate: The Biological Environment*. New York: John Willey and Sons

[37] Geiger R, Aron RH, Paul T. 1961. *The Climate Near The Ground.* Ed ke-5. Cambridge: Harvard University Press.

[38] Wei X, Hahmann AN, Dickinson RE, Yang ZL, Zeng X, Schaudt KJ, Scaaf CB and Strugnell N. 2001. Comparison of albedos computed by land surface models and evaluation against remotely sensed data. Journal of Geophysical Research. Vol,**106**, No.D18:20687-20702

[39] Peng J, Fan W, Xu X, Wang L, Liu Q, Li J, and Zhao P. 2015. Estimating crop albedo in the application of a physical model based on the law of energy conservation and spectral

invariants. *Remote Sens.* 2015, *7*, 15536-15560; doi:10.3390/rs71115536

[40] Yan H, Fan S, Guo C, Hu J, Dong L (2014) Quantifying the Impact of Land Cover Composition on Intra-Urban Air Temperature Variations at a Mid-Latitude City. PLoS ONE **9**(7): e102124. https://doi.org/10.1371/journal.pone.0102124

[41] Hardwick SR, Toumi R, Pfeifer M, Turner EC, Nilus R, Ewers RM. 2015. The relationship between leaf area index and microclimate in tropical forest and oil palm plantation: Forest disturbance drives changes in microclimate. Agricultural and Forest Meteorology, Volume **201**, 2015, Pages 187-195, ISSN 0168-1923, https://doi.org/10.1016/j.agrformet.2014.11.010. (http://www.sciencedirect.com/science/article/pii/S0168192314002780)

[42] Oke TR. 1987. *Boundary Layer Climates*. 2nd ed. Methuen, London, England and New York, NY, USA, 452 pp

[43] Gao J and Wu S. 2014. Simulated effects of land cover conversion on the Surface energy budget in the southwest of China. *Energies* 2014, *7*, 1251-1264; doi:10.3390/en7031251

[44] Sellers WD. 1965. *Physical Climatology*. The University of Chicago Press, Chicago, 60637.

[45] Betts RA. 2001. Biogeophysical impacts of land use on present-day climate: near-surface temperature change and radiative forcing. Atmospheric Science Letters (2001) Volume **1**. doi:10.1006/asle.2001.0023

[46] Soto NS, Cueto RG, Rincón ZH, Benítez SO, Núñez MQ, and Limón NV. 2015. Radiation balance of urban materials and their thermal impact in semi-desert region: Mexicali, México study case. *Atmosphere* 2015, **6**, 1578-1589; doi:10.3390/atmos6101578

[47] Handoko, Abujamin AN, Tania J, Rini H. 1994. *Klimatologi Dasar: Landasan Pemahaman Fisika Atmosfer dan Unsur-unsur Iklim*. Edisi ke-2. Jakarta (ID): Pustaka Jaya

A case study on the impacts of oil palm plantation on butterflies: differences in plantation scale and management implications

Rozza Tri Kwatrina[1]* and Yanto Santosa[2]

[1]Center of Forestry Research and Development, Forestry Research, Development and Innovation Agency, Ministry of Natural Environment and Forestry, Jl. Gunung Batu No. 5, Bogor, Indonesia
[2]Forestry Conservation Department, Faculty of Forestry, Bogor Agricultural Institute, Dramaga Campus, Bogor, Indonesia, 16680

*E-mail: rozzatk317@gmail.com

Abstract. Changes in land use to oil palm plantations have an impact on the biodiversity of tropical butterflies known to have species with a wide ecological range. We research to determine the impact of oil palm plantations on the diversity of butterfly species on two different plantation scales (large scale and small scale of plantation). The results showed that both of the oil palm plantations equally affected the increase in the number and level of species diversity of butterflies as many as 100 - 366% on large-scale oil palm plantations and by 0 - 466% on small-scale oil palm plantation. The development of oil palm plantations also has an impact on changes in species composition which is reflected in the level of similarity in species which is quite low, namely 0 - 0.22 in large oil palm plantations and 0.12 - 0.20 in small-scale plantations. The percentage of species loss in both types of plantations is the same, namely 60%, but there is greater species gain on small-scale oil palm plantations (60 - 400%) than large-scale (166 - 366%). Plantation management that creates habitat heterogeneity can increase the level of butterfly biodiversity on both plantations.

1. Introduction

One of the most pressing global topics now is achieving food security as well as biodiversity conservation. Tropical countries are the regions with the highest potential for conflict, where low food security and high biodiversity occur simultaneously [1]. Based on the conflict risk index, Indonesia was recorded as the seventh highest-ranking country with oil palm (*Elaeis guineensis*) as a strategic commodity.

The main factor as a cause of the high conflict between overlapping agricultural expansion and biodiversity hotspots is land use change. As the largest palm oil exporter in the world, Indonesia has plantations covering an area of 14.03 million hectares in 2018 [2] which has the potential to change the type of origin land cover. Oil palm plantations in Indonesia are developed on a large scale and small scale. Large Scale Plantation (LSP) usually have an area of more than 25 ha and have several age classes of plants, and are managed by the state, national private, or foreign private. Small-scale plantations/smallholder plantation (SP) generally have an area of less than 25 ha, only have one age class of plants, and are managed by communities or community groups. Both types of plantations have the potential to have an impact on biodiversity loss.

Butterflies are one of the biodiversity commonly found in oil palm plantations and have the potential as an environmental bio-indicator [3]. Butterflies also have a role as pollinators, host of parasites, and the prey for predators [4], [5]. Its species richness was also affected by environmental factors; climatology and topography, vegetation structure, and human disturbance [6]. Several studies have shown that butterfly species can survive in oil palm plantation habitats with various age groups of plants, and belong to several feeding guilds [7], [8]. Other studies suggest that butterfly species are sensitive to the effect of converting primary forest to plantations [9]. However, the study on this insect was ştill limited, one of less 'charismatic' group but global dominance among all of the animal communities [10].

The difference in the type of the original land cover and the scale of the plantation affected the impact of land clearing on biodiversity. The oil palm plantations in Indonesia do not fully originate from the conversion of primary forests [11] so that the variety of land cover can have a different impact. This study attempts to look at the impact of large and small oil palm plantations on butterfly species diversity and management implications for species conservation.

2. Methods

The research conducted in July until August 2018 at the oil palm plantation of PT. BPK, which is located in West Kalimantan Province, Indonesia. The study area has six types of a land cover consisting land cover origin before the existence of oil palm plantations (secondary forest) and two types of oil palm plantation (large scale plantation and smallholder plantation). Large Scale Plantation consists of four types of land cover on large-scale oil palm plantations (the old oil palm, mature oil palm, and young oil palm, high conservation value area/HCV), as well as Smallholder Plantation, consists of two smallholdings area.

Young oil palm (YOP) planted in 2015. This area surrounded by oil palm plantations with 2015, 2013, and 2010. The conditions of canopy cover tend to be open with undergrowth dominated by ferns. Mature oil palm (MOP) planted in 2009 and the canopy cover conditions tend to be dense. Old oil palm (OOP) planted in 2001, the canopy cover is quite dense, and the plantation floor covered with ferns. Smallholder plantation one planted in 2000. Apart from oil palm, there are also other types of vegetation such as bananas and cassava. This area borders on community-owned gardens and rubber gardens. Smallholder plantation two planted in 2017 and adjacent to secondary forest and HCV area. This area does not have canopy cover because the age of oil palm is still very young. The plantation floors covered with ferns, and there are water sources/canals around the area. Secondary forest area is located outside the PT BPK area, have water sources in the form of small canals, and bordered by shrubs and oil palm plantation.

Butterfly data collection conducted by using a combination of Line Transect method and Time Search method. A one-kilometer transect was used as an observation track, and record the type and number of mammals species found. Observations carried out simultaneously on all of the habitat type and repeated three times to maximize recording. Measurement of environmental parameters (temperature) also performed in the morning and afternoon in every type of habitat observed.

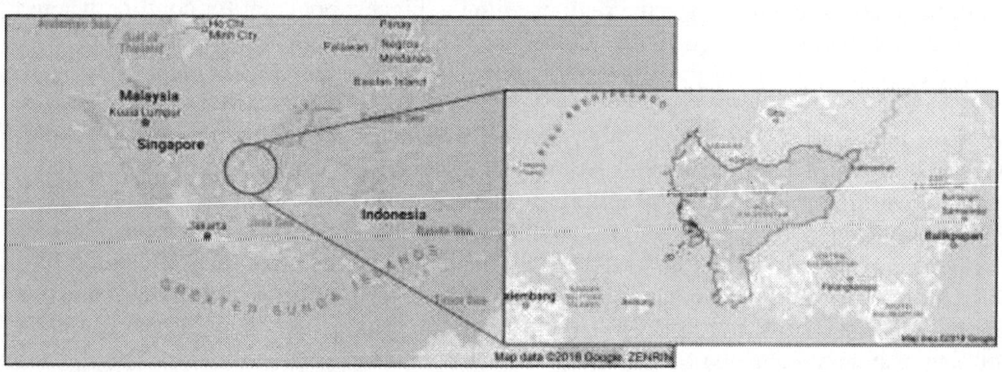

Figure 1. Map of research location in West Kalimantan Province, Indonesia

The data analyzed by quantifying the diversity of species in each type of habitat in oil palm plantations using Margalef Index (DMg), Shannon-Wiener index (H'), Evenness index. The similarity among land cover types calculated using Sorensen similarity index (SI). The composition of species compared between large-scale plantation and smallholder plantation based on the list of species loss, species gains, and species persistence. Loss species is species found before plantation development but not found in the following conditions. Gain species are species that found in the area after plantation development but not found in the previous conditions. Persistence of species is a species found in both conditions before and after plantation development. Data analyzed to estimate the value of species loss and acquisition due to oil palm plantations, namely differences in species gain and species loss. A comparison was made on the number of species, species diversity, and species composition.

3. Results and Discussion

3.1. Comparison of Butterflies Species Diversity

The number and diversity of butterfly species found in large oil palm plantations and smallholder plantations are higher than the original land (secondary forest). The increase in the number and level of species diversity of butterflies in each habitat type indicated by species diversity index, as many as 100 - 366% on large-scale oil palm plantations and by 0 - 466% on small-scale oil palm plantation. A similar trend is also occurring in the total value and diversity of species in the two types of plantations. Index of species diversity in secondary forest as an original land cover type before the development of plantation, large-scale plantation, and smallholder plantation displayed in figure 2.

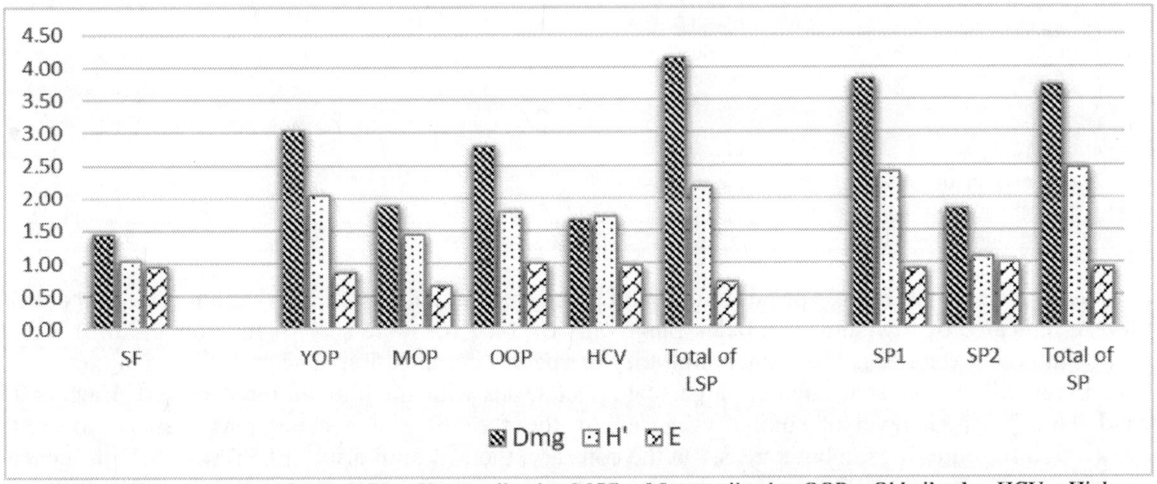

Remarks: SF = secondary forest, YOP = Young oil palm, MOP = Mature oil palm, OOP = Old oil palm, HCV = High conservation value, LSP = Large scale plantation, SP = Smallholder plantation, DMg = Margalef species richness index, H' = Shannon-Wiener species diversity index, E = Evenness index

Figure 2. Comparison of species diversity level in secondary forest, large-scale plantation, and smallholder plantation in PT. BPK West Kalimantan

Totally, 27 butterfly species comprising five families recorded. The families were Hesperiidae, Nymphalidae, Papilionidae, dan Pieridae. Three species were recorded at origin land use (secondary forest), while 6 - 11 species and 3 - 14 species were recorded at large scale plantation, and smallholder plantation, respectively (see table 1). Nymphalidae is the most common butterflies found in all of the habitat types. This trend was similar to the result found in several studies who found Nymphalidae as the biggest group among other butterflies groups [8], [12]. This group known as a diverse habitat range and broad food variety [13].

Table 1. Butterflies species composition at origin land use type and oil palm plantations in PT. BPK West Kalimantan

No.	Species	Family	Secondary Forest	Large Scale Plantation				Smallholder Plantation	
				YOP	MOP	OOP	HCV	SP1	SP2
1	*Pelopidas agna*	Hesperiidae		2	3				
2	*Telicota augias*	Hesperiidae			3				
3	*Acraea violae*	Nymphalidae		1				1	1
4	*Amathusia phidippus*	Nymphalidae	1		38	1	6	1	1
5	*Cethosia hypsea*	Nymphalidae	2					1	
6	*Elymnias cumaea*	Nymphalidae		1					
7	*Elymnias hicetas*	Nymphalidae					4		
8	*Elymnias hypermnestra*	Nymphalidae		6	13	1		3	
9	*Hypolimnas bolina*	Nymphalidae		1	2			6	
10	*Hypolimnas bolina jachynta*	Nymphalidae		1					
11	*Junonia atlites*	Nymphalidae		1	1		2	1	
12	*Junonia orithya*	Nymphalidae		8	8		3	1	
13	*Medusa ciptera*	Nymphalidae					2		
14	*Neptis hylas*	Nymphalidae						2	
15	*Orsotriaena medus*	Nymphalidae						1	1
16	*Pandita sinope*	Nymphalidae					3		
17	*Vindula dejone*	Nymphalidae						1	
18	*Ypthima pandocus*	Nymphalidae						5	
19	*Graphium agamemnon*	Papilionidae				1			
20	*Papilio demoleus*	Papilionidae				1			
21	*Appias olferna*	Pieridae		2	1			3	
22	*Catopsilia phyranthe*	Pieridae			1				
23	*Catopsilia scylla*	Pieridae		3					
24	*Delias hyparate*	Pieridae	1						
25	*Eurema hecabe*	Pieridae				1			
26	*Eurema sari*	Pieridae		1				3	
27	*Leptosia nina*	Pieridae				1		1	
Total of species number (S)			3	11	9	6	6	14	3
Total of individual number (N)			4	25	64	6	20	30	3

We found that butterflies species in oil palm plantation have low similarity with those in secondary forest, indicated by low similarity index range only 0 - 0.22 for large-scale plantation and 0.21 – 0.20 for smallholder (table 2). The higher similarity of species composition is seen in the young age class and the middle age class located on large-scale plantations with smallholder plantations 1. Each is 0.7 and 0.6. This high level of similarity caused by the condition of a better plant canopy to create protection for butterflies in both areas. On the contrary, the low similarity of LSP with SP2 influenced by the age of plants in SP2 (one year), so that differences in vegetation cover on the garden floor can affect species composition. It confirmed by other research [3] that mention the number of species is influenced by the cover of the tree canopy and the intensity of sunlight.

Table 2. Index of similarity of each land cover type at large scale plantation, smallholder plantation, and secondary forest in PT. BPK West Kalimantan

Land cover	SF	YOP	MOP	OOP	HCV	SP1	SP2
SF	1.00	**0.00**	**0.22**	**0.22**	**0.22**	**0.20**	**0.12**
YOP		1.00	0.60	0.13	0.44	**0.70**	0.12
MOP			1.00	0.40	0.67	**0.60**	0.12
OOP				1.00	0.22	0.30	0.12
HCV					1.00	0.30	0.35
SP1						1.00	0.35
SP2							1.00

Changes in species composition are reflected in species loss, species gain, and species persistence. The percentage of species loss in both types of plantations is the same, namely 60%, but there is greater species gain on small-scale oil palm plantations (60 - 400%) than large-scale (166 -

366%). table 3 and table 4 show LSP and SP lose only two species. This amount far less than the species gain which amounts to 20 species in LSP and 12 species in SP. 8 species can be found in LSP and SP. The only species of persistence in both types of gardens is *Amathusia phidippus* or Palm King that commonly found in coconut trees. This species not classified as threatened under the IUCN Red Data List, but some research results refer to Palm King as a rare and endangered species on the Indian Peninsula [14].

Table 3. The exchange of species composition in a large-scale plantation in PT BPK West Kalimantan

Species loss	Species gain	Species persistence
Cethosia hypsea	*Pelopidas agna*	*Amathusia phidippus*
Delias hyparate	*Telicota augias*	
	Acraea violae	
	Elymnias cumaea	
	Elymnias hicetas	
	Elymnias hypermnestra	
	Hypolimnas bolina	
	Hypolimnas bolina jachynta	
	Junonia atlites	
	Junonia orithya	
	Medusa ciptera	
	Pandita sinope	
	Graphium agamemnon	
	Papilio demoleus	
	Appias olferna	
	Catopsilia phyranthe	
	Catopsilia Scylla	
	Eurema hecabe	
	Eurema sari	
	Leptosia nina	

Table 4. The exchange of species composition in smallholder plantation in PT BPK West Kalimantan

Species loss	Species gain	Species persistence
Cethosia hypsea	*Acraea violae*	*Amathusia phidippus*
Delias hyparate	*Elymnias hypermnestra*	
	Hypolimnas bolina	
	Junonia atlites	
	Junonia orithya	
	Neptis hylas	
	Orsotriaena medus	
	Vindula dejone	
	Ypthima pandocus	
	Appias olferna	
	Eurema sari	
	Leptosia nina	

3.2. Discussion

3.2.1. Response of butterfly to oil palm plantation with different scale

We found that LSP and SP together showed a positive impact on increasing the number and diversity of species compared to conditions before the development of oil palm plantations (secondary forests). This study shows that no difference impact in plantation scale on butterfly species diversity. Other studies have already shown that oil palm plantation could affect butterfly diversity found in farmland [8], [9], [15]. The species obtained are generally generalist species in the family group Nymphalidae. Nymphalidae is the largest butterfly group and has a wide range of habitat types.

Species of the Papilionidae, *Graphium agamemnon* and *Papilio demoleus* groups were only found in LSP in the old age group (19 years). Both of them known as swallowtail butterfly. Indonesia has 121 species of swallowtails, over 20 per cent of the world total [16]. *Graphium agamemnon* is one of the charismatic species while *Papilio demoleus* is one of widest spread members of the family Papilionidae and have the quick expansion of its range. Habitat alteration is one of threats swallowtail butterfly. This result reveals that the presence of this species in old oil palm age class may indicate that oil palm plantation could be part of its habitat.

Species diversity influenced by physical factors. One physical factor that influences the diversity of butterfly species is temperature [12]. The temperature may affect the ecology and life history of butterflies [17] and furthermore could affect its species richness. In term of temperature, the optimum temperature for the butterfly to actively search for food is between 23-30 ^{0}C [18]. The temperature at large scale plantation range from 24 until 29.5 0C, from 23 – 26 0C at smallholder plantation, and 27 0C at the secondary forest. The age variation of plants in LSP causes a greater temperature range than SP. The presence of young, medium, old, and HCV class plants provide more variation than only one plant age class in SP or natural forest vegetation in SF.

Table 5. Temperature (^{0}C) at origin land use type and oil palm plantations in PT. BPK West Kalimantan

Time	LSP				SP		SF
	YOP	MOP	OOP	HCV	1	2	
Morning	26	24	26	29.5	26	23	27
Afternoon	26	24	25	28.5	23.5	24.5	27

3.2.2. Management implication in the context of wildlife conservation
Conversion of land cover from secondary forest to oil palm plantation can change the level of diversity and composition of species inhabiting an area. Even though the expansion of tropical forests by man will inevitably disturb swallowtail populations, but the extent of the impact depends on the level of disturbance, the extent of the biotope and its ability to recover, and the willingness of people to allow it to do so [16]. One of the challenges in butterfly conservation is creating habitat heterogeneity in monoculture oil palm plantations. Heterogeneity can provide habitat components needed by butterflies so that they can recover ecologically.

Efforts made on LSP and SP can be different because of differences in land size, level of diversity of plant age, and management by humans. The main components of butterflies are water and feed plants and host plants, so the management carried out should support an increase in the number and quality of the components [19]. Characteristics of LSPs that have canals in the garden area are supporting physical factors for butterflies. Large oil palm plantations can managed by maintaining the various age classes by setting the new planting period and site locations while enhancing the quality of forest patch in the High Conservation Value (HCV) area. For smallholder plantation, the land sharing approach could be a strategy in enhancing the diversity and the richness of vegetation. Combine the oil palm with host trees and food crop diversity are another alternative to maintain the ecological environment for butterflies.

4. Conclusion
The development of oil palm plantation has a positive and negative impact on butterfly diversity. The positive impact is gaining species, while the negative impact is loos of species. Both of Large-scale Plantation and Smallholder plantation indicate no differences impact in plantation scale on butterfly species diversity. The species diversity increasing after the development of oil palm plantation derived from the secondary forest area on both of plantation type. The species dominated by Nymphalidae species. Habitat heterogeneity is a key point in management implication for the species diversity conservation.

Acknowledgments
Our sincere thanks go to the chiefs and staffs of PT. BPK in West Kalimantan for supports and facilities that provided during field activities, BPDPKS for financial support, and all of the teams for dedication and cooperation during data collections.

References
[1] Molotoks A, Kuhnert M, Dawson TP, Smith P. 2017. Global Hotspots of Conflict Risk between

Food Security and Biodiversity Conservation. *Land* (6)67. https://portals.iucn.org/library/sites/library /files/ documents/RD-1985-002.pdf

[2] [Dirjenbun]. Direktorat Jendral Perkebunan. 2016. Statistik Perkebunan Indonesia 2015–2017: Kelapa Sawit. http://ditjenbun.pertanian.go.id

[3] Koh LP, Sodhi NS. 2004. Importance of reverse, fragments, and parks for butterfly conservation in a tropical urban landscape. *Ecological Applications* **14** (6): 1695–1708.

[4] Dahelmi. 2000. Inventarisasi tumbuhan inang kupu-kupu Papilionidae di kawasan agar Alam Lembah Harau, Sumatera Barat. Jurnal Matematika IPA **9**:19–21. http://repository.unand.ac.id/id/eprint/3534.

[5] Atmowi, Buchori TD, Manuwoto S, Suryobroto B, Hidayat P. 2007. Diversity of pollinator insects in relation of seed set of Mustard (Brassica rapa L.: Cruciferae). *Hayati Journal of Bioscience*. **14**:155 – 161. https://doi.org/10.4308/hjb.14.4.155.

[6] Stefanescu C, Herrando S, Páramo F. 2004. Butterfly species richness in the north-west Mediterranean Basin: the role of natural and human-induced factors. *Journal of Biogeography,* **31**(6): 905–915. doi:10.1111/j.1365-2699.2004.01088.x

[7] Santosa Y, Purnamasari I. 2017. Variation of butterfly diversity in different ages palm oil plantation in Kampar, Riau. Prosiding Seminar Nasional Masyarakat Biodiversitas Indonesia **3**(2):278–285. https://doi.org/10.13057/psnmbi/m030220.

[8] Kwatrina RT, Santosa Y, Bismark M, Santoso N. 2018. Ecological Impacts of Oil-Palm Plantation on Butterfly and Bird Species Diversity. *Journal of Tropical Forest management* Vol. **24** (1): 23-31. EISSN: 2089-2063. ISSN: 2087-0469. DOI: 10.7226/jtfm.24.1.23.

[9] Miller DG, Lane J, Senock R. 2011. Butterflies as potential bioindicators of primaryrainforest and oil palm plantation habitats on New Britain, Papua New Guinea. *Pacific Conservation Biology* **17**(2):149-159. doi: 10.1071/PC110149.

[10] Dunn RR. 2005. Modern insect extinctions, the neglect majority. *Conserv Biol* **19**:1030-1036.

[11] Kwatrina RT, Santosa Y, Sunkar A. 2018. History of Status and History of Land Use of Some Private Oil Palm Plantations in Central Kalimantan Province. Paper on Focus Grup Discussion "Oil Palm and Deforestation of Tropical Forest??". IPB International Convention Center, Bogor, Indonesia, 12 April 2018.

[12] Panjaitan R, Atmowidi T, Peggiem D. 2016. "Effect of Temperature on Butterfly Community (Lepidoptera) at Gunung Meja Recreational Forest Area," *KnE Social Sciences*:19–26. DOI 10.18502/kss.v1i1.430

[13] Majumder J, Lodh R, Agarwala BK. 2013. Butterfly species richness and diversity in the Trishna wildlife sanctuary in South Asia. Journal of Insect Science **13** (79). https://doi.org/ 10.1673/031.013.7901

[14] Mathew G, Pulikkal K. 2009. Biology of the Palm King Amathusia phidippus, an extremely rare and endangered butterfly of Peninsula India. *The Journal of The Bombay Natural History Society*. Vol. **106**(1):199-120. https://biodiversitylibrary.org/page/48367187.

[15] Azhar B, Chong LP, Zakaria M, Hassan N, Arif M. 2014. Effects of monoculture and polyculture practices in oil palm smallholdings on tropical farmland birds. *Basic & Scientific Article* ISSN: 2087-0469

[16] Collins NM, Morris MG. 1985. Threatened Swallowtail Butterflies of the World The lUCN Red Data Book. International Union For Conservation of Nature And Natural Resources. Gland. Switzerland and Cambridge, U.K

[17] Karlsson B, Wiklund C. 2004. Butterfly life history and temperature adaptations; dry open habitats select for increased fecundity and longevity. *Journal of Animal Ecology*, **74**(1): 99–104. doi:10.1111/j.1365-2656.2004.00902.x

[18] Freitas AVL, Benson WW, Filho OJM, de Carvalho RM. Territoriality by the dawn's early light: the Neotropical owl butterfly Caligo idomenaeus (Nymphalidae: Brassolinae), *J Res Lepid*, **34**: 14–20, (1997).

[19] Hawkins BA, Porter EE. 2003. Does Herbivore Diversity Depend on Plant Diversity? The Case of California Butterflies. *The American Naturalist*, **161**(1): 40–49. doi:10.1086/345479

IOP Conf. Series: Earth and Environmental Science **336** (2019) 012025 doi:10.1088/1755-1315/336/1/012025

The loss, gain, and diversity of butterfly species due to the development of PT PKWE oil palm plantation, West Kalimantan Province

Lin Nuriah Ginoga[1]*, Yanto Santosa[1], Andi Rezki Mutmainnah[1]

[1]Department of Forest Resources Conservation and Ecotourism, Faculty of Forestry, Bogor Agricultural University, Bogor, 16680, Indonesia

*E-mail: ginogalina@gmail.com

Abstract. An increase in oil palm plantation area of ± 9.3 million hectares within 20 years (1997 – 2006) raises concerns of the international communities on the loss of diversity, including butterflies. Butterflies have multiple ecological functions, including as pollinator and ecosystem balancer, hence research about the impact of oil palm plantation development on the loss, gain, and diversity is important to conduct. The research was conducted in March 2018 in PKWE Plantation, West Kalimantan Province, by using Pollard transect method and trap in seven land covers. The land covers consist of two original land covers before made into oil palm plantation, in the form of shrubs and secondary forest, and five land covers after made into oil palm plantation, in the form of old, medium, young oil palm plantation, High Conservation Value (NKT) area, and yard. The total butterfly species found amounted to 49 species, 315 individuals, four families: Papilionidae (8 species), Pieridae (8 species), Nymphalidae (31 species), and Lycanidae (2 species). The highest number of species and diversity were discovered in the High Conservation Value (HCV) area (S=26; Dmg=6.24), while the lowest ones were located in a yard area (S=11; Dmg=2.09). Based on conservation status, there were a butterfly species which is protected namely *Troides helena*. Regarding the loss and gain of butterfly species, oil palm plantation development had caused a much lower loss than gain. With the loss of five species and gain of 27 species compared to the original land cover in the form of secondary forest, and the loss of three species and gain of 26 species compared to the original land cover in the form of shrubs.

1. Introduction

Oil palm plantation area in Indonesia grew significantly by ± 9.3 million ha, from 2.911.296 ha in 1997 to 12.307.677 ha in 2017 [1]. Oil palm plantation as a source of alternative energy provides a positive contribution to the national economy and as people's livelihood [2]. However, several experts often accuse the conversion of forest area into oil palm plantation as one of the main factors causing the occurrence of deforestation and the loss of biodiversity in various tropical countries [2], [3], [4], [5], [6], [7]. Reference [5] emphasized that the conversion of primary or secondary forest into oil palm plantation have an impact that can damage species diversity including butterfly. Reference [8] stated that primary and secondary forest converted into oil palm plantation lowered the butterfly species richness by around 79-83%.

Butterflies need a suitable habitat for their life. They can live in various habitats ranging from primary forest, secondary forest, agricultural area, plantation, to house yard, with an altitude of 0 - 2 000 meters above sea level. Sources of water, sunlight, temperature, and humidity have important roles in supporting the life of butterflies [9], [10]. The optimum temperature for butterfly is 28 – 35° C [11], while the humidity favored by them is less than 85% [9]. Butterflies utilize plants as their place of life, which are food plant and host plant. Butterflies categorized as herbivore who eat plants. Their food when they are still larvae is part of the plant (vegetation) and their food after becoming an adult is flower nectar [9].

Butterflies are part of biodiversity which must be conserved from extinction and species diversity decline. Butterflies have important values, such as the value of ecology (pollinator), endemism, conservation, education, culture, aesthetics, and economy [12]. Butterfly, as a pollinator, helps in maintaining biodiversity and ecosystem balance. Researches concerning butterfly species diversity in oil palm plantation had been conducted several times, but most of them were conducted in Sumatera area like the study carried out by reference [13] in West Sumatera, in Bengkulu [14], in Riau [15], [16], while in Kalimantan, there have not been many researches conducted. Research regarding the impact of oil palm plantation on the loss, gain, and diversity of butterfly species also has not been undertaken much. Thus this research is important to carry out.

This research aims to identify butterfly species diversity, compare the loss and gain of butterfly species in original land cover before made into oil palm plantation, land cover after converted into oil palm plantation and study the conservation status of the butterflies found. The results of this research can be utilized as a reference and consideration in formulating and developing policies on land management in oil palm industry by keep maintaining wildlife conservation, especially butterflies.

2. Materials and methods

2.1. Location and time

This research was conducted in PT PKWE, West Kalimantan Province, in seven land covers. The two land covers were original land cover before made into oil palm plantation, namely protection forest (HL) and shrubs (SB). The other five land covers were land cover after made into oil palm plantation, namely oldest oil palm plantation (ST) established in 2006, medium oil palm plantation (SS) established in 2011, young oil palm plantation (SM) established in 2016, High Conservation Value (NKT) area, and yard (PKR). Figure 1 presents the research location map. This research was performed in March 2018.

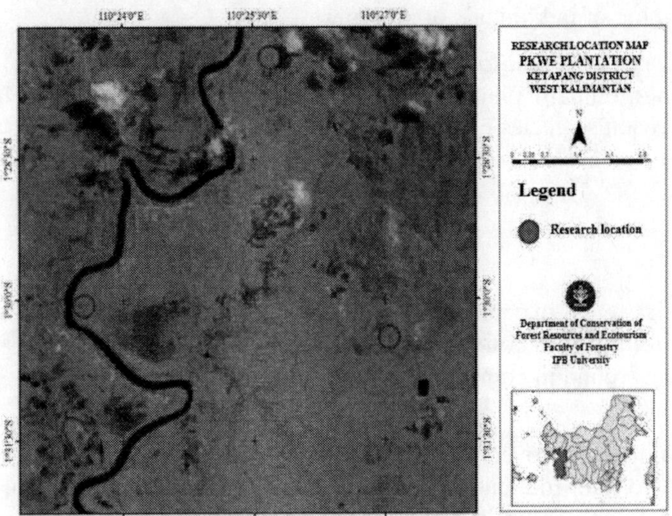

Figure 1. Research location map

2.2. *Tools and materials*
The tools and materials used included Landsat imagery, tally sheet, stationery, GPS Garmin 62s, camera, area map for data collection, butterfly hand net, butterfly trap, tape measure, rope, book of field guide and butterfly species identification titled *Identification guide for butterflies of West Java* (Schulze), *Practical Guide to The Butterflies of Bogor Botanic Garden* [17], and *The Illustrated Encyclopedia of the Butterfly World* [18]. The tools and materials for butterfly preservation were syringe, alcohol, Styrofoam, papilot paper, straight pin, adhesive tape, and camphor.

2.3. *Data collection method*
The data were collected using strip transect method [19]. The length of the transect line was made 1 km long and 150 m wide, and one repetition was carried out in the adjacent line so that the total length of the line became 2 km. The data collection repetition was undertaken three times (3 observation days) simultaneously in all types of land cover in the morning at 09.00-11.00 WIB according to the active time of the butterfly. The observation was performed at a constant speed by using butterfly hand net to catch the butterflies directly. Butterfly trap was also utilized which was installed at the height of ± 1m as many as six traps in each observation line.

2.4. *Data analysis*
The data was analyzed in a quantitative and descriptive qualitative manner, encompassing the number of species, species richness index, species similarity index, species loss and gain ratio, and butterfly conservation status.

Number of Species (S)
The number of species is the number of all types of butterfly found in PKWE plantation.

Species Richness Index (Dmg)
Species richness index serves to show the number of species in a certain type of habitat using Margalef's index [20], with the formula as follows:

$$Dmg = \frac{(S-1)}{\ln N}$$

Information:
Dmg : Species richness index
S : Number of species
N : Total number of individuals of all types

Sorensen's Species/Community Similarity Index
To find out the impact of oil palm plantation on the composition, the number of loss and gain of butterfly species, Sorensen's species/community similarity index was applied [20] as follows:

$$IS = \frac{2c}{a+b}$$

Information:
a: Number of species found in community a
b: Number of species found in community b
c: Number of species found in community a and b

Conservation Status
Butterfly conservation status covers extinction status, trade arrangements, and butterfly species protection according to the International Union for Conservation of Nature (IUCN) Red List, Convention on International Trade in Endangered Species of Wild Fauna and Flora (CITES), and Government Regulation (PP) No. 7 Year 1999 concerning Plant and Animal Species Preservation.

3. Results and discussion

3.1. General condition of the research location

Protection Forest (HL)

Protection forest is a government-owned land, which was adjacent to the oldest oil palm plantation, community-owned field, and several community's residences. This area had a river flow with a closed riparian. It had the characteristics of hilly and a relatively open canopy cover in the hilly area, an area that was still in the form of shrubs, and an area bordering the community fields. The average temperature in this area was 29° C with an average humidity of 83%. The lowest temperature recorded was 28° C and the highest was 32° C with the lowest humidity of 70% and the highest of 91%.

Shrubs (SB)

Shrubs had a relatively open canopy with the floor covered by high and dense shrubs. This area was adjacent to the youngest oil palm plantation and access road. Water often flooded several locations especially during rainy season due to the uneven topography. There was a road building activity next to this area. The average temperature in this area was 27° C with an average humidity of 91%. The lowest temperature recorded was 25° C and the highest was 9° C with the lowest humidity of 74% and the highest of 91%.

The Oldest Oil Palm Plantation (ST)

The oldest oil palm plantation was adjacent to the collision road (CR) which motor vehicles usually passed. The northern part of this area bordered the buffer area before entering the protected forest area. The canopy cover condition was relatively dense with the plantation floor was overgrown with understories. In every two planting points, there was a trench in the form of water flow or waste storage place. The average temperature in this area was 29° C with an average humidity of 86%. The lowest temperature recorded was 26° C while the highest was 29° C with the lowest humidity of 59% and the highest of 91%.

Medium Oil Palm (SS)

Medium oil palm plantation area had a dense canopy cover with the plantation floor was overgrown with a wide range of understory species. This area bordered the access road and NKT area. Several locations in this area were often flooded especially after rain. The average temperature in this area was 29° C with an average humidity of 83%. The lowest temperature recorded was 27° C while the highest was 30° C with the lowest humidity of 77% and the highest of 91%.

The Youngest Oil Palm Plantation (SM)

The youngest oil palm plantation area directly bordered shrubs area. The canopy cover in this area was exceptionally open due to the young oil palm with a height of around 2-3 m. This area was often flooded with water from the trench and boundary canal between blocks. The average temperature in this area was 29° C with an average humidity of 83%. The lowest temperature recorded was 28° C while the highest was 32° C with the lowest humidity of 77% and the highest of 91%.

High Conservation Value Area (NKT)

NKT area was in the form of riparian area, which extends along the river. This area was adjacent to the medium oil palm plantation, had a relatively dense canopy cover in the area of tree vegetation, and was open in river course. The average temperature in this area was 28° C with an average humidity of 83%. The lowest temperature recorded was 26° C while the highest was 30° C with the lowest humidity of 76% and the highest of 91%.

Yard Area (PKR)

Yard area is a community residence area. This area is an open area in which people often plant various types of flowering plants, such as Ixora flower *(Ixora coccinea)*, rosy periwinkle (*Catharanthus roseus*), and *pucuk merah* (*Syzygium oleana*). This area was an area with the highest level of human disturbance as there are house and school buildings. The average temperature in this area was 29° C with an average humidity of 83%. The lowest temperature recorded was 28° C while the highest was 32° C with the lowest humidity of 77% and the highest of 91%.

Overall, there found 49 families among 120 species of understories in PKWE plantation. The family that had the highest number of species and density value was Poaceae family with 16 species. Figure 2 portrays the variation of understory species diversity in various types of land covers.

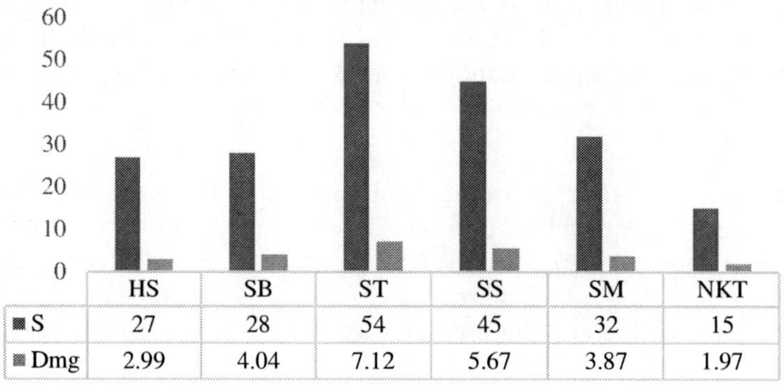

Information: SM = Young oil palm; SS = Medium oil palm; ST = Old oil palm; NKT = High Conservation Value area; SB = Shrubs; HS = Protected Forest

Figure 2. The comparison between the number of species and understory species richness in various types of land covers in PKWE plantation

3.2. Butterfly species diversity in various types of land cover

Butterfly species found in PT PKWE oil palm plantation varied in each type of land cover as presented in the following table 1:

Table 1. Butterfly species diversity in various types of land covers in PKWE plantation

No.	Species	Family	Number of individuals per land cover							Σ Individuals
			SM	SS	ST	NKT	SB	HS	PKR	
1.	*Neptis hylas*	Nymphalidae	7	2	1	3	14	5	1	33
2.	*Euploea mulciber*	Nymphalidae	0	12	0	2	2	4	0	20
3.	*Junonia atlites*	Nymphalidae	5	3	5	1	3	1	1	19
4.	*Orsotriaena medus*	Nymphalidae	12	0	0	0	0	6	0	18
5.	*Eurema hecabe*	Pieridae	2	2	2	5	3	2	1	17
6.	*Parthenos sylvia*	Nymphalidae	13	0	0	1	0	2	0	16
7.	*Catopsilia pyranthe*	Pieridae	5	9	0	0	0	0	0	14
8.	*Mycalesis horsfieldi*	Nymphalidae	3	5	0	1	1	4	0	14
9.	*Mycalesis anapita*	Nymphalidae	7	0	1	4	0	0	0	12
10.	*Eurema sari*	Pieridae	2	1	2	0	1	2	3	11
11.	Others (39 species)									141
	Number of total species		17	16	14	26	19	21	6	49
	Number of total individuals		71	45	32	55	41	60	11	315

Information: SM = Young oil palm; SS = Medium oil palm; ST = Old oil palm; NKT = High Conservation Value area; SB = Shrubs; HL = Protection Forest; PKR = Yard

This research found 49 butterfly species from 315 individuals and four families, namely Papilionidae (7 species), Pieridae (8 species), Nymphalidae (31 species) which is family with the highest number of species, and Lycanidae (2 species) which is family with the lowest number of species. There was no Hesperidae family in this research. NKT area had the highest number of species as many as 26 species, while yard area had the lowest number of species (6 species). The high number of butterfly species in NKT was directly proportional to its species richness value. NKT area has the highest species richness value (Dmg=6.24). Figure 3 shows a comparison between the number of species and species richness in each type of land cover. These results are different with the previous researches conducted by reference [15] that stated that the highest diversity was in the youngest oil palm plantation and reference [16] that revealed that the highest diversity was in the secondary forest area. Environmental condition in each research location is different in the aspect of vegetation, geology, and ecology, is the cause of these differences.

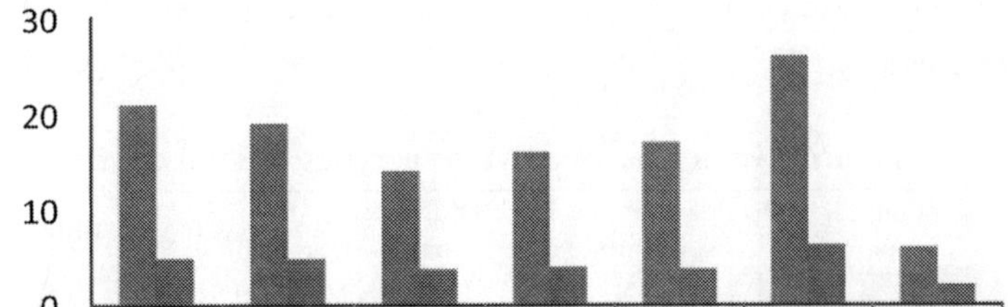

	HS	SB	ST	SS	SM	NKT	PKR
■S	21	19	14	16	17	26	6
■Dmg	4.88	4.85	3.75	3.94	3.75	6.24	2.09

Figure 3. Comparison between the number of species (S) and species richness (Dmg) in various types of land cover

The high level of butterfly species diversity and richness in NKT area caused by the habitat condition, which can support the life of butterflies. In NKT area there are trees and understories which some of them are butterfly feed plants, namely *Piper sp.* as caterpillar feed, *Dillenia suffruticosa* (Griff.) Martelli and *Macaranga triloba* (Thunb.) Müll.Arg. as butterfly feed, and *Bellucia pentamera* Naudin as caterpillar and butterfly feed. Reference [21] stated that butterfly larvae love young betel leaves (*Piper sp.*). The other conditions are the relatively dense canopy cover in forest area and open around the river, abundant water source comes from the river and the moist soil which supports the life of butterflies in that area. This is supported by the research carried out by reference [22] which showed that riparian habitat had high potential as the habitat of butterflies, owing to the availability of water source, open vegetation, and high sunlight intensity. Likewise, reference [23] mentioned that the characteristics of optimal habitat for butterflies are sufficient to light factor, clean or unpolluted air and water as the material required for the humidity of the environment where butterflies live.

The lowest number of butterfly species was found in the yard area, namely two species, as well as the lowest species richness value (Dmg = 2.09), compared to the other types of land cover. These results are different with the result of previous research conducted by reference [15] that declared that the yard area adjacent to the oldest oil palm plantation caused the high number of butterfly species as there were flowering plants grown in people's yard and the yard borders the secondary forest. That condition is different with the condition in the yard area of PKWE plantation that although there were flowering plants, some of them were not in flowering condition, and there was no water source around the yard. Besides, the high level of human disturbance also caused a low level of butterfly diversity in the yard. That condition caused butterflies to prefer oil palm plantation area, NKT area, and protection forest, which were more favorable for their life. Reference [24] stated that the change of habitat and lack of food generated the occurrence of butterfly migration to the area far from human disturbance.

3.3. Butterfly species similarity in various types of land cover

In general, the analysis of butterfly community similarity in various types of land cover in PKWE plantation generated a low species similarity. The highest species similarity value existed in the oldest oil palm plantation and shrubs area with the value of 0.55. The high value, compared to the other land covers, was because there were nine similar species found in both locations. Those similar butterfly species were *Eurema sari, Leptosia nina, Eurema hecabe, Neptis hylas, Junonia atlites, Amathusia phidippus, Hypolimnas bolina, Parantica agloides,* and *Parantica aspasia.* The species similarity between both locations was influenced by the characteristics of habitat in each location. According to reference [25], the high value of the similarity coefficient caused the species found to be similar, or the composition is almost similar in both different location. This showed that the higher the species similarity value, the higher the similarity of habitat characteristics, vice versa. Both shrubs and old oil palm plantation area had much water. In the oldest oil palm plantation, there were two types of trench, which were lengthwise in each two planting points as water flow and waste storage place. Meanwhile, in shrubs area, the puddle existed because of an uneven topography. The following table 2 presents the community similarity values:

Table 2. Species similarity values between various types of land covers in PKWE plantation

Types of land cover	Youngest oil palm	Medium oil palm	Oldest oil palm	NKT	Shrubs	Secondary forest	Yard
Youngest oil palm		0.48	0.39	0.51	0.50	0.42	0.43
Medium oil palm			0.33	0.38	0.51	0.38	0.45
Oldest oil palm				0.40	**0.55**	0.46	0.40
NKT					0.49	0.47	**0.19**
Shrubs						0.45	0.40
Protection forest							0.37
Yard							

The lowest species similarity value existed in yard area and NKT area with the same number of species, which was three species (IS=0.19), namely *Eurema hecabe, Neptis hylas,* and *Junonia atlites*. The species similarity value was low because both areas had distinct habitat characteristics. Yard area is an open area with buildings, rocky roads, and open water source are rarely found, low amount of understories but there are several flowering plants. NKT area is a riparian forest, closed in forest part and open area around the river flow has many understories, stakes, and trees. NKT was an area with the highest number of butterfly species, namely 26 species, while the yard is an area with the lowest number of butterfly species, namely 6 species so that the difference of number of species between the two locations reached 20 butterfly species. That supported the low level of species similarity in both types of land cover, in line with reference [26] that distinct habitat had a low species similarity.

The most dominating butterfly species in PKWE plantation was *Neptis hylas*, which amounted to 33 of the total 315 butterfly individuals found. This butterfly species can also be found in all types of land cover, which showed its high adaptation power with different land covers. *Neptis hylas* is one of the species that are members of Nymphalidae family, which is adaptable. Reference [27] stated that Nymphalidae is a butterfly family who had the highest number of species and is cosmopolitan, has a widespread family distribution in many areas in the world, has a great survival ability in multiple types of habitat as they are *polifag*. *Polifag* is a group of animal living and eating in various types of plants from various families [28]. The *polifag* nature causes these species do not rely on solely one feed plant so that it is more adaptive in many kinds of environmental conditions. According to reference [29], the host plants of Nymphalidae family were are Annonaceae, Asteraceae, Moraceae, Rubiaceae dan Anacardiaceae. Also, reference [15] mentioned that the feed plants of Nymphalidae family butterflies are of Asteraceae, Melastomataceae, Solanaceae, and Poaceae family. The host plants of Nymphalidae family found in the location were *Ageratum conyzoides* (L.) L., *Mikania micrantha* Kunth, *Climedia hirta* (L.) D. Don, *Uncaria acida* (Hunter) Roxb., and *Paspalum conjugatum* P.J. Bergius.

Several types of butterfly existed in all types of land cover, but there were also species that existed only in one type of land cover. There were three butterfly species that exist in all types of land cover, namely *Eurema hecabe, Neptis hylas,* and *Junonia atlites.* The butterfly species that could be found only in young oil palm plantation was *Ideopsis vulgaris,* in medium oil palm plantation were *Graphium antiphates, Papilio polytes, Gandaca harina, Euploea tulliolus,* and *Mycalesis perseus,* in old oil palm plantation was *Mycalesis janardana*, in NKT area were *Eurema alitha*, *Hypolimnas misippus*, *Cupha erymanthis*, and *Faunis canens*, in shrubs area were *Moduza procris* and *Ideopsis gaura*, and in secondary forest were *Danaus melanippus* and *Junonia orithya*. However, there were no certain species only found in yard area. The difference of butterfly species between types of land cover was due to the difference of habitat characteristics supporting the life of butterflies, both from the factor of geology, ecology, and larvae and adult butterfly feed plant distribution [10].

3.4. *The percentage of loss and gain of butterfly species diversity due to oil palm plantation development*

The change of land cover due to oil palm plantation development had an impact on the change of biodiversity [30]. The existing change can cause a loss or gain of new butterfly species in that area. The loss of butterfly species is a loss of butterfly species, which originally found in original land in the form of HS and SB before made into oil palm plantation (ST, SS, SM, areal NKT, and PKR). On the other hand, species gain is the finding of butterfly species in oil palm plantation, which is not found in original land. The percentages of loss and gain of butterfly species due to the change of secondary forest into oil palm plantation are portrayed in figure 2, while the percentages of loss and gain of butterfly species due to the change of shrubs land into presented in figure 4.

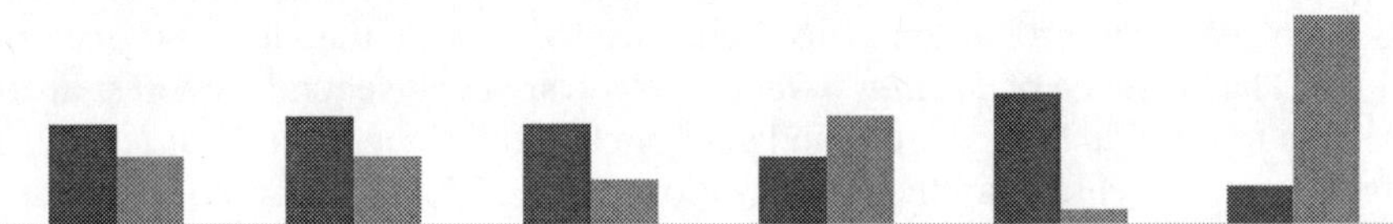

	ST	SS	SM	NKT	PKR	SK
■ Lose	61.90%	66.67%	61.90%	42.11%	80.95%	23.81%
■ Gain	42.86%	42.86%	28.57%	66.67%	9.52%	128.57%

Information: ST = Old oil palm, SS = Medium oil palm, SM = Young oil palm, NKT = High Conservation Value area, PKR = Yard, SK = After converted to oil palm plantation

Figure 4. The loss and gain of butterfly species in the original land of secondary forest

	ST	SS	SM	NKT	PKR	SK
■ Lose	52.63%	52.63%	52.63%	42.11%	73.68%	15.79%
■ Gain	42.11%	36.84%	26.31%	78.95%	5.25%	136.84%

Information: ST = Old oil palm, SS = Medium oil palm, SM = Young oil palm, NKT= High Conservation Value area, PKR = Yard, SK = After converted to oil palm plantation

Figure 5. The loss and gain of butterfly species in the original land of shrubs

Figure 4 and figure 5 show that the change of secondary forest and shrubs into oil palm plantation in four land covers, namely ST, SS, SM, and PKR caused a higher species loss than gain. It was different with NKT area, which had a lower species loss than gain. The highest loss happened in yard area, while the highest gain happened in NKT area. Overall, oil palm plantation development both in the original land of secondary forest and shrubs had caused a higher species gain than species loss.

3.5. *Butterfly conservation status*

Wildlife conservation status can be categorized based on three references, namely IUCN Red List, CITES, and Government Regulation (PP) No. 7 Year 1999 (table 3).

Table 3. Conservation status of butterflies found in research location

No.	Species	Family	Conservation status		
			IUCN	CITES	PP NO.7/1999
1	*Troides helena*	Papilionidae	-	Appendix II	Protected
2	*Eurema alitha*	Pieridae	LC	-	-
3	*Mycalesis janardana*	Nymphalidae	LC	-	-
4	*Thaumantis noureddin*	Nymphalidae	LC	-	-
5	*Ideopsis vulgaris*	Nymphalidae	LC	-	-
6	*Tanaecia pelea*	Nymphalidae	LC	-	-

This research found one protected butterfly species, namely *Troides helena,* which was found in two types of land covers, namely NKT area and secondary forest area. This butterfly species protected in Indonesia since 1980, by the Decree (SK) of Minister of Agriculture No. 576/Kpts/Um/8/1980 and No. 716/Kpts/Um/8/1980. This species also belongs to the list of species with the CITES status of Appendix II so that it is prohibited from being traded except that it is originated from the captive breeding [31]. The presence of *Troides helena* in NKT area and secondary forest area was because both areas were able to provide feed sources and suitable habitat. Other than *Troides helena,* there were five other butterfly species which had the conservation status of Least Concern (LC) based on the category of IUCN's conservation status. Butterfly species, which had the LC status based on IUCN mean that they had been evaluated, but not categorized as threatened or nearly threatened, so that the status is low risk. In other words, they have no urgent conservation dependency.

4. Conclusion

NKT area is the land cover with the highest number of species and species richness with the value of S=26 and Dmg=6.24 respectively. The lowest number of species existed in the yard area with the lowest species richness value as well, which were S=6 and Dmg=2.09 respectively. NKT area was the habitat that was most favorable for butterfly species existence.

The highest community similarity was obtained between the oldest oil palm plantation area and shrubs area (IS=0.55), while the lowest community similarity was obtained between NKT area and yard area (IS=0.19). The most dominating butterfly species in the research location was *Neptis hylas. Troides helena was* a protected species found in this research.

The change of secondary forest into oil palm plantation resulted in butterfly species loss by 23.81% and butterfly species gain by 128.57%, while the change of shrubs land into oil palm plantation resulted in butterfly species loss by 15.79% and butterfly species gain by 136.84%.

References

[1] [Ditjenbun] Directorate General of Plantation. 2016. *Statistics of Indonesian Plantation 2015-2017 (in Indonesian)*. Jakarta (ID): Direktorat Jendral Perkebunan

[2] Sheil D, Casson A, Meijaard E, Noordwijk MV, Gaskell J, Groves JS, Wertz K, Kanninen M. 2009. *The Impacts and Opportunities of Oil Palm in Southeast Asia*. Bogor (ID): CIFOR

[3] Donald PF. 2004. Changes in bird communities following conversion of lowland forest to oil palm and rubber plantations in southern Thailand. *Bird Conservation International*. **16**: 71-82

[4] Fitzherbert EB, Struebig MJ, Morel A.2008. How will oil palm expansion affect biodiversity?. *Trends in Ecology & Evolution.* **23**: 538–545

[5] Koh LP, Wilcove DS. 2008. Is oil palm agriculture really destroying tropical biodiversity?. *Conserv Lett*. Vol **1**: 60-64

[6] Danielsen F, Beukema H, Burgess ND, Parish F, Bruhl CA, Donald PF, Murdiyarso D, Phalan B, Reijnders L, Struebig M, Fitzherbert EB. 2009. Biofuel plantations on forested lands: double jeopardy for biodiversity and climate. Conserv Biol **23**:348–358

[7] Obidzinski K, Andriani R, Komarudin H, Andrianto A. 2012. Environmental and social impacts of oil palm plantations and their implications for biofuel lumber production in Indonesia. *Ecology and Society*. **17**(1): 25
[8] Hamer KC, Newton RJ, Edwards FA, Benedick S, Bottrell SH, Edwards DP. 2015. Impacts of selective logging on insectivorous birds in Borneo: the importance of trophic position, body size, and foraging height. *Biol. Conserv*. **188**: 82–88
[9] Sihombing. 1999. Satwa Harapan I: The Introduction of Cultivation Science and Technology (in Indonesian). Bogor (ID): Pustaka Wirausaha Muda
[10] Amir M, Woro AN, Sih K. 2003. The Insects of Halimun Mount National Park, West Java. Bogor (ID): Biodiversity Conservation Project
[11] Watanabe MIT. 2003. Thermoregulation and flying habits of Japanese sulphur Butterfly coliaserate (Lepidoptera: Peridae) in open habitat. Entomol Sci. **6**:111-118
[12] Lestari DF, Putri RDA, Ridwan M, Purwaningsih AD. 2015. Butterfly diversity (insecta: Lepidoptera) in Alas Bromo Forest Tourism, North Lawu BKPH, Karanganyar, Central Tengah (in Indonesian). *Prom Sem Nas Masy Biovid Indon.* **1**(6): 1284-1288
[13] Naboya R, Moerfiah, Wiedarti S. 2010. Butterfly Species in Oil Palm and Rubber Plantation in Ragusa Region, Dharmasraya District, West Sumatera (in Indonesian). Bogor (ID): Pakuan University
[14] Ririn KA. 2014. Inventory of insect species in oil palm flowers in PT Agri Andalas (Persero) oil palm plantation Pasar Ngalam, Air Periukan Sub-district, Seluma District (in Indonesian) [Skripsi]. Bengkulu (ID): Bengkulu University
[15] Wahyuni I. 2016. Butterfly diversity in various types of land cover in PT Adimulia Agrolestari oil palm plantation, Kampar, Riau (in Indonesian) [undergraduate thesis]. Bogor (ID): Bogor Agricultural University
[16] Nurjannah S. 2016. The role of area with high conservation value for biodiversity conservation in oil palm plantation in Riau Province (in Indonesian). *Risalah Kebijakan Pertanian dan Lingkungan*. **3**(1): 68-77
[17] Peggie D, Amir M. 2006. *Practical Guide to the Butterflies of Bogor Botanic Garden. LIPI.* Bogor (ID): Zoology Division of Biology Research Center
[18] Smart P. 1975. The Illustrated Encyclopedia of the Butterfly World. London (EN): Salamander Books Ltd
[19] Pellet J, Bried JT, Parietti D, Gander A, Heer PO. 2012. Monitoring Butterfly Abundance: Beyond Pollard Walks. *Journal Pone.* **7**(7): 11-16
[20] Magurran AE. 1988. *Ecological Diversity and Its Measurement*. New Jersey (USA): Princeton University Press
[21] Vane-Wright RI, de Jong R. 2003. The butterflies of Sulawesi: annotated checklist for a critical island fauna. *Zoo Verh Leiden.* **343**: 3-267
[22] Mustari AH, Pramana Y, Nurlinda R. 2013. Butterfly diversity in Bantimurung Bulusaraung National Park (in Indoensian). *Media Konservasi*. **18**(2): 63-68
[23] Aidid L. 1991. The study of butterfly captive breeding in Bantimurung National Park, Maros District, South Sulawesi Province (in Indonesian) [undergraduate thesis]. Bogor (ID): Bogor Agricultural University
[24] Ngatimin S, Tamrin A, Nasruddin A, Gassa A. 2017. Introduction of butterfly species for students of SD 129 INPRES, Bantimurung Sub-district, Maros District, South Sulawesi (in Indonesian). *Jurnal Hutan dan Masyarakat.* **9**(1): 30-36
[25] Meidilaga. 2013. Butterfly diversity, captive breeding design and its development as tourism object in Walat Mount Education Forest Sukabumi (in Indonesian) [undergraduate thesis]. Bogor (ID): Bogor Agricultural University
[26] Koneril R, Saroyo. 2012. Butterfly (Lepidoptera) distribution and diversity in Old Manado Mount, Laur Bunaken National Park Area, North Sumatera (in Indonesian). *Jurnal Bumi Lestari*. **12**(2):357-365

[27] Tabadepu H, Damayanti B, Bandung S. 2008. Butterfly record from Salak Mount, Indonesia. *J.Entomol Indonesia.* **5**(1):10-16
[28] Sreekumar PG, Balakrishnan M. (2001). Habitat and altitude preferences of butterflies in Aralam Wildlife Sanctuary, Kerala. *Tropical Ecology.* **42**(2): 277-281.
[29] Dendang B. 2009. Butterfly diversity in Selabintana Resort, Gunung Gede Pangrango National Park, West Java (in Indonesian). *Jurnal Penelitian Hutan dan Konservasi Alam.* **4**(1): 25-36.
[30] Muin A. 2013. Conservation-oriented oil palm plantation business (in Indonesian) [dissertation]. Bogor (ID): Bogor Agricultural University
[31] Simbolon K, Iswari A.1990. *Butterfly Species Protected by Law in Indonesia (in Indonesia)*. Jakarta: Directorate General of Forest Protection and Nature Conservation (PHPA) Ministry of Forestry RI

How does oil palm plantation impact bird species diversity? a case study from PKWE Estate, West Kalimantan

Catharina Yudea[1], Yanto Santosa[2]

[1]Bachelor of Forest Resources Conservation and Ecotourism Department, Faculty of Forestry, Bogor Agricultural University
[2]Department of Forest Resources Conservation and Ecotourism, Faculty of Forestry, Bogor Agricultural University

*E-mail: cthyudea@gmail.com

Abstract. The prospect of Indonesian palm oil industry nowadays tends positive and shows an increasing trend. This matter also caused many negative responses to the presence of environmental damage and biodiversity loss, including bird species diversity. Since there are only a few researches related to oil palm plantation impact on bird species diversity, this research is necessary to enrich the data related. Data retrieval was carried out on March 2018, in PKWE Estate, West Kalimantan, using strip transect method in six different types of habitat/land cover. Two types of land cover which are secondary forest and shrubs considered as land cover before the establishment of oil palm plantation (baseline), while the others are habitat types that exist in oil palm plantation. The result shows that both baselines have the highest number of species among all habitat types, with a total of 26 species were found in each baseline. However, as a unity, turn out that the level of bird species diversity in oil palm plantation area is higher than the baseline area, where 33 species of bird were found (Dmg= 6.98). The existence of oil palm plantation also has changed the composition of bird species, the highest similarity index with both baselines is showed by old growth of oil palm. The change of species composition caused ten birds species lost and 17 species were gained on the secondary forest cover, meanwhile, 11 species lost and 18 species were gained in shrubs.

1. Introduction

Palm oil is known as the most widely consumed vegetable oil in the world that used in many packaged products sold in the market. At present, the palm oil mill industry being dominated by the Southeast Asia region, where Indonesia is the world's biggest producer of palm oil [1]. The prospect of Indonesian palm oil industry seems positive and keep shows an increasing trend. According to the FAO and IndexMundi [2], there were 18 million ha planted with oil palm in 2013, with a total production volume of 55 million tons of CPO during the same year, in 2015 the production increased to 62 million tons and this increased projected to continuously improve. This has an impact on the increasing level of land requirements resulting in land cover conversion, while the production of palm oil continues to grow, its expansion also increased rapidly [3]. The establishment of oil palm plantation certainly has changed the type of land cover. Based on several references oil palm rapid expansion claimed as one of the major drivers for the land-use change in Indonesia and has been spotlighted internationally of this dramatic land-use change [4], [5]. This conversion activity causes habitat destruction, degradation and fragmentation, and changes in species that are sensitive to migratory species. If this continues, biodiversity will threatened. Biodiversity that continues to decline shows enormous costs if many irreplaceable species begin to become extinct. Habitat loss is one of the important factors that caused a global biodiversity crisis, and the existence of land management can affect biodiversity needs. Several studies have shown that changes in land use-land cover are one of the main causes of biodiversity loss [6], [7].

This matter also caused negative responses on oil palm plantation establishment, said that the expansion connected to the environmental damage, destruction of habitat, and biodiversity loss, including the loss of bird species diversity. NGOs and researchers often declared that the conversion of forest to oil palm has led to a major threat to biodiversity and environmental problems [8], [9]. Indonesia is one of the most cost-efficient countries in the world for oil palm agriculture due to the environment suitability [10] and also a home of many wildlife species because located in a hotspot biodiversity area [11], including birds. Bird is one of the biodiversity that commonly found in oil palm plantation. Its species presence was affected by environmental factors; climatology and topography, vegetation structure, and human disturbance. It has an important role in nature and has high sensitivity on the surroundings changes. That is why birds often used widely for environmental changes indicators and included in evaluation studies for overall biodiversity conservation [12]. However, this is also made a concern on bird conservation, threatened by the habitat disturbances and loss, especially by the establishment of oil palm plantation. Therefore, we conduct this research with an intention to estimate the impact of oil palm plantations on bird species diversity.

2. Methods

2.1. Study site

Data retrieval was held on March 2018, in PKWE Estate, Ketapang District (0°19'26,51'' LS – 3°4'16,59'' LS and 109° 47'36,55'' BT – 111°21'37,36'' BT), West Kalimantan Province. We chose six plots for counting birds, divided in six sites, secondary forest, shrubs, High Conservation Value area (HCV), young, intermediate, and old growth of oil palm plantation. The oil palm plantations area consists of several age classes. Each age classes have different types and characteristics of land cover. The first class is young palm plantation, where the observation plot took place in the palm plantation planted in 2016 or two years old at the time of the study and located adjacent to shrubs land cover. The young palm characterized by 2-3 m tall plants, no canopy covers and groundcover with shrubs and grasses. The second class is middle-aged oil palm plantation, the plot took in palm plantation grown in 2011 or seven years old when the research was conducted. Most have fairly wide canopy covers and groundcover with fern, shrubs and grasses. The third is old palm plantation, the observation plot took in palm plantation grown in 2006 or aged 12 years. Oil palm in this category is characterized by closed canopy cover, the groundcover only slightly covered with fern, shrubs and grasses. The HCV area is an area that looks like secondary forest inside the oil palm plantation area. This land cover is quite dense dominated by big trees, surrounded by community of the oil palm plantations and parts of riparian areas.

2.2. Data collection

This study defines the habitat into two types, i.e., habitat before the oil palm plantation establishment and habitat after the oil palm plantation establishment. We used Landsat image analysis of each site within one to two years before plantation land clearance to determine the type of land cover prior. Based on the satellite imagery interpretation, secondary forest and shrubs representing the baseline or the land cover before the establishment of oil palm plantation, and the rest of land cover are habitat after the oil palm plantation establishment. In each type of land cover, one observation track was made considering the relatively small area and based on the assumption that bird species in each of the same habitat types were homogeneous.

The inventory of birds was carried out with strip transect method (Figure 1), with total length of the track is 2 km, by making a straight line as far as one kilometre to the north/south/west/east, then turn approximately 150 m (distance between lines), and go back as far as one kilometre in the opposite direction to the start line. This inventory is carried out simultaneously on each type of land cover, in the morning (06.00-08.00 WIB) and afternoon (16.00-18.00 WIB) for each transect, repeated for three times in each transect to maximize the number of species recorded. Observer walks along the track by noting the type species and a number of birds seen or heard, the time encounter and the place where birds are found.

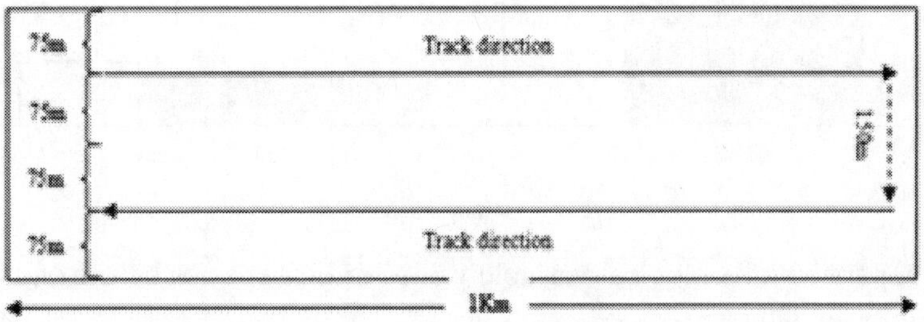

Figure 1. The illustration of strip transect method

2.3. Data analysis
Impacts of oil palm plantations on bird diversity were determined by comparing the current biodiversity found on each plantation covers, with those found on the current baseline (with an assumption that the current baseline resembles the previous land cover before the plantation establishment). To measure and compare the diversity of birds in all habitat conditions number of species that found on each land covers in PKWE estate is counted. Also Margalef species Richness index [13] was used to simply count the number of different species in a given area. The evenness index was used to determine the evenness of the species in each land cover and the similarity index was used to determine the similarity of bird species between the baseline and each of oil palm plantation covers [14], and used as the basis for assessing the impact of oil palm plantation establishment on the diversity of birds, number of bird diversity loss and gain.

3. Results and discussion
3.1. Comparison of bird species diversity
Based on observation data in all types of land cover, 49 species of bird were recorded. Compared to all existing land cover types, secondary forest and shrubs cover have the highest bird encounter with total 26 species were found in each cover. Although both land covers have the same number of species, but the value of species richness indices shows a differ, the secondary forest has a higher value (6.01) rather than shrubs cover (5.48). The difference is due evenness factor, where evenness indices on shrubs is lower than secondary forest, this indicates a high dominance or abundance in certain species on the site. Species richness not only takes into account the number of species, but also the number of individuals, so the abundance of a species will affect the value of the index [13]. Meanwhile, the lowest bird species encounter and species richness indices was found on old growth oil palm, only 4 species of bird were found, with species richness value of 1.44. Overall the results of observations on the bird diversity in the study sites are shown in table 1 and figure 2.

Table 1. Comparison of ecological variable values of bird species diversity in each land cover of PKWE estate

	OG	IG	YG	HCV	Secondary forest	Shrubs
Number of species	4	9	16	15	26	26
Species richness	1.44	3.12	3.99	4	6.01	5.48
Evenness indices	0.95	0.95	0.9	0.87	0.88	0.85

*where: YG= young growth oil palm, IG= intermediate growth oil palm, OG= old growth oil palm, HCV= high conservation value area

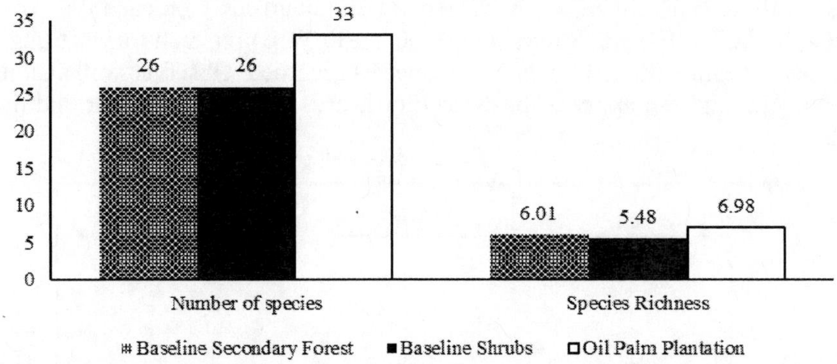

Figure 2. Comparison of ecological variable values of bird species before and after oil palm Establishment

This study has a different result with another similar study, conducted by [15], where the results represent that bird diversity affected by plantation age, in the early establishment of oil palm plantation, the diversity of birds tends to be lower and show an increasing number along with the older age of plantation. This study shows the opposite result, the oldest growth oil palm has the lowest species diversity instead, intermediate growth oil palm has a higher number, and the young growth oil palm has the highest number. In old growth oil palm cover, the state of the plantation's undergrowth was less dense than the other types of the plantation, this is due to the harvesting activities in the surrounding block area. In addition to the harvesting activities, the area was frequently visited by humans, so the environment tends to have a high level of human disturbance [16]. Remarked by reference [17], [18], human activities and presence caused a disturbance that can affect species diversity in a given area or habitat. A better habitat is a place where far from human disturbances and there is enough source of food, this kind area is likely to have many bird species [19]. Both baselines (secondary forest and shrubs) have the highest number of species because of almost no disturbance from humans.

Shrub is a good habitat for some species of birds to perch and look for food on the shrubs or bushes area. The denser the cover of the undergrowth, birds will be easier in hiding and looking for food, this condition can increase bird diversity. As been stated in reference [20] when the undergrowth removed experimentally, the bird species richness was declined. As stated by reference [21] and [22], the increase in ground vegetation cover can marginally increase bird and butterfly diversity in plantations. Among all habitat types in oil palm, young growth oil palm cover has a high number of species because in this area the undergrowth is denser. Other factors known to affect the responses of birds to oil palm include the age and size of the plantation trees, percentage of canopy cover [23], the presence of epiphytic ferns on oil palm trees, and human disturbance such as weeding or harvesting [18]. The results obtained show that the way of managing and maintaining oil palm plantations greatly affects the diversity of bird species.

As a whole management unit, oil palm plantation has a higher number of species and species richness indices compared to the two baselines, where 33 species of birds recorded with richness indices value of 6.98. Bird densities, domination, species occurrence, and turnover rate depend mainly on area and structure variables [24]. The characteristics and habitat types affect the diversity of bird species, the more diverse habitat the higher bird species diversity tends to get [25]. The difference on the oil palm plantation which does not only consist of oil palm monoculture plantation but also HCV areas whose resembles a secondary forest, while each baseline formed only from one characteristic habitat, has caused a decrease in the number of species and species diversity value.

3.2. Comparison of bird species composition

Each type of land cover has a different number of species and species composition. Changes in land cover that occur not only caused a differ in the value of the diversity of bird species but also changed the composition of the bird species. An indication of differences in species composition can be found based on the composition and feeding guild of bird species before and after the establishment of oil palm in table 2 and comparison of the Sorensen similarity indices that can be seen in figure 3.

Table 2. Composition and feeding guild of bird species before and after the establishment of oil palm plantation

No	Scientific name	Famili	Feeding guilds	Area before oil palm plantation		Oil palm plantation area			
				SF	SB	OG	IG	YG	HCV
1	*Microhierax fringillarius*	Falconidae	Insectivorous, carnivorous	V	V				
2	*Ixobrychus cinnamomeus*	Ardeidae	Carnivorous				V	V	
3	*Lanius schach*	Laniidae	Insectivorous		V			V	
4	*Psittacula longicauda*	Psittacidae	Frugivorous, granivorous		V			V	
5	*Lonchura fuscans*	Estrildidae	Granivorous	V					
6	*Centropus bengalensis*	Cuculidae	Insectivorous	V	V		V		
7	*Centropus sinensis*	Cuculidae	Insectivorous	V		V		V	
8	*Chalcoparia singalensis*	Nectariniidae	Nectivorous	V	V				
9	*Dicaeum cruentatum*	Dicaeidae	Insectivorous	V					
10	*Ardea purpurea*	Ardeidae	Carnivorous		V				
11	*Halcyon smyrnensis*	Alcedinidae	Insectivorous	V				V	
12	*Todiramphus chloris*	Alcedinidae	Carnivorous				V		
13	*Orthotomus ruficeps*	Sylviidae	Insectivorous	V			V		
14	*Aegithina tiphia*	Aegithinidae	Insectivorous		V				
15	*Pycnonotus aurigaster*	Pycnonotidae	Omnivorous		V				
16	*Ichthyophaga ichthyaetus*	Accipitridae	Carnivorous		V			V	
17	*Butastur indicus*	Accipitridae	Carnivorous		V				
18	*Elanus caeruleus*	Accipitridae	Carnivorous	V					
19	*Corvus enca*	Corvidae	Omnivorous						V
20	*Hemipus hirundinaceus*	Campephagidae	Insectivorous	V	V				
21	*Copsychus saularis*	Muscicapidae	Insectivorous						V
22	*Anthracoceros malayanus*	Bucerotidae	Frugivorous, carnivorous						V
23	*Anthracoceros albirostris*	Bucerotidae	Frugivorous, carnivorous						V
24	*Amaurornis phoenicurus*	Rallidae	Carnivorous		V		V	V	
25	*Acridotheres cristatellus*	Sturnidae	Omnivorous		V				V
26	*Rhipidura javanica*	Rhipiduridae	Insectivorous	V	V				V
27	*Merops viridis*	Meropidae	Insectivorous	V	V				
28	*Egretta eulophotes*	Ardeidae	Carnivorous					V	
29	*Bubulcus ibis*	Ardeidae	Carnivorous		V		V	V	
30	*Hirundo tahitica*	Hirundinidae	Insectivorous	V		V			V
31	*Pycnonotus plumosus*	Pycnonotidae	Omnivorous	V					
32	*Pycnonotus goiavier*	Pycnonotidae	Omnivorous	V	V		V	V	V
33	*Pycnonotus simplex*	Pycnonotidae	Omnivorous		V				
34	*Pycnonotus brunneus*	Pycnonotidae	Omnivorous	V	V		V		
35	*Chrysophlegma miniaceum*	Picidae	Insectivorous	V					V
36	*Prinia flaviventris*	Cisticolidae	Insectivorous	V	V			V	
37	*Ducula aenea*	Columbidae	Frugivorous	V					V
38	*Treron vernans*	Columbidae	Frugivorous	V	V			V	V
39	*Treron olax*	Columbidae	Frugivorous		V				
40	*Alcedo meninting*	Alcedinidae	Carnivorous						V
41	*Loriculus galgulus*	Psittacidae	Frugivorous	V					
42	*Dicrurus paradiseus*	Dicruridae	Insectivorous						V
43	*Dicrurus macrocercus*	Dicruridae	Insectivorous		V				
44	*Psilopogon australis*	Megalaimidae	Frugivorous	V					
45	*Spilopelia chinensis*	Columbidae	Granivorous	V	V	V	V	V	V
46	*Eurystomus orientalis*	Coraciidae	Insectivorous	V	V			V	
47	*Gracula religiosa*	Sturnidae	Omnivorous	V					V
48	*Collocalia esculenta*	Apodidae	Insectivorous	V	V	V		V	
49	*Ixobrychus flavicollis*	Ardeidae	Carnivorous					V	

*where: SF= secondary Forest, SB= shrubs, YG= young growth oil palm, IG= intermediate growth oil palm, OG= old growth, HCV= high conservation value area

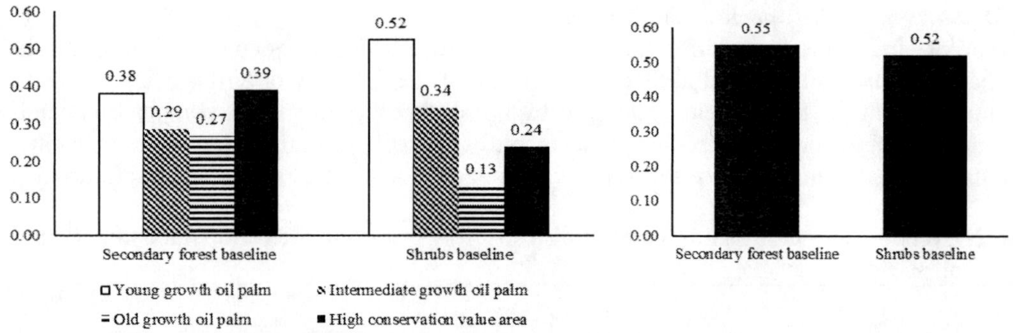

Figure 3. Comparison of Sorensen similarity indices between (a) each type of land covers in oil palm plantation with the baseline (b) overall oil palm plantation and the baseline

Table 2 shows that predators and carnivores can be more commonly found in SB areal, while in other types of land cover there are fewer or even only one predator species. One type of feed such as small birds exists in all locations. Vegetation structure can influence the ability to detect prey. AS SB is an open habitat, make predators tend to perch and find prey more easily. Besides that, some locations within the SB area are submerged by water, which can be a place to find prey for several species of carnivores.

Based on figure 3 we can see that bird communities differed greatly between the two baselines (secondary forest and shrubs) compared with all types of covers in oil palm plantation. From 33 species recorded in the oil palm plantation area, Insectivorous birds dominated the number of bird species (42.42%), followed by carnivorous bird (24.24%), omnivorous and frugivorous bird (15.15%), and granivorous (3.03%). No nectivorous bird was found in oil palm plantation. The most significant species composition changes in both baselines occurred in old-growth oil palm, marked by the lowest value of similarity indices, 0.27 in the secondary forest and 0.13 in shrubs. Shown by figure 3 that young growth oil palm and shrubs have the highest similarity indices with value > 0.5 and became the highest similarity indices with the shrubs baseline, meanwhile, the highest similarity with the secondary forest is HCV area with a value of only 0.39. The rate of species similarity indices affected by the characteristic in both habitats, the more similar both habitats, the higher species similarity will obtain [26].

HCV area in oil palm plantations is an area that intended for the protection of riparian and all waterways and restoration of buffer zones. This area function toward providing a relatively healthy habitat for wildlife species, this area has a similar habitat characteristic with secondary forest, there is still dense trees cover, which is not found in other types of oil palm plantation cover. Young growth oil palm has the highest similarity with shrubs, there are entirely no tree types on both land covers, the height of the oil palm is still short, no canopy tree is formed, and most of the vegetation that exists is in the form of undergrowth or shrubs. On the other side, old growth oil palm has the most significant difference in both baselines, due to the lack of habitat characteristics. Its habitat condition has been disturbed, only slightly covered by undergrowth and shrubs (low density of undergrowth) so that the acquisition of the species is the least. With the existence of the undergrowth or shrubs in this area, there is still a similarity between old growth oil palm and the baseline.

Overall, the comparison of similarity indices between oil palm plantation with each baseline show almost the same value in the range 0.52-0.55. This indicates that land cover changes in the form of secondary forest or even shrubs cover have a pretty significant difference, where almost half of the total bird species, with 16 species of birds in the secondary forest are similar, and 15 species in shrubs are similar. Different species in each baseline lead the assumption of the loss and gain of bird species. Figure 4 shows the assumption of loss and gain from both baselines with each land cover types that exist in an oil palm plantation, meanwhile, table 2 and 3 shows result between overall oil palm plantation and the baseline.

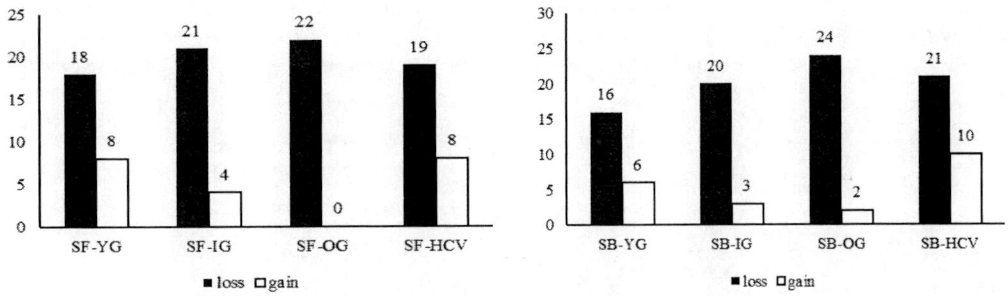

(a) (b)

*where: SF= secondary Forest, SB= shrubs, YG= young growth oil palm, IG= intermediate growth oil palm, OG= old growth oil palm, HCV= high conservation value area

Figure 4. Comparison assumption of bird species loss and gain in each type of land covers (a) from secondary forest baseline (b) from shrubs baseline

The comparison of loss and gain estimation from both baseline (figure 4a and 4b), shows that each type covers of oil palm plantation have a higher number in the species lost than the species gained. Old growth oil palm cover has the highest number of species lost, also the lowest number of species gained from even both baseline type. From the secondary forest baseline, 22 species are lost and there are no species gained, from the shrubs baseline 24 are lost and 2 are gained. This represents the lowest value in similarity indices between old growth oil palm cover with the baseline. This loss and gain data show that the new appearance or loss of certain species of bird is caused by the selection of habitat based on the compatibility of the bird with the habitat and the availability of resources for the birds to sustain their life [17].

Birds found in oil palm cover type (young, intermediate and old growth oil palm) tend to be simple communities of generalists and non-forest species and they tend to dominate [27], frugivorous birds also rarely recorded in the oil palm, due to the lack of fruiting trees and dominated only oil palm monoculture. Vegetation on the baseline found to be more diverse. Vegetation plays an important role in bird survival and supports birdlife, including species diversity, structure, population density, and density of the canopy [28]. It provides the availability of food, cover, and other psychological conditions [17]. The more diverse and abundant of the vegetation compositions, including the abundance of epiphytes and fruits, the larger number of bird species [29], [30]. In the secondary forest, the main vegetation types are ranging from the undergrowth, seedlings, saplings, and tall trees that are not owned by other land covers, the presence of trees cause canopy stratification also more diverse in this land cover. Vertical distribution of leaf or canopy stratification diversity also affects the diversity of bird species [31]. This makes the species found on the baseline more diverse, thus causing each oil palm covers to lose much more. While the list of bird species that considered lost and gained from the baseline into oil palm plantation is shown in table 3, table 4 and figure 5.

Table 3. List of bird species contributed to the dissimilarity of species composition between oil palm plantations and the baseline

Secondary forest as the baseline				Shrubs as the baseline			
Founded only in oil palm plantation		Founded only in secondary forest		Founded only in oil palm plantation		Founded only in shrubs	
1	*Ixobrychus cinnamomeus*	1	*Psilopogon australis*	1	*Ixobrychus cinnamomeus*	1	*Microhierax fringillarius*
2	*Ixobrychus flavicollis*	2	*Microhierax fringillarius*	2	*Ixobrychus flavicollis*	2	*Treron olax*
3	*Bubulcus ibis*	3	*Elanus caeruleus*	3	*Egretta garzetta*	3	*Butastur indicus*
4	*Egretta garzetta*	4	*Chalcoparia singalensis*	4	*Alcedo meninting*	4	*Dicrurus macrocercus*
5	*Todiramphus chloris*	5	*Hemipus hirundinaceus*	5	*Halcyon smyrnensis*	5	*Pycnonotus aurigaster*
6	*Amaurornis phoenicurus*	6	*Merops viridis*	6	*Todiramphus chloris*	6	*Pycnonotus simplex*
7	*Alcedo meninting*	7	*Lonchura fuscans*	7	*Corvus enca*	7	*Merops viridis*

8	*Corvus enca*	8	*Pynonotus plumosus*	8	*Centropus chinensis*	8	*Ardea purpurea*
9	*Acridotheres javanicus*	9	*Loriculus galgulus*	9	*Copsychus saularis*	9	*Hemipus hirundinaceus*
10	*Lanius schach*	10	*Dicaeum cruentatum*	10	*Dricurus paradiseus*	10	*Aegithina tiphia*
11	*Copsychus saularis*			11	*Hirundo tahitica*	11	*Chalcoparia singalensis*
12	*Dicrurus paradiseus*			12	*Ducula aenea*		
13	*Psittacula longicauda*			13	*Chrysophlegma miniaceum*		
14	*Ichthyophaga ichthyaetus*			14	*Orthotomus ruficeps*		
15	*Anthracoceros albirostris*			15	*Gracula religiosa*		
16	*Anthracoceros malayanus*			16	*Anthracoceros albirostris*		
				17	*Anthracoceros malayanus*		

Table 4. List of bird species contributed to the dissimilarity of species composition between oil palm plantations (without HCV) and the baseline

Secondary forest as the baseline				Shrubs as the baseline			
Oil palm without HCV		Secondary forest		Oil palm without HCV		Shrubs	
1	*Ixobrychus cinnamomeus*	*1*	*Microhierax fringillarius*	*1*	*Ixobrychus cinnamomeus*	*1*	*Microhierax fringillarius*
2	*Ixobrychus flavicollis*	*2*	*Lonchura fuscans*	*2*	*Ixobrychus flavicollis*	*2*	*Chalcoparia singalensis*
3	*Lanius schach*	*3*	*Chalcoparia singalensis*	*3*	*Centropus chinensis*	*3*	*Ardea purpurea*
4	*Psittacula longicauda*	*4*	*Dicaeum cruentatum*	*4*	*Halcyon smyrnensis*	*4*	*Aegithina tiphia*
5	*Todiramphus chloris*	*5*	*Elanus caeruleus*	*5*	*Todiramphus chloris*	*5*	*Pycnonotus aurigaster*
6	*Ichthyophaga ichthyaetus*	*6*	*Hemipus hirundinaceus*	*6*	*Hirundo tahitica*	*6*	*Butastur indicus*
7	*Amaurornis phoenicurus*	*7*	*Rhipidura javanica*	*7*	*Egretta garzetta*	*7*	*Hemipus hirundinaceus*
8	*Egretta garzetta*	*8*	*Merops viridis*	*8*	*Orthotomus ruficeps*	*8*	*Acridotheres cristatellus*
9	*Bubulcus ibis*	*9*	*Pycnonotus plumosus*			*9*	*Rhipidura javanica*
		10	*Chrysophlegma miniaceum*			*10*	*Merops viridis*
		11	*Ducula aenea*			*11*	*Pycnonotus simplex*
		12	*Loriculus galgulus*			*12*	*Treron olax*
		13	*Psilopogon australis*			*13*	*Dicrurus macrocercus*
		14	*Gracula religiosa*				

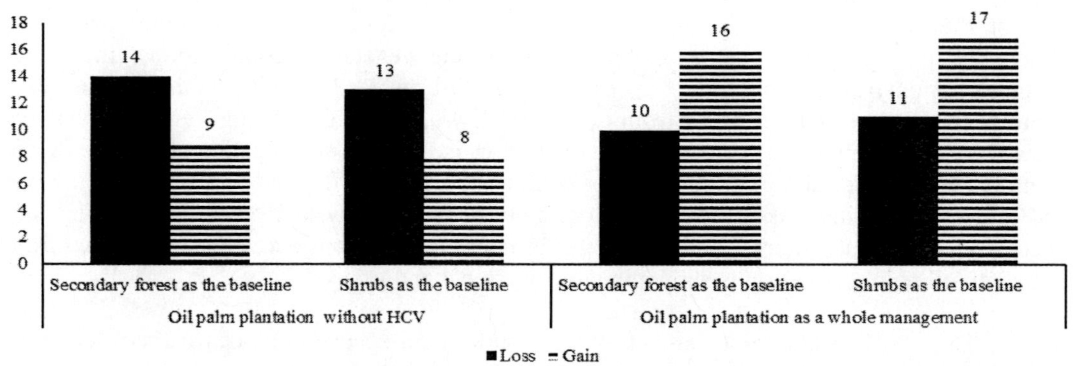

Figure 5. Comparison assumption of bird species loss and gain between whole oil palm plantations and plantation area without HCV with the baseline

All species that considered lost indicate that those species cannot fit in oil palm plantation habitat, meanwhile gained species or similar species that also recorded in oil palm plantation indicate that those species are capable of using oil palm plantations for their habitat, also indicates that oil palm plantation can be hospitable for some bird species. The total number of the gained species from both land cover show almost the same bird species composition. The bird community that found in the oil palm plantation is influenced by the relatively heterogeneous landscape, both within and surrounding the oil palm plantation. HCV area in the oil palm plantations enrich heterogeneity, this area has a similar habitat characteristic to secondary forests. The crucial role of HCV area in the oil palm plantation shown by table 3 and figure 5, which indicate that without this area the presence oil palm plantation has caused the number of bird species lost is higher than the species that gained. This area has various types of trees, forest edge or border typical, therefore forest species birds can be found and recorded in oil palm plantation area, just like *Anthracoceros albirostris* and *Anthracoceros malayanus.*

Meanwhile, in oil palm cover, trenches, shrubs and marshes become the typical characteristic of this habitat. Areas that close to water sources, just like swamp area has their own habitat characteristics and favored by certain types of birds, such as *Alcedo meninting, Ergetta garzetta, Bubulcus ibis, Ixobrychus cinnamomeus, Ixobrychus flavicollis.* Those species founded only in oil palm plantation areas, they found nesting on platforms of reeds in shrub, feed on insects, fish and amphibians [32]. The presence of oil palm plantation also increased and provided insects [15], this caused insectivorous birds gained, such as *Lanius schach, Copsychus saularis, Centropus chinensis.* The presence of insectivorous birds in oil palm plantation bring positive impacts, they can contribute in controlling the pest [33]. Some of the bird's families that have been found only in oil palm cover, among others are alcedinidae and ardeidae. Its species, among others, *Halcyon smyrnensis, Todiramphus chloris, Ixobrychus cinnamomeus, Ixobrychus flavicollis* are species that common and have large range [32]. This shows that most species appear to be able adapt in the oil palm habitat, many species of birds are capable of using oil palm plantations but indicate that habitats supporting communities in oil palm plantation are dominated by a common and widespread species, but the presence of HCV area inside the plantation can bring more birds diversity. Reference [34], [35] remarked the importance of the surrounding forest area and by adjoining, higher species of birds obtained than plantations that lacked such connectivity.

4. Conclusion

The results indicate that conversion from secondary forest and shrubs oil plantation results in the change of bird number of species and bird species composition. Overall oil palm plantation has increased the bird species diversity, from 26 species on each baseline into 33 species. However, if we look based on the comparison of the baseline with all types of land covers on oil palm plantation, the result shows a decrease of species diversity in all oil palm land covers. The highest decreased from both baselines is shown by old growth oil palm.

The most significant change of species composition in each baseline, both shown by old growth oil palm with the lowest similarity value of 0.27 with secondary forest, and 0.13 with shrubs cover. The land cover change from the secondary forest into oil palm plantations led to the gain of 16 bird species and the loss of 10 species, while the change from shrubs led to the gain of 17 species and the loss of 11 species. Insectivorous bird is the most found species. Oil palm habitat tends to bring species that considered as species of lower conservation concern and with extensive ranges. The existence of HCV area in oil palm plantations is crucial because it brings more diverse, so that oil palm plantation areal can has higher diversity.

Acknowledgments

Further thanks to PT Smart tbk for funding this study, also for providing the permissions and facilities during the study period.

References

[1] McCarthy J, Zen Z (2010). Regulating the oil palm boom: assessing the effectiveness of environmental governance approaches to agro-industrial pollution in Indonesia. Law & Policy, **32**: 153-179

[2] Pacheco P, Gnych S, Dermawan A, Komarudin H and Okarda B. 2017. The palm oil global value chain: Implications for economic growth and social and environmental sustainability. Working Paper 220. Bogor, Indonesia: CIFOR. (p. 5-7)

[3] Lubis MF, Lubis I. 2018. Production Analysis of Oil Palm (Elaeis guineensis Jacq.) in Pelalawan, Riau. Bul. Agrohorti **6** (2): 281 – 286
[4] Euler M, Schwarze S, Siregar H, Qaim M. 2016. Oil palm expansion among smallholder farmers in Sumatra, Indonesia. Journal of Agricultural Economics, **67**: 658–676
[5] Gatto M, Wollni M, Qaim M. 2015. Oil palm boom and land-use dynamics in Indonesia. The role of policies and socioeconomic factors. Land Use Policy **46**: 292-303
[6] Alessandra F, Maiorano L, Boitani L. 2007. Changes in land-use/landcover patterns in Italy and their implications for biodiversity conservation. Landscape Ecol, **22**: 617-631
[7] Yadav PK, Kapoor M, Sarma K. 2012. Land use land cover mapping, change detection and conflict analysis of Nagzira-Navegaon Corridor, Central India using geospatial technology. Intl J Remote Sensing GIS **1** (2): 90-98
[8] Amnesty International. 2016. The Great Palm Oil Scandal: Labour Abuses Behind Big Brand Names. London (UK): Amnesty International
[9] Fitzherbert, EB. Struebig, M. J.,Morel, A.,Danielsen, F., Bruehl, C. A., Donald, P. F. &Phalan, B. (2008). How will oil palm expansion affect biodiversity? Trends in Ecology & Evolution **23**, 538–545
[10] Casson A. 2000. The Hesitant Boom: Indonesia's Oil Palm Sub-sector in an Era of Economic Crisis and Political Change. CIFOR Program on the Underlying Causes of Deforestation. Available at http://www.cifor.cgiar.org/ publications/pdf_files/CASSON.pdf (accessed August 2007). Center for International Forestry Research (CIFOR), Bogor, Indonesia
[11] Mittermeier RA, Gil PR, Hoffman M, Pilgrim J, Brooks T, Mittermeier CG, Lamoreux J, da Fonseca GAB. 2004. Hotspots revisited. Cemex, Conservation International, and Agrupacion Sierra Madre, Monterrey, Mexico
[12] Gregory RD, Noble DG, Custance J. 2004. The state of play of farmland birds: population trends and conservation status of lowland farmland birds in the United Kingdom. Ibis **146** (Suppl. 2), 1–13
[13] Magurran A. 1988. Ecological Diversity and its Measurement. Croom Helmed Limited, London
[14] Krebs CJ. 1985. Experimental Analysis of Distribution and Abundance. 3rd ed. Harper & Row Publisher, New York
[15] Yoza D. 2000. Dampak Perkebunan Keiapa Sawit Terhadap Keanekaragaman Jenis Burung di Areał Perkebunan PT. Ramajaya Pramukti, Kabupaten Dati II Kampar Propinsi Dati I Riau [Thesis]. Bogor (ID): Institut Pertanian Bogor
[16] Sheldon FH, Styring A, Hosner PA, 2010. Bird species richness in a Bornean exotic tree plantation: Along-term perspective. Biol.Cons.**143**,399–407
[17] Welty JC, Baptista L. 1988. The Life of Bird. New York (US): Sounders College Publishing
[18] Azhar B, PuanCL, Zakaria M, Hassan N, Arif M. 2014. Effects of monoculture and polyculture practices in oil palm small holdings on tropical farmland birds. BasicAppl.Ecol.**15**(4),336–346
[19] Widodo W. 2009. Komparasi Keragaman Jenis Burung-Burung di Taman Nasional Baluran dan Alas Purwo Pada Beberapa Tipe Habitat. Jurnal Berkala Penelitian Hayati. (14): 113-124
[20] Najera A, Simonetti JA, 2010. Can oil palm plantations become bird friendly? Agrofor.Syst. **80**,203–209
[21] Koh LP. 2008. Can oil palm plantations be made more hospitable for forest butterflies and birds? Journal of Applied Ecology **45** 1002–1009
[22] Azhar B, Lindenmayer DB, Wood J, Fischer J, Manning A, McElhinny C, Zakaria M, 2011. The conservation value of oil palm plantation estates, small holdings and logged peat swamp forest for birds. ForestEcol. Manag. **262**(12),2306–2315
[23] Azhar B, Lindenmayer DB, Wood J, Fischer J, Manning A, McElhinny C, Zakaria M, 2013. The influence of agricultural system, stand structural complexity and landscape context on foraging birds in oil palm landscapes. Ibis **155**(2),297–312
[24] Tworek S. 2002. Different bird strategies and their responses to habitat changes in an

agricultural landscape. Ecological Research **17**, 339–359
[25] Johnsingh AJT, Joshua J. 1994. Avifauna in three vegetation types on Mundanthurai Plateu. Journal of Tropical Ecology, **10**:323-335
[26] Sihotang DF, Patana P, Jumilawaty E. 2013. Identification diversity of bird species in restoration area at Sei Betung Resort, Gunung Leuser National Park. Peronema Forestry Science Journal, volume **2** (2):59-66
[27] Edwards FA, Edwards DP, Hamer KC, Davies RG. 2013. Impacts of logging and conversion of rainforest to oil palm on the functional diversity of birds in Sunda land. Ibis **155**(2),313–326
[28] Dickson JG, Conner RN, Fleet RR, Croll JC, Jacson JA .1979. The Role of Insectivorous Birds in Forest Ecosystems. New York (US): Academic Press
[29] Ricklefs RE. 1978. Ecology. New York (US): Chiron Press
[30] Winkler H, Preluetherener M. 2001. Behaviour and ecology of bird in tropical rain forest canopy. Plant Ecology, **153**: 193-202
[31] Parker GG, Brown MJ. 2000. Forest canopy stratification-is it useful?. The American Naturalist, **155**(4): 473-484
[32] Ali Salim; Daniel J C (1983). The book of Indian Birds, Twelfth Centenary edition. New Delhi: Bombay Natural History Society/Oxford University Press
[33] Koh LP. 2008. Birds defend oil palms from herbivorous insects. Ecol.Appl.**18**,821–825
[34] Pulliam, H.R. (1988) Sources, sinks, and population regulation. The American Naturalist, **132**, 652–661
[35] Raman, T.R.S. (2006) Effects of habitat structure and adjacent habitats on birds in tropical rainforest fragments and shaded plantations in the Western Ghats, India. Biodiversity and Conservation, **15**, 1577–1607

Impact of oil palm plantation on mammal and herpetofauna species diversity

Yanto Santosa[1]*, Sani Sari Sri Rejeki[2]

[1]Wildlife Ecology and Management Division, Department of Forest Resources Conservation and Ecotourism, Faculty of Forestry, Bogor Agricultural Univsersity, Bogor, West Java, Indonesia

*E-mail: yantohaurjaya@yahoo.co.id

Abstract. Since 2008, Indonesia has succeeded in becoming the largest palm oil producer in the world. However, this success has drawn some accusations that oil palm plantations come causing deforestation and decreasing biodiversity. The research has been carried out in the topic of "the origin of the status and type of land cover and variations in the wildlife species diversity in oil palm plantations" in Province of Riau, South Sumatra, Central Kalimantan and West Kalimantan. The results of the interpretation of Landsat imagery coverage of 1 year before being oil palm plantation and interviews with resource persons showed that the type of land cover of the 8 study locations was 5 in the form of secondary forest, 2 bushes and the rest in the form of rubber plantations. The status of the land as a whole APL or customary land. Furthermore, the results of the comparison of the number and composition of species of mammals and herpetofauna before and after being oil palm plantations show that the impact of oil palm plantations on these two variables of diversity species according to the origin of land cover types. For the origin of the type of land cover in the form of shrubs or rubber plantations, the impact of oil palm plantations is positive because it increases the number of species both mammals and herpetofauna (except the number of species of herpetofauna on KJNP estate plantations). As for the origin of the type of land cover in the form of secondary forest, the existence of oil palm plantations increase the number of mamalia in 2 locations and herpetofauna in 4 locations The presence of oil palm plantations is thought to have caused changes in the composition of species with a range of 50 to 100% for mammals and 23 to 76% for herpetofauna.

1. Introduction

The expansion of oil palm plantations in developing countries is carried out as one of the economic development policies [1]. In 2017, palm oil has become the main product and play a role in the Indonesian economy from the plantation sector [2]. The area of oil palm plantation in Indonesia reaches 12.31 million ha in 2017 [3]. The expansion of oil palm plantation is thought to be a trigger for deforestation because it has rapid expansive properties in a short time [4]. This is also reinforced by a statement [5] that the conversion of forest areas in Indonesia is the largest allocated for oil palm plantations.

Deforestation has caused a diversity of wildlife species such as mammals and herpetofauna to decline. Mammals and herpetofauna are important components in maintaining ecosystem balance.

Mammals play a role in pest control, seed spreaders, predators, and pollinators for plants [6]. In addition, mammals also act as level I and II consumers and top predators in the food chain [7]. Meanwhile, Herpetofauna is a combination of reptiles and amphibians which are used as bioindicators of environmental change. This is because herpetofauna is very susceptible to environmental changes [8] and all aspects of life are affected by the climate in these habitats [9].

This research was carried out because there was still a lack of knowledge about the impact of the establishment of oil palm plantations on mammals and herpetofauna in various land coverings. The purpose of this study was to identify palm oil plantations, species of palm plantations, and plant oil palm oil. Estimate the impact of oil palm plantations on species of mammals and herpetofauna and Estimate the impact of oil palm plantations on loss/gain in species of mammals and herpetofauna.

2. Materials and methods

The impact of oil palm plantation mammals and herpetofauna diversity was known by comparing the types covered in land cover before and after oil palm plantation (OPP) establishment. Landsat imagery was interpretation to find out the history of land cover change for two years coverage before OPP establishment since it is not possible to find the exact land cover type as those before the conversion to oil palm. The present land cover type is assumed to represent the past land cover before the oil palm plantations establishment. In the same line, it is assumed that the mammal and herpetofauna diversity would also be similar.

2.1. Study area

The study was conducted on 8 large-scale oil palm plantations (PSB) located in the Riau (PT KPR, PT SAR, PT AMA, PT GDH and KJNPE), South Sumatera (PT WMA), Central Kalimantan (PT BLP) and West Kalimantan (PKWE) Province on two types area (baseline area and oil palm plantations area). Data collection was conducted out in 2016-2018.

2.2. Study methods

Animal inventory was carried out simultaneously on each type of land cover present in each habitat. The inventory of mammals was done using the list transect method with three repetitions. There 15 camera traps were used to record observation indirectly. The observation was done in the morning (06:00-08:00 WIB) and afternoon (15:30-17:30 WIB). Herpetofauna inventory was done by using Visual Encounter Survey method for 3 hours starting from 19:00 to 22:00 with three repetitions.

2.3. Data analysis

To find out the impact of oil palm plantation on mammals and herpetofauna species diversity, the calculation of number of species, evenness index and species richness index are calculated as follows:

- Species richness index (Dmg)
  Margalef Index (species richness) used to determine the species richness of each species in the community that found in the field, the formula is:

$$Dmg = \frac{S-1}{\ln N}$$

  Information:
  Dmg = species richness index
  S = number of species
  N = total number of individuals of all species

- Evenness index (E)
  Evenness Index has a function to determine the evenness of each type in every community.

$$E = \frac{H'}{\ln(S)}$$

Information:
E = evenness index (a value between 0-1)
H ' = diversity
ln = natural logarithm
S = number of species

Whereas to determine the impact of oil palm plantation on species composition, a number of mammals and herpetofauna diversity loss and gain in all location, the following formula is used:

- Dissimilarity index (ID)
  Dissimilarity index is used to determine the dissimilarity between the observation location based on the species mammals and herpetofauna found

$$ID = 1 - IS, \text{ where } IS = \frac{2c}{(A+b)}$$

Information:
ID = Dissimilarity indeks
IS = similarity index
a = the number of species that are only found in community A
b = the number of species that are only found in community B
c = the number of species found in community A and B

3. Results and discussion

3.1. History of land cover types and status

The type of land cover of the 8 study locations was no longer in the form of primary forest, but 5 of them were in the form of secondary forest (PT KPR, PT SAR, PT AMA, PT GDH and PT BLP), 2 shrubs (KJNPE and PKWE) and the rest in the form of rubber plantations (PT WMA). Table 1 presents data on the type and status of land cover before the oil palm plantation establishment.

Table 1. Land cover history

Study location		Land cover types before oil palm plantation established	Land status before oil palm plantation established
Riau	PT. KPR	Secondary forest	Others land use
	PT. SAR	Secondary forest	Others land use and Transmigration area
	PT. AMA	Secondary forest	Others land use
	PT. GDH	Secondary forest	Others land use
	KJNP Estate	Shrub	Others land use
South Sumatera	PT. WMA	Rubber plantation	Others land use
Central Kalimantan	PT. BLP	Secondary forest	Others land use
West Kalimantan	PKW Estate	Shrub	Others land use

Besides that, the status of the land as a whole is not a forest area but APL, customary land and transmigration area. Transmigration policy (1905-1940/ Dutch colonial period; and 1969- present) – around 8.94 million land (Most of the part of the forest) were conversed into crops plantation, especially in order to increase rice production. Meanwhile, Forest use rights (HPH) using its systematic logging as it was started since the 1970s (1985 – 1997 as the peak) with degradation rate around 1.26 million per year which made it as a major acceleration of the deforestation as it turned out to be an access for the newcomers. The policy about increasing deforested forest production which allowing investors to open the land (especially for the degraded land) to open oil palm plantation and industrial forest started in the late 1980s).

3.2. Impact on diversity of species

There are 11 species of mammals found in the baseline area of each location. The highest number of species mammals is found in PT KPR with five species, and the lowest is in PT KJNP (**S = 0**). PT

KPR also has the highest species richness index compared to the other locations (Dmg = 2.23). Three locations have zero of species richness index, namely PT GDH, KJNPE and PKWE. All species of mammals are shown in table 2.

Table 2. List of mammals species before OPP establishment

No	Scientific name	PT KPR	PT SAR	PT AMA	PT GDH	PT BLP	KJNPE	PKWE	PT WMA
1	*Callosciurus notatus*	2	5	0	0	2	0	3	5
2	*Macaca fascicularis*	1	0	6	25	0	0	0	0
3	*Macaca nemestrina*	0	1	0	0	0	0	0	0
4	*Muntiacus muntjak*	1	0	0	0	0	0	0	0
5	*Pongo pygmaeus*	0	0	0	0	1	0	0	0
6	*Tracypithecus auratus*	0	0	10	0	0	0	0	0
7	*Prionailurus bengalensis*	1	0	0	0	0	0	0	0
8	*Rattus tiomanicus*	0	0	0	0	0	0	0	1
9	*Sus scrofa*	1	1	0	0	0	0	0	0
10	*Tupaia minor*	0	0	0	0	1	0	0	0
11	*Tupaia splendidula*	0	2	1	0	0	0	0	0
	Number of individuals	**6**	**9**	**17**	**25**	**4**	**0**	**3**	**6**
	Number of species (S)	**5**	**4**	**3**	**1**	**3**	**0**	**1**	**2**

The species of mammals found in the OPP area are more than in the baseline area. There are 12 species of mammals are found in the OPP area. PT WMA and PT BLP have the most number of species, namely six species. While PT GDH has the least number of mammals which are two species, the highest species richness index was found in PT BLP (Dmg = 1.47) and the lowest in PT GDH (Dmg = 0.51). The species of mammals found can be seen in table 3.

Table 3. List of mammals species after OPP establishment

No	Scientific Name	PT KPR	PT SAR	PT AMA	PT GDH	PT BLP	KJNPE	PKWE	PT WMA
1	*Aonyx cinereus*	0	0	0	0	0	0	8	1
2	*Callosciurus notatus*	2	0	0	5	14	1	7	6
3	*Callosciurus prevostii*	0	0	0	0	1	0	0	0
4	*Macaca fascicularis*	0	0	0	0	0	36	26	28
5	*Macaca nemestrina*	0	1	0	0	0	0	0	0
6	*Paradoxurus hermaphroditus*	1	0	3	0	0	0	0	0
7	*Presbytis cristata*	0	0	0	0	0	6	0	1
8	*Prionailurus bengalensis*	0	4	0	2	1	0	1	1
9	*Rattus tiomanicus*	0	9	11	0	12	0	4	17
10	*Sus barbatus*	0	0	0	0	1	0	0	0
11	*Sus scrofa*	1	8	1	0	0	0	0	0
12	*Tupaia minor*	0	0	0	0	1	0	0	0
Number of individuals		**4**	**22**	**15**	**7**	**30**	**43**	**46**	**54**
Number of species (S)		**3**	**4**	**3**	**2**	**6**	**3**	**5**	**6**

For the origin of land cover types in the form of shrubs or rubber plantations, the impact of oil palm plantations is positive because it increases the number of mammal species. Whereas for the origin of land cover types in the form of secondary forests, the existence of oil palm plantations has caused unchanged (2 locations), increased (2 locations) and decreased (1 location) number of mammals. Seen from species richness index, the highest change in value is found in PKWE with Dmg = +1.04. Changes in the number and value of mammal species can be seen in table 4.

Table 4. Comparison of ecological variable value of mammal species diversity

Types of Land Cover Before Oil Palm Plantations Establish	Study Location		Impact Of Oil Palm Plantations on Mammals Species Diversity
Secondary forest	Riau	PT. KPR	S= Decreasing 2 species (from 5 to 3 species) Dmg= - 0,79
		PT. SAR	S= Does not changes (4 species) Dmg= - 0.39
		PT. AMA	S= Does not changes (3 species) Dmg= + 0.03
		PT. GDH	S= Increasing 1 species (from 1 to 2 species) Dmg= + 0.51
	Central Kalimantan	PT. BLP	S= Increasing 3 species (from 3 to 6 species) Dmg= +0.03
Shrub	Riau	KJNP Estate	S= Increasing 3 species (from 0 to 3 species) Dmg= +0.53
	West Kalimantan	PKW Estate	S=Increasing 4 species (from 1 to 5 species) Dmg= +1.04
Rubber plantations	South Sumatera	PT. WMA	S=Increasing 4 species (from 2 to 6 species) Dmg= + 0.70

There are two areas that have declining species richness index (PT KPR and PT SAR) (table 4). Both locations have a baseline in the form of a secondary forest. Similar to the research of [7] who also found that changes from secondary forest to oil palm plantations had a negative impact than shrub or rubber plantation areas. One of the factors that influence the decline in the species richness index is the difference in habitat structure. As stated by [10] that the high diversity of mammal species influenced by the high diversity of vegetation species. Secondary forest areas have more diverse species of vegetation and more complex strata than OPP area. This is similar to [11] statement that forested areas have a high diversity of vegetation. Variations in strata and the number of species of vegetation greatly affect the presence of mammals because mammals need trees that can be used as a place of refuge and become a source of food. According to [12] each stratum can support the life of a certain species of wildlife. [13] stated an increase in the number of different habitats led to an increase in species diversity.

The expansion of this large-scale crop monoculture has provided an unnatural environment in which many animals have rapidly adapted and become pests. One of them is the rat that is adaptable, aggressive, highly competitive, fierce and difficult to control [14]. Based on Tables 2 and 3, *Rattus tiomanicus* is gained after OPP established in 4 areas (SAR, PT AMA, PT BLP and PKWE). Research by [15] also found that *Rattus tiomanicus* dominated 68% in 28 oil palm plantations that researched. [16] explained that habitat change had occurred in *Rattus tiomanicus* which is currently rarely found in mature forests. Habitation of this biome appears to have effectively 'pre-adapted' the species for life in oil palm plantation [17]. [18] also stated that rats tend to increase in the plantation area. One of the factors that make *Rattus tiomanicus* increase in oil palm plantations is the abundance of feed. [19] explained that the uncontrolled population of *Rattus tiomanicus* reached 600 individuals per hectare and caused damage to oil palm plantations. The study of [20] states that many adult *Rattus tiomanicus* foreage oil palm fruit. Another study of Rattus tiomanicus in an oil palm plantation also revealed that on average about 96% of their stomach contents was made up of oil palm fruit [21]. Besides, *Rattus tiomanicus* also forage for insect, molluscs and other invertebrates among the ground vegetations of oil palm plantation fields [22]. [23] stated that rodent relations and the abundance of oil palm fruit were considered as numerical responses, which meant that the more the number of oil palm fruits, the more the number of rats. Furthermore, the study also presents correlative evidence that at least 1 species rats in oil palm plantations increase in abundance with food availability in a number of circumstances.

Based on table 5, the total number of herpetofauna found before OPP was 29 species (16 amphibians and 13 reptiles). PT BLP has the highest number of species (S = 12) and the highest species richness index (Dmg = 3.88). The lowest number of species and species richness index in PT GDH (S = 0; Dmg = 0).

Table 5. List of herpetofauna species before OPP establishment

No	Scientific name	PT KPR	PT SAR	PT AMA	PT GDH	PT BLP	KJNPE	PKWE	PT WMA
Amphibi									
1	*Ingerophrynus biporcatus*	0	0	0	0	1	0	0	1
2	*Pseudobufo subasper*	0	0	0	0	1	0	0	0
3	*Fejervarya cancrivora*	0	0	1	0	1	4	0	3
4	*Fejervarya limnocharis*	0	0	1	0	5	7	2	2
5	*Microhyla achatina*	1	0	0	0	0	0	0	0
6	*Microhyla heymonsi*	1	0	0	0	0	0	0	0
7	*Microhyla sp*	0	0	0	0	0	4	0	0
8	*Amnirana nicobariensis*	0	0	0	0	0	1	0	1
9	*Chalcorana raniceps*	0	1	0	0	0	0	0	0
10	*Huia sumatrana*	0	0	0	0	0	1	0	0
11	*Hylarana erythraea*	3	0	2	0	0	1	0	4
12	*Limnonectes paramacrodon*	0	0	0	0	0	0	0	1
13	*Limnonectes sp.*	0	0	0	0	3	0	0	0
14	*Pulchrana baramica*	0	0	0	0	1	1	0	0
15	*Pulchrana glandulosa*	0	0	1	0	0	1	5	0
16	*Polypedates leucomystax*	0	1	1	0	2	4	0	0
17	*Polypedates macrotis*	0	0	0	0	1	0	0	0
Reptile									
18	*Boiga dendrophila*	0	0	1	0	0	0	0	0
19	*Dendrelaphis caudolineatus*	0	1	0	0	0	0	0	0
20	*Dendrelaphis pictus*	0	0	0	0	0	0	0	1
21	*Bungarus pasciatus*	0	0	0	0	0	0	0	1
22	*Cyrtodactilus marmoratus*	0	0	0	0	0	6	0	0
23	*Hemidactylus frenatus*	0	0	0	0	0	1	0	1
24	*Dasia vittata*	0	0	0	0	1	0	0	0
25	*Eutropis multifasciata*	0	0	0	0	2	0	0	0
26	*Eutropis rudis*	0	0	0	0	2	0	1	2
27	*Eutropis sp*	0	0	0	0	0	2	0	0
28	*Takydromus sexlineatus*	0	0	0	0	1	0	0	0
29	*Varanus salvator*	0	0	0	0	1	0	0	0
	Number of individuals	**5**	**3**	**7**	**0**	**22**	**33**	**8**	**17**
	Number of species	**3**	**3**	**7**	**0**	**12**	**12**	**3**	**10**

There is number of herpetofauna found in the OPP area than baseline area. As seen in table 6, there are 16 species of reptiles and 24 species of amphibians. The highest number of species was found in KJNPE, and PT WMA (S = 19) and the highest species richness index was found at PT WMA (Dmg = 3.83). PT AMA has the least number of species (S = 12) and the lowest species richness index (Dmg = 2.62).

Table 6. List of herpetofauna species after OPP establishment

No	Scientific name	PT KPR	PT SAR	PT AMA	PT GDH	PT BLP	KJNPE	PKWE	PT WMA
Amphibi									
1	*Duttaphrynus melanostictus*	2	6	10	3	0	5	0	4
2	*Ingerophrynus biporcatus*	1	1	0	1	7	0	0	5
3	*Phrynoidis aspera*	7	0	0	1	0	1	0	0
4	*Pseudobufo subasper*	0	0	0	1	0	0	0	0
5	*Fejervarya cancrivora*	0	2	4	0	19	6	4	34
6	*Fejervarya limnocharis*	9	8	21	44	12	18	30	21
7	*Leptobrachium abbotii*	0	0	0	0	0	1	1	0

No	Scientific name	PT KPR	PT SAR	PT AMA	PT GDH	PT BLP	KJNPE	PKWE	PT WMA
8	*Leptobrachium hasseltii*	0	0	0	0	0	0	0	2
9	*Leptobrachium nigrops*	1	0	0	0	0	0	0	0
10	*Kaloula pulchra*	0	1	0	2	0	9	0	0
11	*Microhyla achatina*	4	2	9	4	0	0	0	0
12	*Microhyla heymonsi*	6	9	1	9	0	31	0	0
13	*Polypedates leucomystax*	4	10	13	7	9	2	7	6
14	*Polypedates macrotis*	0	0	0	0	2	0	0	0
15	*Amnirana nicobariensis*	2	0	1	2	3	5	8	3
16	*Chalcorana chalconota*	0	4	3	0	0	2	0	2
17	*Chalcorana raniceps*	0	1	0	0	0	0	0	0
18	*Huia sumatrana*	0	0	0	0	0	19	0	0
19	*Hylarana erythraea*	4	0	0	6	16	3	21	13
20	*Limnonectes malesianus*	0	1	1	0	0	0	0	0
21	*Limnonectes paramacrodon*	0	0	0	0	0	0	1	1
22	*Limnonectes sp.*	0	0	0	0	9	12	0	0
23	*Pulchrana baramica*	1	0	2	4	0	5	1	4
24	*Pulchrana glandulosa*	0	0	1	0	0	0	9	6
Reptile									
25	*Dendrelaphis caudolineatus*	0	0	1	0	0	0	0	0
26	*Dendrelaphis pictus*	0	1	0	1	0	0	2	2
27	*Oligodon purpurascens*	0	0	0	0	0	0	1	0
28	*Bungarus candidus*	0	0	0	0	0	0	0	1
29	*Bungarus pasciatus*	0	0	0	0	0	0	0	1
30	*Naja sumatrana*	0	0	0	1	0	0	0	1
31	*Cyrtodactilus marmoratus*	0	0	0	0	0	2	0	0
32	*Hemidactylus frenatus*	1	1	0	0	6	2	0	1
33	*Hemidactylus platyurus*	1	0	0	1	0	0	0	0
34	*Homalopsis buccata*	0	0	0	1	0	0	0	0
35	*Malayopython reticulatus*	0	2	0	0	0	1	0	0
36	*Eutropis multifasciata*	0	0	0	0	3	0	0	2
37	*Eutropis rudis*	0	0	0	0	5	0	4	0
38	*Eutropis sp*	0	0	0	0	0	1	0	0
39	*Takydromus sexlineatus*	0	0	0	0	0	1	9	1
40	*Varanus salvator*	0	1	0	0	1	0	0	0
	Number of individuals	**43**	**50**	**67**	**88**	**92**	**126**	**98**	**110**
	Number of species	**13**	**15**	**12**	**16**	**14**	**19**	**13**	**19**

Similar to changes in the number and species richness index of mammal species, the amount and species richness index of herpetofauna also have positive changes in the area of Shrub and Rubber plantation. Unlike the case with the secondary forest area, the comparison between positive and negative changes are balance (2 location positive and 2 location negative). Seen in table 7, the highest change in value is found in PT KPR with Dmg = +1.95.

Table 7. Comparison of ecological variable value of herpetofauna species diversity

Types of Land Cover Before Oil Palm Plantations Establish	Study Location		Impact of Oil Palm Plantations on Herpetofauna Species Diversity
Secondary forest	Riau	PT. KPR	S= Increasing 9 species (from 4 to 13 species) Dmg= +1.95
		PT. SAR	S= Increasing 11 species (from 3 to 14 species) Dmg= +1.76
		PT. AMA	S= Increasing 10 species (from 3 to 13 species) Dmg= -0.47
	Central Kalimantan	PT. BLP	S= Increasing 3 species (from 12 to 15) Dmg= -1.01

Types of Land Cover Before Oil Palm Plantations Establish	Study Location		Impact of Oil Palm Plantations on Herpetofauna Species Diversity
Shrub	Riau	KJNP Estate	S= Decreasing 7 species (from 19 to 12 species) Dmg= +0.58
	West Kalimantan	PKW Estate	S= Increasing 10 species (from 3 to 13 species) Dmg= +1.66
Rubber plantations	South Sumatera	PT. WMA	S= Increasing 8 species (from 11 to 19 species) Dmg= +0.65

The change from rubber plantation and shrub to OPP has a positive impact, namely an increase in the value of species diversity. PT AMA and PT BLP (secondary forest) have a decrease in the species richness index, even though the number of species increases (Table 7). This can be caused by the influence of the number of individuals. Similarly, the study of [24] also found a decrease in species richness index when the number of species increased. [24] also noted that the margalef index was not only influenced by the number of species but also by the number of individuals. Reference [25] explains that the margalef index has high sensitivity and the ability to respond to the difference in species so that the dominance of species will affect the values of the index. The existence of sources of water and canopy cover are the factors that most influence the presence of herpetofauna, because both of these affect the humidity in a habitat. According to [26] the existence of water sources is a limiting factor for animals that are dependent on water, such as herpetofauna. Secondary forest areas in PT AMA and PT BLP have very abundant water sources because they are close to the river flow. It is different from the condition of secondary forests in PT KPR and PT SAR which only have water sources from the trench.

3.3. Impact on community of species

The presence of oil palm plantations has a change in species composition with a range of 50 to 100% for mammals and 23 to 76% for herpetofauna. PT AMA, PT GDH and KJNPE had the highest of dissimilarity value of 100%, meaning that all species found in the baseline area were not found in the OPP area. While the highest dissimilarity value for herpetofauna is found in PT SAR, all dissimilarity values can be seen in table 8.

Table 8. Dissimilarity value of mammals and herpetofauna before and after OPP establishment

Types of land cover before oil palm plantations established.	Study location		Impact of Oil Palm Plantation on Mammal and Herpetofauna Species Composition (Dissimilarity)	
			Mammal	Herpetofauna
Secondary forest	Riau	PT. KPR	50 %	53 %
		PT. SAR	50 %	76 %
		PT. AMA	100 %	43 %
		PT. GDH	100 %	-
	Central Kalimantan	PT. BLP	56 %	26 %
Shrub	Riau	KJNP Estate	100%	23 %
	West Kalimantan	PKW Estate	67%	62.5 %
Rubber plantations	South Sumatera	PT. WMA	50%	27 %

The location of secondary forest in PT KPR and rubber plantation of PT WMA has a distance that is very close or adjacent to the OPP area. This is thought to be one of the factors affecting the similarity between the two habitats. [27] said that the similarity index is influenced by the distance between habitat. Besides distance, [28] explain the same vegetation composition and other environmental factors also affect the similarity index. In addition, the presence of HCV areas can also increase the number of mammal and herpetofauna species. Based on [29] who found that similarity between baseline and OPP was influenced by the presence of HCV that was part of the OPP. PT WMA has a similarity value of 50%, and 4 of 6 species of mammals are found in the HCV area. The presence of HCV in oil palm plantations has an important role in maintaining and increasing the diversity of mammal and herpetofauna species. [27] said that HCV is a place of refuge especially for arboreal animals such as primates. As stated in [7] found the HCV area had a high variation of vegetation and that the vegetation was used as a shelter and feed source for mammals. Similar to the location of KJNPE which has the lowest dissimilarity value of 23% or 77% types of herpetofauna can be found in the shrub and OPP area. In general, the number of species found in the baseline area can still be found in the OPP area, and even there are some new species found in the OPP area. This proves that the OPP area can still be a suitable habitat for herpetofauna.

3.4. Impact on loss/gain of species

Figure 1 presents data on the number of species loss and gain due to the conversion of secondary forest to oil palm plantations. The highest species of losses are found in PT KPR and PT AMA with a total loss of 3 species mammals, while the highest number of gains is in PT BLP with a total gain of 4 species. In PT KPR, there are three loss species (*Macaca fascicularis, Muntiacus muntjak* and *Prionailurus bengalensis*). Meanwhile, the species loss in PT AMA are *Macaca fascicularis, Trachypithecus auratus* and *Tupaia spendidula.* Similar to the research of [30] who found *Mutiacus muntjak* only in wooded areas and did not find it in the OPP area. [31] explained that *Mutiacus muntjak* is one species of mammal that is negatively affected by forest fragmentation or conversion. This mammals is a diurnal active [32], so that *M. muntjak* activity must be disrupted by the activities of employees working in oil palm plantations. Naturally, wildlife will avoid if there human activity [33].

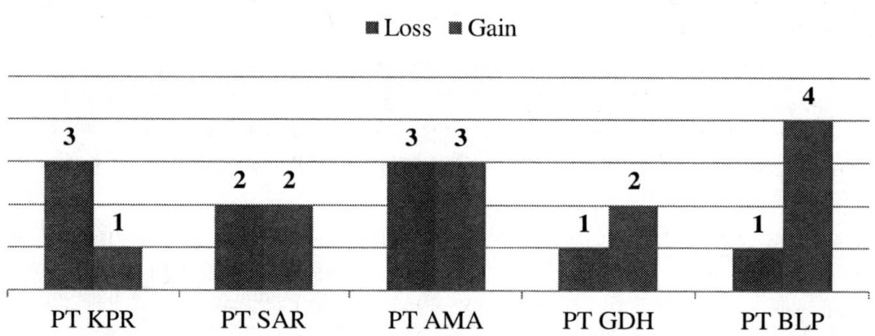

Figure 1. Comparison of mammal species loss and gain percentage from secondary forest

In addition, the species loss is caused in the secondary forest area has a variety of vegetation and more food sources. Reference [34] cites that low biodiversity in oil palm plantation area is caused by monoculture and there are not major components of forest vegetation. Research [35] proved that mammals of primates such as *Macaca fascicularis* were more commonly found in secondary forests because they had abundant feed sources. *Artocarpus interget* is one species that dominated in secondary forest PT KPR and PT AMA. According to the results of [36] cempedak is one of 9 types of primate feed.

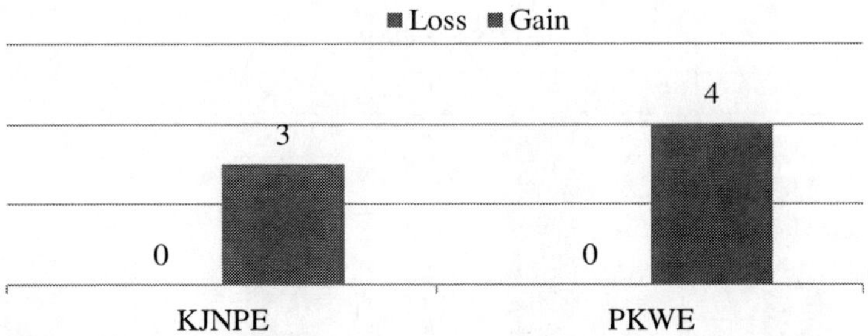

Figure 2. Comparison of mammals species loss and gain percentage from shrubs

The conversion from shrub into OPP does not cause loss of mammal species (figure 2). In KJNPE, there are three species gain, namely *Callosciurus nottatuts*, *M. fascicularis* and *T. auratus*. While in PKWE there are four species gain, namely *Prionailurus bengalensis, M. fascicularis, Rattus tiomanicus* and *Aonyx cinerus*. Plantain squirrel or *Callosciurus nottatus* is a species of mammal that has high adaptability, wide distribution and tolerance to habitat changes [7] so that this species can be found in the OPP area. Furthermore, the discovery of *P. bengalensis* was caused because *R.tiomanicus* was found in the OPP area which is the main food *P. bengalensis*. This type of mammal is often used as a biocontrol in rat pests on oil palm plantations [37]. [38] said carnivorous species like a *P. bengalensis* have a role a balancing ecosystem mainly as a predator of a small animal such as *R. timonanicus* and *C. nottatus*.

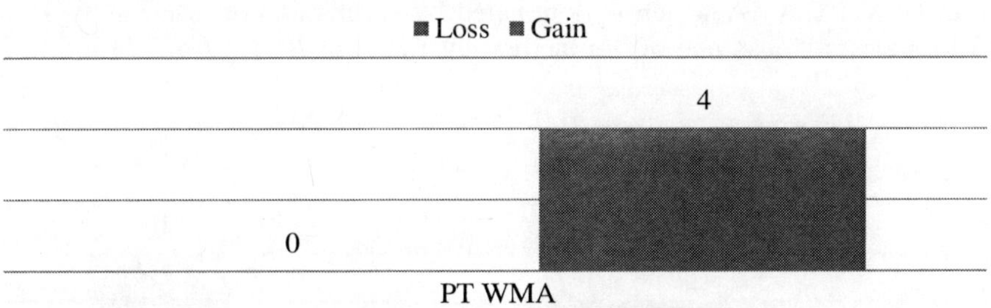

Figure 3. Comparison of mammals species loss and gain percentage from rubber plantation

As with shrub, the conversion from rubber plantation to OPP does not cause loss of species (figure 3). There are three species of mammals found in OPP PT WMA (*P. bengalensis, T. auratus, M. fascicularis* and *Aonyx cinerus*). Like PKWE, *Aonyx cinerus* was also found in the PT WMA OPP area. Large and deep canals surround the PT WMA OPP area. This canal contains peat water which lives in various kinds of fish such as *Clarias nieuhofii, Mystus sp, Channa striata* and *Channa pleurophthalma*. [39] stated the food composition of small claw otters is Pisces and Mollusca as the main food, Insecta and Amphibians as the second food, Crustacea and Mammals as complementary foods. The study of [40] and [41] also found that *A.cinereus* was only found in oil palm plantation areas and nesting inside of the canal.

The effect of conversion from secondary forest to OPP area tends to have a positive impact. All locations that have higher number of species gain than the number of species loss (figure 4). PT SAR has the highest number of species gain which is 12 species. While the lowest is PT BLP with four species gain. PT KPR and PT BLP have not a loss species. Different from PT SAR and PT AMA which have one species loss.

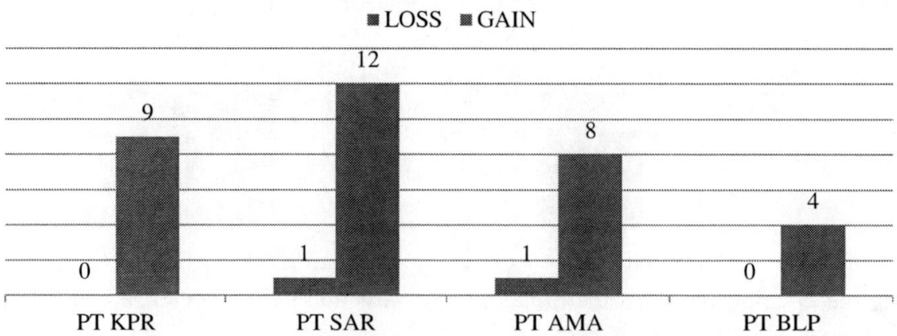

Figure 4. Comparison of herpetofauna species loss and gain percentage from secondary forest

The secondary forest of PT SAR only found three species of herpetofauna, while the OPP area had 15 species of herpetofauna. The small amount of herpetofauna found in the secondary forest of PT KPR because this area has a water source with a low pH which is 4. According to [42] amphibians will grow optimally if the support has a pH of 6-8. In contrast to [43] states that Anura can tolerate a pH range of water from 4.3 to 7.5. Physical factors such as pH influence anura growth [44] The only loss species in OPP PT AMA is *Boiga dendrophila.* Gold ringed cat snakes or *B. dendrophila* are an arboreal snake. This snake commonly found in habitat dominated by trees with lots of twigs [45]. This is related to the conditions of secondary forest of PT AMA which are dominated by Dipterocarpacea and Anacardiaceae (*Shorea pauciflora, Shorea teysmanniana* and *Gluta renghas*). It is different with the condition in HCV PT AMA which is dominated by shrubs and does not have strata such as the secondary forest area. This was one of the factors not found in *B. dendrophila* in the PT AMA OPP area.

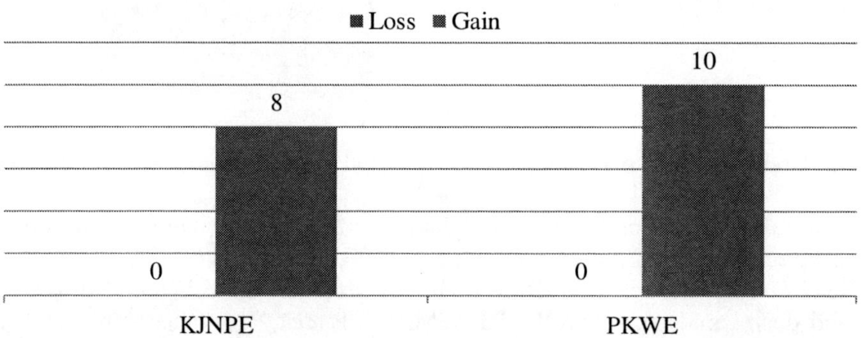

Figure 5. Comparison of herpetofauna species loss and gain percentage from shrub

Similar to the secondary forest area, the shrub area also has a high number of species gain (8 species in the KJNPE and ten species in the PKWE). The factors that affect loss and gain species in KJNPE and PKWE are canopy cover conditions. Shrub has a canopy cover, and OPP has a canopy cover. This can affect the humidity and temperature in the area. The same thing was also explained by [46] that canopy cover affects the temperature and humidity of a habitat. Herpetofauna is a poikilotherm whose body temperature depends on the temperature of the environment [47]. Research by [48] states that herpetofauna prefers habitat with a closed canopy because it has high humidity. [43] in his research obtained the optimal humidity range for herpetofauna between 84% and 99%.

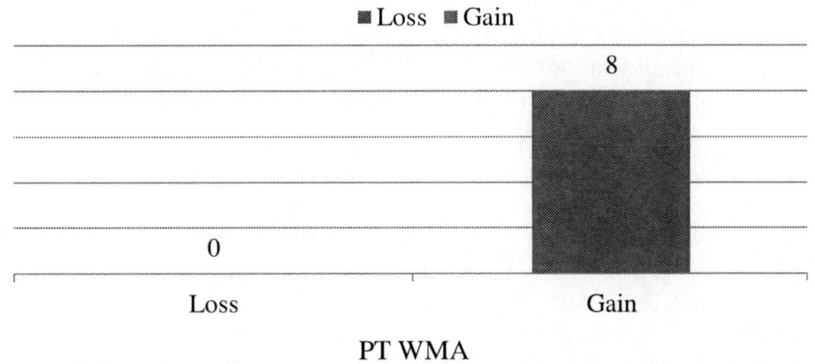

Figure 6. Comparison of herpetofauna species loss and gain percentage from shrub

Conversion from rubber plantation to oil palm plantations also have a positive impact. PT WMA has eight species gain and no one species loss due to land conversion. Species gain are *Duttaphrynus melanostictus, Pulchrana baramica, Chalcorana chalconota, Polypedates leucomystax, Leptobrachium hasseltii, Naja sumatrana and Cancdidus bungarus.* The number of species gains of amphibians is many because the OPP area has more water sources than the rubber plantation area. Anura is always associated with water and anura requires water to lay eggs or breed [49]. In addition, [50] amphibians are always associated with water, because water can maintain changes in body temperature.

4. Conclusion

All the oil palm plantations being studied was converted from the secondary forest with non-forest area status. The presence of oil palm plantations does not always cause a decline in the diversity of species of mammals and herpetofauna. Oil palm plantations in all study areas resulted in a change in the species composition of both mammals and herpetofauna.

Acknowledgments

Our sincere thanks go to the chief and staff of oil palm plantation company in our study (PT KPR, PT SAR, PT AMA, PT GDH, KJNPE, PKWE and PT WMA) for permission, support and facilities that had been provided during field activities.

References

[1] Azzahra F, Dharmawan AH, Pandjaitan. 2017. Perempuan dan resiliensi nafkah rumahtangga petani sawit: analisis dampak ekspansi perkebunan kelapa sawit di provinsi jambi. *Jurnal Sosiologi Pedesaan* 23-35

[2] GAPKI. 2017. *Reflection of palm oil industry 2016 prospect 2017*. Jakarta (ID): Indonesia Palm Oil Entrepeneurs Association (GAPKI)

[3] Direktorat jendral perkebunan. 2016. *Statistik Perkebunan Indonesia 2015-2017*. Jakarta (ID): Kementerian Pertanian Republik Indonesia

[4] Buckland H. 2005. The oil for ape scandal: How palm oil is threatening orang-utan survival. London

[5] Sheil D, Casson A, Meijaard E, Van Noordwijk M, Gaskell J, Sunderland-Groves J, Kanninen M. 2009, The impacts and opportunities of oil palm in Southeast Asia. Africa. http:// doi.org/10.17528/cifor/002792

[6] Novriyanti N. 2016. Take control of forest pests [Internet]. [downloaded 2018 Nov 25]. Available on: https://www.researhgate.net/publication/307966325_ kendalikan_hama_hutan

[7] Santosa Y, Dalimunthe Y, Purnama Sari I. 2017. Impact of oil palm plantation on mammal and herpetofauna diversity: case study in South Sumatra Province. *International Journal Of*

Environmental And Ecological Engineering. **11** (9)
[8] Kusrini MD, LF Skerratt, S Garland, L Berger, W Endarwin. 2008. Chytridiomucosis in frog Mount Gede Pangrangro Indonesia. *Diseases of Aquatic Organism*: 187-194
[9] Qurniawan TF. 2015. Model of microclimatic influence on fluctuation of herpetofauna diversity in camps area. *Jurnal Teknosains*. **4**(2): 172-178
[10] Kartono AP. 2015.Diversity and abudanceof mammala in PT Sukses Tani Nusasubur oil palm estate, East Kalimantam. *Media Konservasi* **20**(2): 85-92
[11] Muin A. 2013. Pengusaha Perkebunan Kelapa sawit berwawasan konservasi [Disertation]. Bogor(ID): Bogor Agricultural University
[12] Alikodra HS. 2002. *Pengelolaan Satwa Liar Jilid I.* Bogor(ID): Yayasan Penerbitan Fakultas Kehutanan IPB
[13] MacArtur RH and MacArtur JW. 1961. On bird species diversity. *Ecology* 42:594-598
[14] Baker S, Singleton G, Smith R. 2007. The nature of the beast: Using biological processes in vertebrate pest management. In: McDonald DW, Service K, eds. Key Topics in Conservation Biology. Blackwell Publishing, Malden, pp. 173–85
[15] Hafidzi MN and Saayon MK. 2001. Status of rat Infestation and recent control strategies in oil palm plantations in penisular malaysia. *Pertanikan J. Trop. Agric.Sci*. **24** (2) : 109-114
[16] Harrison J. L. (1957) Habitat of some Malayan rats. Proc. Zool. Sot. Land. 28. 1-21
[17] Wood B. J. (1994) Rodents in agriculture and forestry. In Rodent Pests and Their Control, eds A. P. Buckle and R. H. Smith, pp. 45-83. CAB International, Oxon, UK
[18] Kamim ABM. 2018. Perebutan ruang kehidupan dan gangguan terhadap animal rights : studi atas konflik satwa-manusia sebagai implikasi dari ekspansi perkebunan sawit di indoneisa. Multidisipliner mahasiswa indonesia. **1**(2): 199-217
[19] Wood BJ. 2001. Rat control in oil palm plantation and rice field. *Pesticide outlook*. 71-74
[20] Lukito PA and Sudrajat. 2017. Pengaruh kerusakan buah kelapa sawit terhadap kandungan free acid dan rendemen CPO di kebun Talisayan I Berau. Jurnal Agrolestari. **5** (1): 37-44
[21] Wood BJ and Liau SS. 1984. A long-term study of Rattus tiomanicus populations in an oil palm plantation in Johore, Malaysia: II. Recovery from control and economic aspects. *Journal of Applied Ecology* . **21** (1) : 465–472
[22] Buckle AP, Chia TH, Fenn MGP and Visvalingam M. 1997. Ranging behaviour and habitat utilisation of Malayan wood rat (*Rattus tiomanicus*) in an oil palm plantation in johore, Malaysia. *Crop Protection.***16**(5) : 467-473
[23] Puan CL, Goldizen AW, Zakaria, Hafidzi MN and Baxter GS. 2011. Relationship among rat number, abudance of oil palm fruit and damage levels to fruit in an oil palm plantation. Integrative zoology. **6**(2)
[24] Kwatrina RT, Santosa Y, Bismark M, Santoso N. 2018. The impact of oil palm plantation establishment on the habitat type, species diversity and feeding guild of mammals and herpetofauna. *Biodiversitas* **19** (4):1213-1219
[25] Magurran AE. 1988. *Ecological Diversity and Its Measurement*. New Jersey (US): Princeton University
[26] Odum. 1971. *Fundamental of Ecology*. Tokyo (JP): W.B Saunders Company Ltd
[27] Nugroho SS and Santosa Y. 2018. The estimation loss and gain of mammals species diversity due to oil palm plantation: a case study of PT BMPE, Riau. Presented in International Conference on Natural Resources Management (NRM) of Consortium Studies on Smallholder Palm oil (CSSPO) 2018, Serawak 10th july
[28] Kendeigh SC. 1980. *Ecology with special reference to animal and men*. New Delhi (IN) : Prentice Hall of India Private limited
[29] Santosa Y, Sunkar A, Purnamasari I, Yohanna. 2017. Sejarah Asal Usul Status , Riwayat Penggunaan lahan dan Keanekaragaman Hayati Kebun Kelapa Sawit PT Waimusi Agroindah. Bogor (ID): Bogor Agricultural University
[30] Arief H. 2013. Diversity of animals and protection status in oil palm plantation area sungai terik

estate PT Nusa Lestari, Seruyan Paser, East Kalimantan. *Media Konservasi* **18** (2): 78-82

[31] Hidayat W, Abdullah, Khairil. 2018. Estimasi populasi gajah sumatera berdasarkan metode defekasi di kawasan hutan peunaron kabupaten aceh timur. *Jurnal Edubio Tropika* **6**(1) : 1-72

[32] Payne J, Framcis CM, Philips K., Kartikasari SN. 2000. *Panduan Lapang Mamalia di Kalimantan, Sabah, Serawak, Brunei Darussalam*. Jakarta(ID): Wildlife Conservation Society-Indonesia Program

[33] Corbat GB and Hill JY. 1992. *Mammals of Indo-Malayan Region: A Sysmatic Review*. Oxford(UK): Oxford University Press

[34] Danielsen F, Beukema H, Burgess ND, Parish F, Brulh CH, Donal PF, Murdiyarso D, Pahlan B, Reijnders L. Struebig M and Fitzherbet EB. 2009. Biofuel plantation on forested land : double jeopardy for biodiversity and climate. *Conservation Biology* **23**(2): 248-258

[35] Chivers DJ. 1980. Malayan Forest Primate: 10 Years on Studi in TropicalRrainforest. New York(US): Plenum Press

[36] Sumarni S. 2016. Studi jenis tumbuhan pakan kelasi (*Presbitis rubicunda*) pada kawasan hutan Wisata Baning Kabupaten Sintang. *Piper* No.3

[37] Putra W. 2016. Pola aktivitas harian macan akar (Prionailurus bengalensis) pada musim hujan di kawasan perkebunan PT Surya Sawit Sejati Kalimantan Tengah (essay). Padang (ID): Andalas University

[38] Maryanto I, Noerdjito M, Partomihardjo T. 2012. Mount Slamet Ecology: Geology, Climatology, Biodiversity and Social Dynamic. Jakarta (ID): LIPI Press

[39] Ananta SME. 2017. Komposisi makanan berang-berang cakar kecil (*Aonyx cinereus*) pada perbedaan fitur lokasi penemuan fese dikecamatan lubuk alung kabupaten padang pariaman (essay). Padang (ID): Andalas University

[40] Abdullah, S Ali, Ar-Rasyid UH. 2015. Habitat Characteristic of small – clawed Otter (*Aonyx cinereus*) in Ujong Nga, Samatiga, West Aceh. *Jurnal Natural* **15** (1):2015

[41] Purba HS. 2017. Perbandingan keanekaragaman jenis mamalia antara lahan pasca terbakar dan tidak terbakar di perkebunan sawit di PT Waimusi Agroindah Sumatera Selatan (essay). Bogor(ID): Bogor Agricultural University

[42] Hartanto R. 2003. Biology Kodok Lembu atau Bullfrog (*Rana Catesbeiana*). Semarang (ID): Diponegoro University

[43] Darmawan. (2008). Keanekaragaman Amfibi di Berbagai Tipe Habitat: Studi di Eks-HPH PT Rimba Karya Indah Kab. Bungo Prov. Jambi [essay]. Bogor(ID): Bogor Agricultural University

[44] Fuyi Y, Qingchun S, Jinglin L, Guoshuang C. 2004. Effects of Salinity and Alkalinity on Tadpole and Abnormal Young Frog of Rana chensinensis. *Journal of Northeast Forestry University* 1-11.

[45] Rodda HG, Fritts TH, McCoid MJ, Campbell EW. 1999. An Overview of The Biology of the Brwon Treesnake (Boiga irregularis) a Costly Introduced Pest on Pacific Island. New York(US): Cornell University Press

[46] Putra PT. 2011. Evaluasi kenyamaan pada beberpa taman kota di Jakarta Pusat (essay). Bogor(ID): Bogor Agricultural University

[47] Juniarmi R, Nurdin J, Zakaria IJ. 2014. Population density and lizard distribution (*Mabuya multifasciata*) of small island in Padang city. *Jurnal Biologi Universitas Andalan*. **3**(1): 51-56

[48] Halliday T and Adler K. 2000. *The Encyclopedia of Reptiles and Amphibians*. New York(US): Fact on file inc.

[49] Iskandar DT. 1998. *Amfibi Jawa dan Bali LIPI-Seri Panduan Lapang Bogor*. Bogor (ID) : Puslitbang Biologi Lipi.

[50] Yanuarefa MF, Hariyanto G, Utami J. 2012. *Panduan Lapang Herpetofauna (Amfibi dan Reptil) Taman Nasional Alas Purwo*. Banyuwangi (ID): Balai Taman Nasional Alas Purwo

IOP Conf. Series: Earth and Environmental Science **336** (2019) 012028 doi:10.1088/1755-1315/336/1/012028

Does oil palm plantation establishment cause loss/gain of bird species? (Case study: KGP oil palm plantation, West Kalimantan)

Diah Eres Dwi Karuniyanti[1], Yanto Santosa[1]

[1]Forest Resource Conservation and Ecotourism Department, Faculty of Forestry, Bogor Agricultural University, Bogor, Indonesia

*Email: diah.eres@gmail.com

Abstract. Oil palm plantation expansion in Indonesia always been associated as the cause of deforestation and loss of biodiversity. This research was conducted aiming to verify that allegation. The study was conducted from July to August 2018 and carried out in an independent oil palm plantation company, KGP in West Kalimantan. To identify the loss and gain of the bird species, an observation about the bird diversity was conducted. Five transects were laid out in each land covers (4 in the oil palm plantation and 1 in the shrub [baseline]). The analysis showed that the bird species found in the shrub was slightly smaller (22 species) then the species found in the oil palm plantation land covers (27 species). A total of 3 species were found only in shrubland cover which later is assumed as the loss. On the other side, 18 species were only found in the oil palm plantation assumed as a gain. These could be happening as 60% of the bird species composition found in shrub were almost the same as those in the High Conservation Value (HCV) land covers which belong to oil palm plantation land cover. Hence, the existence of oil palm plantation did cause a loss to the bird species but also happen to gain some other new species.

1. Introduction

Oil palm plantation has been covering large scale of areas in Indonesia plantation history [1]. This has been led oil palm plantation as one of the most promising sources of Indonesia's economic growth as it has been contributed being a source of income for 2.2 million farm households [2] which lead to more land expansion of some plantation companies.

Although this growth giving a positive effect on our economic sector, the expansion itself has been assumed to cause some matters such as deforestation [3]. Land covers changes can cause some problems related to environmental stabilities. The conversion can cause soil fertility decrease, frequent floods, local extinction and even flora and fauna diversity loss [4], [5] which happened especially in bird species diversity. As birds are very sensitive to changes in the ecosystem which makes them as a strong bio-indicator for environmental changes resulting from land use changes [6]. Hence, it is necessary to identify the bird species diversity loss or gain by comparing the oil palm plantation land covers es to its origin land cover.

2. Methods

2.1. Study area
The study was conducted in the West Kalimantan Province, Indonesia, particularly in the Ketapang district. This province lies in the 2°08 LU - 3°05 LS and between 108°0 BT - 114°10 BT and has rainfall average of 397.0 mm maximum and 34.3 mm minimum. Whereas the rainy season falls on October – March and dry season falls on April – September. This research was focusing on wildlife diversity loss and gain as a cause of the oil palm plantation establishment. The study area comprised of oil palm plantation and shrub land covers. The oil palm plantation land cover divided into 4 land covers which were young-aged, middle-aged, old-aged oil palm plantation and high conservation value forest (HCVF).

2.2. Procedures
Data were collected through July to August 2018 using line transect method. The transects were 1 km length and 100 m width. Each land covers were placed one transect. To record the bird diversity, the transects were visited in the morning from 6.00 to 9.00 and from 16.00 to 18.00 in the afternoon. Each transect were repeated for 3 times, but not in the rainy day. While to identify the bird species we used [7] field guide. Camera and binoculars were also used to make the observation easier.

There were 5 transects in total, whereas 4 transects lied in the oil palm plantation management land and 1 transect lied outside the plantation, the shrubland cover. The palm oil plantation was classified into 3 age classes: young (1-8 years old), middle (9-10 years old), old (more than 10 years 0ld). Physically, each class was recognized by its height and front coverage. Young-aged oil palm plantation tends to have an open canopy covers. While shrubland cover was covered by undergrowth and few of vegetation species. The bird diversity survey was using visual and acoustic to identify species, count the numbers and determine the locations of the birds. Meanwhile, to describe the characteristic of the HCV, we assessed the vegetation using single plots measurement 113x100 m. The plants were classified into the growth stage of tree, sapling, pole, and seedling.

2.3. Data analysis
The total number of species that was actually recorded then being analyzed using Margalef Index (Dmg) to determine the richness of each species in each land cover. This index has the ability to respond to differences in species and is highly sensitive [8].

$$Dmg = \frac{S-1}{ln\ (N)} \tag{1}$$

While to analyze the gain or loss of the species, we estimated the species proportion in the area (Faculty of Forestry IPB – SMARTRI 2018). In some cases, the survival of some certain species will increase their species richness and the homogenization of the species origin in a long period of time (Mc Cune and Vellend, 2013). Therefore, this relation can be applied to estimate the species proportion which will survive in a certain period of time in an area when the habitat experiencing proportion decrease from its own origin area [9]:

$$S_{survived}/S_{initial} = \left(A_{survived}/A_{initial}\right)Z \tag{2}$$

3. Results and discussions

3.1. Bird species diversity amongst land covers
There are 30 bird species recorded from all land covers. The highest number of species found in the HCV land cover which has 17 species (table 1). While the lowest number of species found in the middle-aged palm oil land cover. The highest number of individuals found in the oil palm plantation

land cover was Long-tailed parakeet (*Psittacula longicauda*). This happened as lots of this species of bird flying across the land covers area, as they were on their way out for foraging and back to their own nests. However, the highest individuals number found in the bushes land cover was yellow-vented bulbul (*Pycnonotus goiavier*). This species found in all over land cover areas.

Yellow-vented bulbul is one of the insectivorous bird species that have a high tolerance to environmental change as it can found in every land cover [6]. The study conducted by [6] shows the same result as this species found being the most abundant species amongst the land covers. These insectivore birds could perform the role of natural pest control in the oil palm plantation land area [10]. Meanwhile, other land covers showed a variety of species which had the highest abundant. In the young-aged oil palm land cover, Long-tailed parakeet showing the highest number individuals compared to other bird species found in the land covers. While in the middle-aged oil palm plantation, the highest species abundant is Pacific swallow (*Hirundo tahitica*). The bird species found in the old-aged oil palm plantation showed that none of the species being dominant. Almost all of the species shared the same number of individuals. This condition also happened in the HCV forest, almost all of the species shared the same number of individuals, although there was one species which had a slightly higher number of species which is yellow-vented bulbul. This species categorized as a generalist species [11] which is not dependent on forest existence [6], [12].

However, the differences between each land cover resulting in richness varieties of richness value. The highest value of richness resulted in HCV land cover (Dmg= 4.16). Compared to other bushland cover, that value is different significantly. However, the bushland cover has higher richness value compared to other land covers other than HCV land cover. The lowest value of richness found in the middle-aged land cover which has more than 1 value gap from bushland cover as shown in figure 1. The variation in height of oil palm also affecting the existence of the bird species as it created a microhabitat which suited well for bird-nesting places, foraging or physiology activities for particular species [13].

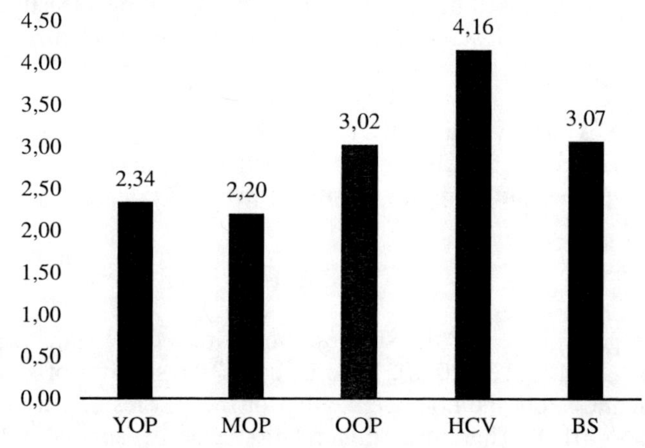

YOP: Young-aged Oil Palm Plantation; MOP: Middle-aged Oil Palm Plantation; OOP: Old-aged Oil Palm Plantation; HCV: High Conservation Value Forest, BS: Bushes

Figure 1. Species richness index values in each land covers

Nevertheless, compared to bushland cover, HCV forest has more heterogenous bird species. Almost all of the bird species found in the bushland covers were insectivorous species. While the bird species found in the HCV forest were varied from predator such as Black-winged kite (*Elanus Caeruleus*), frugivore and insectivore bird species. This study shows the same result as [14] which also showed that forest area has more heterogeneous species. The existence of the heterogenous bird species can be affected by the vegetation condition in the land cover itself [15]. As HCV forest has a more heterogenous vegetation composition compared to bushland cover. Bushland cover was dominated by undergrowth and only few vegetation found in the land cover. The abundance of epiphytes and fruits [16], openness of forest floor, composition of tree species could also affect the diversity of birds [15]. Overall, bush land cover species composition has the highest similarity to the HCV forest compared to other land covers.

Table 1. Bird species list found in the whole sample of the study area

No.	Common Name	Scientific Name	Land Covers				
			Bushes	**YOP**	**MOP**	**OOP**	**HCV**
1	White-breasted Woodswallow	*Artamusleucorhynchus*		v			
2	Ashy Tailorbird	*Orthotomusruficeps*	v		v	V	v
3	Barn Swallow	*Hirundorustica*				V	
4	Black-winged Kite	*Elanus caeruleus*					v
5	Brown-throated Sunbird	*Anthreptesmalacensis*	v				v
6	Common Emerald Dove	*Chalcophapsindica*		v		V	v
7	Crimson Sunbird	*Aethopygasiparaja*					v
8	Dusky Munia	*Lonchurafuscans*				V	
9	Green Iora	*Aegithinaviridissima*					v
10	Large Green Pigeon	*Treroncapellei*		v			
11	Lesser Coucal	*Centropusbengalensis*		v	v	V	
12	Little Spiderhunter	*Arachnotheralongirostra*					v
13	Long-tailed Parakeet	*Psittaculalongicauda*		v			
14	Long-tailed Shrike	*Laniusschach*	v				v
15	Olive-backed Sunbird	*Cinnyrisjugularis*	v	v			v
16	Olive-winged Bulbul	*Pycnonotusplumosus*	v				v
17	Orange-bellied Flowerpecker	*Dicaeumtrigonostigma*					v
18	Pacific Swallow	*Hirundotahitica*			v	v	
19	Pied Fantail	*Rhipidurajavanica*	v				
20	Pink-necked Green Pigeon	*Treronvernans*	v				
21	Rufous-tailed Tailorbird	*Orthotomussericeus*		v	v	v	V
22	Rusty-breasted Cuckoo	*Cacomantissepulcralis*			v		
23	Scarlet Minivet	*Pericrocotusflammeus*					V
24	Sooty-headed Bulbul	*Pycnonotusaurigaster*	v	v		v	V
25	Spotted Dove	*Streptopeliachinensis*	v	v	v	v	V
26	Swallows	*Callocalia*sp.		v		v	
27	White-breasted Waterhen	*Amaurornisphoenicurus*			v	v	
28	White-throated Kingfisher	*Halcyon smyrensis*	v				
29	Yellow-bellied Prinia	*Priniaflaviventris*	v			v	V
30	Yellow-vented Bulbul	*Pycnonotusgoiavier*	v	v	v		V
	Species Total		**13**	**11**	**7**	**12**	**17**

The similarities happened as there are some sources which support some species found in both areas. The species such as the sunbirds species which dieting on nectar. However, in other land covers, only a few species found in both bush and other land covers which allow more difference between the land covers. The value of the similarity index as shown in figure 2 tells that the establishment of oil palm plantation changes the composition of the bird species. Only 30% - 33% of the species found in oil palm land covers are similar to the species found in bushland cover.

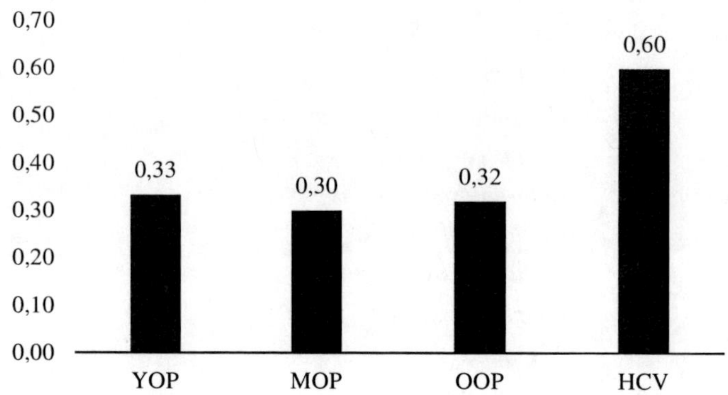

Figure 2. Index similarity between each land cover and bushland cover

3.2. Bird species gain and loss

The varieties of species through all over the land covers led to a composition difference between each land cover. The composition of bird species change comprises both species gain and loss as shown in table 2. Some land covers experiencing higher loss than gain species. The proportion of loss and gain (figure 3) showing that young-aged and middle-aged oil palm plantation has a higher loss species proportion compared to other land covers. While old-aged oil palm plantation shows the same proportion of gain and loss species. However, HCV forest land cover experiencing higher gain species than the loss species.

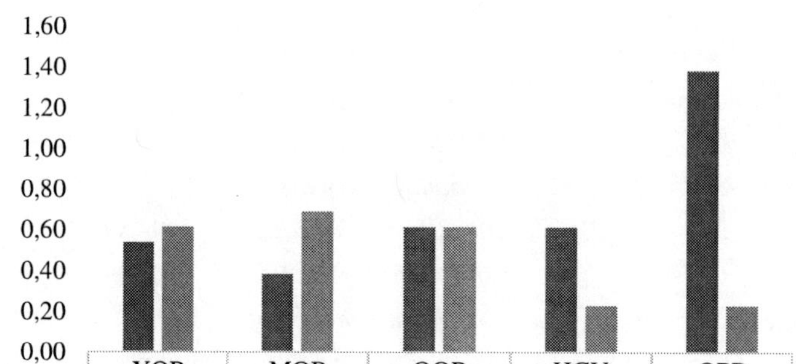

	YOP	MOP	OOP	HCV	OPP
Proportion gain	0,54	0,38	0,62	0,62	1,38
Proportion loss	0,62	0,69	0,62	0,23	0,23

Proportion gain Proportion loss

Figure 3. Gain and loss of bird species proportion

Table 2. Species gain and loss in each land covers

Young-aged Oil Palm Plantation		Medium-aged Oil Palm Plantation	
Loss	**Gain**	**Loss**	**Gain**
Long-tailed Shrike	Long-tailed Parakeet	Long-tailed Shrike	Lesser Coucal
Brown-throated Sunbird	Lesser Coucal	Brown-throated Sunbird	Rufous-tailed Tailorbird
White-throated Kingfisher	Rufous-tailed Tailorbird	Olive-backed Sunbird	White-breasted Waterhen
Ashy Tailorbird	Common Emerald Dove	White-throated Kingfisher	Pacific Swallow
Pied Fantail	White-breasted wood swallow	Sooty-headed Bulbul	Rusty-breasted Cuckoo
Olive-winged Bulbul	Large Green Pigeon	Pied Fantail	
Yellow-bellied Prinia	Swallows	Olive-winged Bulbul	
Pink-necked Green Pigeon		Yellow-bellied Prinia	
		Pink-necked Green Pigeon	
Old-aged Oil Palm Plantation		**High Conservation Value Forest**	
Loss	**Gain**	**Loss**	**Gain**
Long-tailed Shrike	Dusky Munia	White-throated Kingfisher	Crimson Sunbird
Brown-throated Sunbird	Lesser Coucal	Pied Fantail	Orange-bellied Flowerpecker
Olive-backed Sunbird	Rufous-tailed Tailorbird	Pink-necked Green Pigeon	Rufous-tailed Tailorbird
White-throated Kingfisher	Common Emerald Dove		Green Iora
Pied Fantail	White-breasted Waterhen		Common Emerald Dove
Olive-winged Bulbul	Barn Swallow		Black-winged Kite
Yellow-vented Bulbul	Pacific Swallow		Little Spiderhunter
Pink-necked Green Pigeon	Swallows		Scarlet Minivet

Oil Palm Plantation		
Loss	**Gain**	
Pied Fantail	White-breasted wood swallow	Lesser Coucal
Pink-necked Green Pigeon	Barn Swallow	Little Spiderhunter
White-throated Kingfisher	Black-winged Kite	Long-tailed Parakeet
	Common Emerald Dove	Orange-bellied Flowerpecker
	Crimson Sunbird	Pacific Swallow
	Dusky Munia	Rufous-tailed Tailorbird
	Green Iora	Rusty-breasted Cuckoo
	Large Green Pigeon	Scarlet Minivet
		Swallows
		White-breasted Waterhen

Nevertheless, the species gain from all land covers shows a higher gained species than the loss species (table 2). Most of the loss species are insectivore bird, however, there are some other bird species such as frugivore, nectarivore, and predator. The loss of certain bird species will result in species diversity and the loss of the role played by the species. Ashy tailorbird is one of the species which give a positive impact on oil palm plantation as it helps the oil palm during their reproduction as their pollinator [16]. However, this species loss in the young-aged oil palm will give some effects to the plantation itself where the pollination will run slower as there is no more natural supporter.

Nevertheless, the loss of this species does not happen in other land covers. This condition might happen as there is less supply of its diet. As stated in [6] that the presence of bird species is affected by the source of the species diet. However, most of the bird species which are classified as the loss species come from nectarivore species. [13] stated that this bird species are hard to be seen in the oil palm plantation as there is less source of its diet. While the predator bird which happens to be lost is White-throated kingfisher. However, the loss of this species will not affect greatly to the environment as the plantation area is not a swamp area.

Though the establishment of this plantation resulting in some loss species, this establishment also gain some new species. There were 18 species gained from all over the land covers. Most of the gained species were also insectivorous bird species. Higher species richness and gain were boosted by the presence of HCV forest as part of the plantation land cover. Due to some bird species are depended heavily on the presence of forest such as pigeons, the heterogeneity of the habitat needed [14]. Therefore, with the support of HCV forest, oil palm plantation may perform essential roles in conserving the biodiversity [14].

4. Conclusion

The oil palm plantation area experiencing more gain than the loss of bird species (gain: 138%; loss: 23%).Each land cover happened to have a variety of gain and loss percentage. In the young-aged oil palm plantation encounter higher loss than gain (gain: 54%; loss:62%), which also happened in the middle-aged oil palm plantation (gain: 38%; loss: 69%). While the old-aged oil palm plantation experiencing a balance loss and gain percentage in the number of 62%. The HCV forest experiencing a higher gain (62%) compared to its loss, 23%. Therefore, this establishment of the oil palm plantation provided new habitat for some particular bird species, though the richness of the bird species showing a variation. The oil palm plantation land cover which includes HCV forest giving essential support to the diversity and ecological role of each species.

References

[1] Santosa Y, Sugiharti W. 2017. Impact of fire on bird diversity in oil palm plantation: Case study in South Sumatra Province. *Int. Journal of Environmental and Ecological Engineering.* **11**(9):105-108
[2] Directorate Generale of Plantations. 2014. Tree Crop Estate Statistic of Indonesia 2013-2015. Central Bureau of Statistic, Jakarta
[3] Clay J. 2004. *World Agriculture and Environment.* Washington (US): World Wild Fund
[4] Lamb D, Erskine PD, Parrotta JA. 2005. Restoration of degraded tropical forest landscapes. Science **310**: 1628-32
[5] Ruswenti E, Novarino W, Rizaldi. 2014. Jenis-jenis burung di perkebunan kelapa sawit PT. AndalasWahana Berjaya (AWB), Kabupaten Dharmasraya, Sumatera Barat. Jurnal Biologi Universitas Andalas. Vol **3**(3) : 219-226
[6] Erniwati, Zuhud EAM, Santosa Y, Anas I. 2016. Bird diversity at smallholding and large scale palm oil plantation in different age class at Riau province. *Biodiversitas.* **17**(2):791-798
[7] MacKinnon J, Philipps K. 1993. Field Guide for Bird in Sumatera, Java, Bali and Kalimantan. Puslitbang Biologi-LIPI. Jakarta
[8] Magguran AE. 1988. *Statistical Ecology.* New York: Jhon Wiley and Sons
[9] Limm SL, Askins RA. 1995. Forest losses predict bird extinctions in Eastern North-America.

Proceedings of the National Academy of Sciences. **92**(20):9343-9347

[10] Koh, L.P., D.S. Wilcove, 2008. Is palm oil agriculture really destroying tropical biodiversity?. Conservations Letter 1, pp. 60-64

[11] Wells DR. 2007. *The birds of the Thai-Malay Peninsula*. Volume II. Passerines. Christoper Helm. London (UK)

[12] Wee YC. 2009. Observation on the behavior of the Yellow-vented Bulbul, *Pycnonotusgoiavier*, in a humid equatorial environment. *J.zoological Society of London.* **157**: 25-45

[13] Edwards FA, Edwards DP, Larsen TH, Hsu WW, Benedick S, Chung AYC, VunKhen C, Wilcove DS, Hamer KC. 2013. Does logging and forest conversion to oil palm agriculture alter functional diversity in a biodiversity hotspot? Anim Conserv **17**: 163-173

[14] Kwatrina RT, Santosa Y, Bismark M, Santoso N. 2018. Ecological impacts of oil-palm plantation on butterfly and bird species diversty. *Jur. Man. Hutan Tropika.* **24**(1): 23-31

[15] Clough Y, Kruess A, Tscharntke T. 2007. Local and landscape factors in differently managed arable fields affect the insect herbivore community of a non-crop plant species. Appl Ecol **44**: 22-28

[16] Orians GH. 1969. The Number of Birds Species in Some Tropical Forest. New York (US): Saunders College

[17] Ellwood MD, Foster WA. 2004. Doubling the estimate of invertebrate biomass in a rainforest canopy. *Nature.* **429** (6991): 549-551

[18] Yue S, Brodie JF, and Bernard H 2015 Oil palm plantations fail to support mammal diversity. *Ecol Appl* **25**: 2285-2292

Loss and gain of the bird species after the establishment of oil palm plantation in South Sumatra

Erniwati[1] and Yanto Santosa[2]

[1]Department of Forestry, Faculty of Agriculture, Universitas Bengkulu. Jl. W.R. Supratman, Kota Bengkulu 38371A, Bengkulu, Indonesia. Tel.: +62-73621170.
[2]Ecology and Wildlife Management Division, Department of Forest Resource Conservation and Ecotourism, Faculty of Forestry, Bogor Agricultural University. Jl. Raya Dramaga, Bogor 16680, West Java, Indonesia.

*E-mail: erniwati@unib.ac.id

Abstract. The development of oil palm has been documented as one of factor caused deforestation leading to the loss of biodiversity, however, to what extent the impacts of oil palm conversion on biodiversity in South Sumatra remain little-known. To answer this issue, we studied the diversity of bird in the area before and after oil palm established by calculating species number, richness, evenness, similarity, and composition. Secondary forest, shrubs and rubber plantation were identified as area before oil palm establishment through Landsat image analysis, while in the established oil palm plantation area, we classified into the young, mature and old stand of oil palm. We recorded 49 bird species, 25 families and 298 individuals. Our result showed that conversion secondary forest, shrubs and rubber plantation to oil palm plantation resulted in the variation of losing, gaining, and persisting of bird species at both habitats. Losses of bird species varied about 16%-38%, but the loss of bird species was more drastic in shrubs. Gaining of bird species also varied about 12.5% to 45% where conversion rubber plantation to oil palm impacted the highest bird species procuration. Meanwhile, bird species persistence ranged between 38%-50%. Secondary forest as the habitat condition before the establishment of oil palm had the highest bird diversity (33 species), while the palm oil plantations had the lowest bird species (11 species). The most of bird species lost were insectivorous birds. However, bird species gain was also dominated by insectivorous birds. Declines in insectivorous bird species that are important for an ecosystem process service in this study area may not significantly decline in that process service because of the other species in the similar functionally group increase in the response. Calculation of loss and gain of biodiversity as a result of land use change is necessary as a strategy for biodiversity conservation in an agricultural landscape.

1. Introduction

Rapid losses of species as a consequence of human activity have been continuing at both global and local scales [1]. At the local level, the dramatic effect of the loss of species occurs when the previously lost species were key species or its functional ecological role was not able to be placed by another species, meanwhile the effects of gain species are probably higher than previous species existed if they have a new ecological role in the new ecosystem [2].

The development of oil palm has been blamed as a major cause of environmental problems such as deforestation and loss of biodiversity in tropical countries [3], [4], [5], [6]. Some studies show that conversion of forests into oil palm plantations reducing wildlife species such as birds, lizards and mammals [4], [7], [8], loss of High Conservation Value species [9], declining invertebrate richness such as ants and beetles [10], butterflies [6] and loss of soil organic matter [11], and having consequences on microclimate and wildlife species that favor sheltered habitats [8]. However, recent studies have shown that well-managed palm oil plantations are capable of supporting diversity to some extent. The abundance of arthropods and butterflies in oil palm plantations adjacent to forests was similar to natural forests, although the species richness was lower than natural forest. [12], [13], [10].

Understanding how the conversion of habitat into oil palm affects the magnitude of bird species extinction would be particularly useful for highly threatened regions such as Sumatra Island. Birds are good bioindicator of ecosystem change leading to biodiversity loss [14], [15]. The bird is important in the ecosystem because of its role as seed dispersal and pollinator [16]. Also, birds can suppress arthropod density, thereby increasing the yield of oil palm plantations [15], [17]. Previous research has found that insectivorous birds can control leaf-eating pests [15] and owls (*Tyto alba*) can control plant pest [18]. In this study, we assessed the relative gains and losses of bird diversity before and after the establishment of oil palm plantation and its impact on ecosystem function.

2. Materials and methods

2.1. *Study area*

The study was conducted in an oil palm plantation with an area of 9700 ha, in Ogan Komiring Ilir District, South Sumatra. Data were collected on different habitat conditions; consisted area before and area after palm plantation establishment. Secondary forest, shrubs and rubber plantation nearby oil palm plantation were identified as a habitat reference before oil palm plantation established. Land use cover before oil palm plantation was analysed by Landsat Imagery. Landsat images of the coverage were acquired for the 3 years before the oil palm was planted. The plantation consisted of different ages of oil palm stands. Therefore we classified into 3 locations, namely: young age of oil palm stand (5 years old), medium/mature age (10 years old), old age (16 years old). In general, oil palm covered by undergrowth plant. The total area of the remnant of the secondary forest is about 22 ha. It was bordered by the river at left and right side. This area has been pointed as High Conservation Value by the oil palm company. Saplings and pole of native tree species such as *Malaleuca leucadendron* were still found at that area. Furthermore, Shrub area was about 580 ha, most of the area was a swampy area. Its vegetation was dominated by cover crop. However, some saplings and poles of *Malaleuca leucadendron* and *Combretocarpus rotundus* were still found. Rubber plantation in research area was 27 years old. Undergrowth was dominated by *Melastoma malabathricum, Imperata cylindrica* and *Mimosa pudica.*

2.2. *Procedure*

2.2.1 *Data collection*

Birds were observed using binoculars and camera. MacKinnon and Philips field guide [19] was used to identify the birds. Data collection was conducted in 2017 using the combination of line transect method and point observation. Birds were observed along a one-km-long transect within 50 m to the left and right side of transect line. In total we had 6 transects, all transect in each research site were visited in the morning from 6.00 to 9.00 and in the afternoon from 16.00 to 18.00. The bird observation were repeated for 3 days at the same plots to maximize the number of bird species recorded. The observation were delayed during the rainy days.

2.2.2 *Data analysis*

Bird species diversity at the area before and after the establishment of oil palm plantation was estimated by the total number of species that was actually recorded in the research sites, the Shannon-Wiener Index (H’), Margalef (Dmg) Index, Evennees index (E) followed [20]. The similarity of bird communities among different sites was estimated using the Sorensen Index. Functional diversity of bird was assessed by categorizing birds according to feeding guild based on [19]. To assess particular bird species lost, we counted the number of species exists before conversion and species lost after that. For the species gain, we assessed the number of species which did not exist before and species was found after conversion, and for the persistence species, we defined as the species existence before and after changing habitat.

3. Results and discussion

3.1. Bird diversity before and after the establishment of oil palm plantation

Our result showed that the conversion of secondary forest, shrubs and rubber plantation caused the decreasing of bird diversity in the research area. The change of bird diversity at habitat after oil palm plantation establishment ranged from 2 species to 17 species. The conversion of secondary forest or High Conservation Value (HCV) into oil palm plantation had the highest bird diversity lost (vary 17 to 22 species at different oil palm stand age), while the conversion of rubber plantation into oil palm had the lowest bird species lost (vary 3 to 5 species). Within the oil palm plantation, different stand age of oil palm responded differently to the habitat change, medium age stands of oil palm plantation accommodated less bird diversity compared to young and old age stand of oil palm (table 1).

The change of habitat into oil palm plantation influenced the abundance of the individual bird. The conversion shrubs into oil palm had the highest decreasing than other habitats before oil palm establishment. It varied about 24% to 63% within the aged stand, where the old stand of oil palm had the highest loss of bird abundance (table 1). Meanwhile, the conversion of rubber plantation into oil palm plantation decreased the number of the individual bird about 11% to 19% at medium and old age stand of oil palm, however at the opposite, young age stand of oil palm, received higher individual bird (64%).

Moreover, evenness index tended to similar at both habitat condition before and habitat condition after oil palm establishment, except at young age stand of oil palm had lower evenness index. It indicated that there are no dominant species occupied the habitat. The change of habitat into oil palm did not influence the distribution of species at each habitat. Most species evenly distributed among habitats both before and after oil palm establishment.

Table 1. Bird diversity before and after the establishment of oil palm plantation in South Sumatra

Species diversity	Area before oil palm establishment			Area after oil palm establishment		
	HCV	Rubber	Shrubs	Young stand	Medium stand	Old stand
Number of species (S)	33	16	30	16	11	13
Number of individu (N)	64	36	78	59	32	29
Species richness (Dmg)	7.68	4.19	6.66	3.68	2.89	3.56
Evenness index (E)	0.93	0.93	0.92	0.86	0.92	0.92

Data source: [21]

Based on Margalef Index, all the habitats before oil palm plantation establishment had the higher species richness (Dmg) which varied from 4.19 to 7.68 (figure 1). The dramatical changes of bird species richness occurred after HCV habitat converted into oil palm. The bird species richness slightly reduced after conversion of rubber plantation area. Our result is consistent with previous research [22], [23], which concluded that the conversion of secondary forest into oil palm plantation in Riau province reduced the bird species richness. However, on the contrary, these results are different from another study in Kalimantan [24] which revealed that the change of habitat after oil palm plantation establishment increased the diversity of birds.

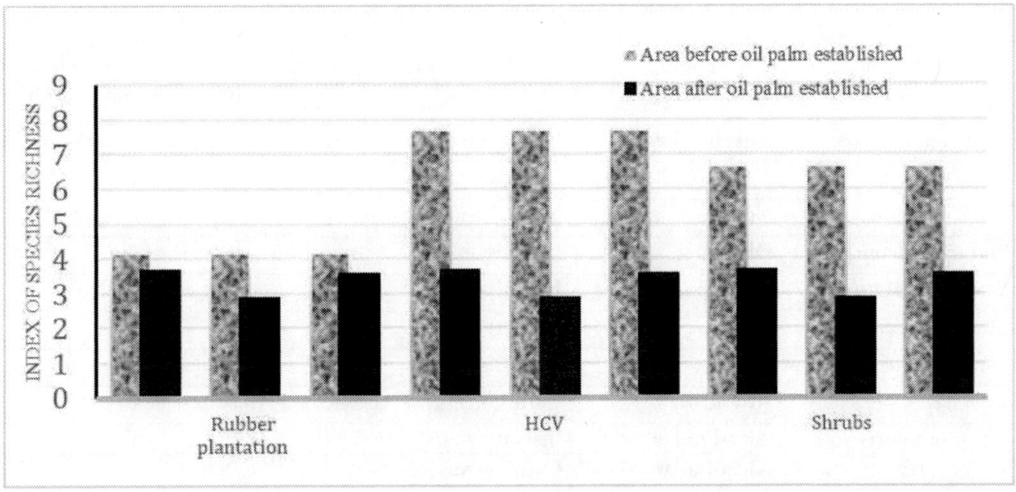

Figure 1. Species richness (dmg) of bird species before and after oil palm plantation establishment in Southern Sumatra

Our study confirmed that the highest loss of bird species richness occurred when secondary forest (HCV) was converted to oil palm plantation. The degradation of the habitat from complex vegetation structure into monoculture system such as oil palm plantation influences the abundance of food availability and variety of niche. The diversity of vegetation is an important factor affecting bird species diversity in the habitat before oil palm plantation established. Forest conversion to palm oil plantation has changed the plant species composition from a heterogeneous to a homogeneous one. Therefore it has altered the food availability and habitat conditions for potential bird species.

3.2. Bird composition, loss and gain of bird before and after oil palm plantation establishment

A total of 49 bird species recorded, 25 families and 298 individuals. The abundant species varied with both habitats, but some dominant bird species were found at both habitat before and after oil palm establishment, namely Sooty-headed Bulbul (*Pycnonotus aurigaster),* Yellow-vented Bulbul (*Pycnonotus goiavier)* which are insectivorous and omnivorous species from family Pycnonotidae, Zebra Dove (*Geopelia striata),* Spotted Dove (*Streptopelia chinensis)* which are granivorous bird from family Columbidae. Those species are highly adapted species which are often found in the open habitat, shrubs, secondary forest and yard [19]. A swampy bird species such as White-breasted Waterhen (*Amaurornis phoenicurus*) also was found at both HCV, Shrubs and oil palm. It is because those habitats close to stream, river or there is a swampy area spot surrounded the habitats.

Table 2. Composition and feeding guild of bird species before and after establishment of oil palm

Scientific name	Family	Feeding guild	Area before establishment			Area after establishment
			Rubber plantation	HCV	Shrubs	Oil palm
Gallus gallus	Phasianidae	Herbivorous, Insectivorous				√
Ixobrychus cinnamomeus	Ardeidae	Carnivorous	√	√	√	√
Lanius schach	Laniidae	Insectivorous		√	√	√
Lonchura punctulata	Estrildidae	Herbivorous		√		√
Lonchura leucogastra	Estrildidae	Herbivorous	√			
Centropus bengalensis	Cuculidae	Insectivorous		√		√
Centropus sinensis	Cuculidae	Carnivorous			√	
Passer montanus	Ploceidae	granivorous			√	

Scientific name	Family	Feeding guild	Area before establishment			Area after establishment
			Rubber plantation	HCV	Shrubs	Oil palm
Anthreptes malacensis	Nectariniidae	Nectivorous	√	√		√
Aethopyga siparaja	Nectariniidae	Nectivorous			√	
Nectarinia jugularis	Nectariniidae	Nectivorous	√	√	√	√
Dicaeum trochileum	Dicaeidae	Insectivorous		√	√	
Caprimulgus macrurus	Caprimulgidae	Insectivorous			√	√
Picoides moluccensis	Picidae	Insectivorous		√	√	
Dendrocopos macei	Picidae	Insectivorous		√	√	
Ardea cinerea	Ardeidae	Carnivorous	√	√	√	√
Halcyon smyrnensis	Alcedinidae	Insectivorous	√	√	√	√
Halcyon chloris	Alcedinidae	Carnivorous	√	√	√	
Cisticola juncidis	Sylviidae	Insectivorous			√	
Orthotomus sepium	Sylviidae	Insectivorous	√	√		√
Orthotomus sericeus	Sylviidae	Insectivorous		√		√
Aegithina tiphia	Aegithinidae	Insectivorous		√		
Pycnonotus aurigaster	Pycnonotidae	Insectivorous	√	√	√	√
Elanus caeruleus	Accipitridae	Carnivorous		√	√	
Haliastur indus	Accipitridae	Carnivorous				√
Hemipus hirundinaceus	Campephagidae	Insectivorous				
Lalage nigra	Campephagidae	Insectivorous		√	√	
Amaurornis phoenicurus	Rallidae	Insectivorous, Granivorous		√	√	√
Artamus leucorhynchus	Artamidae	Insectivorous			√	√
Acridotheres javanicus	Sturnidae	Omnivorous		√		
Rhipidura javanica	Rhipiduridae	Insectivorous		√	√	√
Merops philippinus	Meropidae	Insectivorous	√	√	√	
Pycnonotus plumosus	Pycnonotidae	Omnivorous	√	√		
Pycnonotus goiavier	Pycnonotidae	Omnivorous	√	√	√	√
Anhinga melanogaster	Phalacrocoracidae	Carnivorous			√	
Pelargopsis capensis	Alcedinidae	Carnivorous		√		
Picus miniaceus	Picidae	Insectivorous		√		
Prinia familiaris	Sylviidae	Insectivorous				
Prinia flaviventris	Sylviidae	Insectivorous	√	√	√	√
Geopelia striata	Columbidae	Granivorous	√	√	√	√
Treron curvirostra	Columbidae	Carnivorous			√	
Alcedo coerulescens	Alcedinidae	Insectivorous		√		√
Alcedo meninting	Alcedinidae	Carnivorous		√		√
Gerygone sulphurea	Acanthizidae	Insectivorous.			√	√
Pericrocotus flammeus	Campephagidae	Insectivorous		√		
Streptopelia chinensis	Columbidae	Granivorous, Fruitsvorous	√	√	√	√
Eurystomus orientalis	Coraciidae	Large Insectivorous		√	√	

Scientific name	Family	Feeding guild	Area before establishment			Area after establishment
			Rubber plantation	HCV	Shrubs	Oil palm
Collocalia linchi	Apodidae	Small insectivorous	√		√	√

Our result revealed that conversion secondary forest, shrubs and rubber plantation to oil palm plantation resulted in the variation of losing, gaining, and persisting of bird species at both habitats. Loss of bird species varied about 29%-47%, HCV and Shrubs experienced similar loss even though the loss of bird species were slightly higher. Gaining of bird species also varied about 18% to 88% where conversion rubber plantation to oil palm impacted the highest bird species procuration, while HCV affected the lowest gaining of species after habitat change. Moreover, bird species persistence ranged between 53%-75% (table 3).

Table 3. Number of species loss, gain and persist in the habitat before and after establishment of oil palm

Habitat before converting	Habitat after converting	Loss of bird species	Gain of species	Persist of species
HCV	Oil palm	14 Species (42 %)	6 species (18 %)	20 Species (61 %)
Shrubs	Oil palm	14 Species (47 %)	9 Species (30 %)	16 Species (53%)
Rubber plantation	Oil palm	4 species (29%)	14 (88%)	12 Species (75%)

Insectivorous birds dominated the number of bird species lost after conversion into oil palm followed by carnivorous bird, omnivorous bird, granivorous and frugivorous bird and herbivorous bird. We also found some nectivorous bird in oil palm plantation. The same pattern also occurred for gaining and persistence species (table 4). The large number and variety of individual abundance of insectivorous birds in oil palm plantations related to the abundance of food resources. Thick ground vegetation cover both before and after habitat change may be related to the abundance of arthropods as the food of some bird insectivorous. The abundance of insectivorous such as Yellow-vented Bulbul (*Pycnonotus goiavier*) and Red Junglefowl (*Gallus* gallus) might play an important ecological role in controlling arthropods including insects pest in oil palm plantation [17]. Birds can suppress arthropod density, thereby increasing the yield of oil palm plantations [25].

Some insectivorous birds lost after habitat conversion into oil palm, such as Scarlet-headed Flowerpecker (*Dicaeum trochileum),* Sunda Pygmy Woodpecker (*Picoides moluccensis),* Fulvous-breasted Woodpecker (*Dendrocopos macei),* Common Iora (*Aegithina tiphia),* Pied Triller (*Lalage nigra),* Blue-tailed Bee-eater (*Merops philippinus),* Banded Woodpecker (*Picus miniaceus),* Scarlet Minivet (*Pericrocotus flammeus)* and Oriental Dollarbird (*Eurystomus orientalis),* but their functional role in ecosystem may be replaced by other insectivorous which was gained or was persisted after conversion, such as Red Junglefowl (*Gallus gallus),* Cinnamon Bittern (*Ixobrychus cinnamomeus),* Yellow-bellied Prinia (*Prinia flaviventris).* The ecosystem functioning is decreased as the number of species in a community decreases. However, the profound effect of the loss of species occurs when the previously lost species were key species, or its functional ecological role was not able to be placed by another species [26].

Table 4. Composition of species lost, gain and persist based on feeding guild before and after the establishment of oil palm plantation

Feeding guild	HCV into oil palm			Shrubs into oil palm			Rubber into oil palm		
	Lost	Gain	Persist	Lost	Gain	Persist	Lost	Gain	Persist
Insectivores	9 Sp	5 Sp	8 Sp	4 Sp	7 Sp	10 Sp	1 Sp	10 Sp	5 Sp
Omnivores	2 Sp		2 Sp			1 Sp	1 Sp		
Carnivores	3 Sp		4 Sp	2 Sp	5 Sp	2 Sp	1 Sp	2 Sp	2 Sp

Feeding guild	HCV into oil palm			Shrubs into oil palm			Rubber into oil palm		
	Lost	Gain	Persist	Lost	Gain	Persist	Lost	Gain	Persist
Nectivores			1 Sp		1 Sp	1 Sp			2 Sp
Granivores, Fruitsvores			5 Sp	2 Sp	1 Sp	2 Sp			2 Sp
Herbivores		1 Sp		1 Sp			1 Sp	2 Sp	

Note: Sp = number of species lost, gained or persisted

4. Conclusion

There were some species loss, gains and persist after the conversion of habitat into oil palm plantation. The secondary forest and shrubs experienced higher loss of bird species than rubber plantation. Meanwhile rubber plantation had the highest bird species gain after habitat conversion. The most bird species that lost, gained and persisted after land use change were insectivorous bird. The large number and variety of individual abundance of insectivorous birds both at habitat before and after conversion related to abundant of insect. Bird declining because of changing habitat may disrupt ecosystem processes and service of potential importance to human well being, however in our study declines in insectivorous bird species that are important for a particular ecosystem process service may not necessarily decline in that process service because of the other species in the similar functionally group increase in the response.

Acknowledgments

This study is part of the project 'Land History and Biodiversity in Palm Oil Plantation in Indonesia in South Sumatra' supported by the Indonesian Sawit Funds (BPDPKS). We would like to warmly thank all plantation managers and staff from palm oil plantation company in South Sumatra for their help in this study.

References

[1] Millennium Ecosystem Assessment. 2005. *Ecosystems and Human Well-Being.* Washington: (US): Island Press

[2] Wardle DA, Richard D.Bardgett, Ragan M. Callaway and Wim H. Van der Putten. (2011) Terrestrial Ecosystem Responses to Species Gains and Losses. http://dx.doi.org/10.1126/science.1197479 *Science*. Volume: **332**, pp 1273-1277

[3] Basiron Y. 2007. Palm oil production through sustainable plantations. *Eur J Lipid Science and Technol.* **109**(4): 289-295

[4] Fitzherbert EB, Struebig MJ, Morel A, Danielsen F, Brush CA, Donald PA, Phalan B. 2008. How will oil palm expansion affect biodiversity?. *Trends Ecol Evol.* **23**: 538-545

[5] Koh LP, Wilcove DS. 2007. Cashing in palm oil for conservation. *Nature*. 448: 993-994.

[6] Koh LP, Wilcove DS. 2008. Is palm oil agriculture really destroying tropical biodiversity?. *Conserv Lett.* **1**: 60-64

[7] Danielsen FINN, Beukema H, Burgess ND, Parish F, Bruhl AC, Donald PF, Murdiyarso D, Phalan BEN, Reijnders L, Struebig M., Fitzherbert EB. 2009. Biofuel plantations on forested lands: double jeopardy for biodiversity and climate. *Conserv Biol.* **23**(2):348-358

[8] Peh KSH, Johny DJ, Navjot SS, Susan LHL, Charlotte AM. 2005. Lowland rainforest avifauna and human disturbance: persistence of primary forest birds in selectively logged forests and mixed-rural habitats of southern Peninsular Malaysia. *Biol Conserv.* **123**(4): 489-505.

[9] Aratrakorn S, Thunhikorn S, Donald PF. 2006. Changes in bird communities following conversion of lowland forest to palm oil and rubber plantations in southern Thailand. *Bird Conserv Inter.* **1**: 71-82

[10] Senior MJM, Hamer KC, Bottrell S, Edwards DP, Fayle TM, Lucey JM, Mayhew PJ, Newton R, Peh KSH, Sheldon FH, Stewart C, Styring AR, Thom MDF, Woodcock P, Hill JK. 2013. Trait-dependent declines of species following conversion of rain forest to oil palm plantations. *Biodiv Conserv.* **22**: 253-268

[11] Saxon E, Roquemore S. 2011. Ch. 6: Palm Oil. In Union of Concerned Scientists, The Root of the Problem: What's Driving Tropical Deforestation Today?. Cambridge (CG): Union of Concerned Scientists (pp. 51-63)
[12] Ghazali A, Asmah S, Syafiq M, Yahya M, Aziz N, Peng T, Norhisham AR, Puan CL, Turner EC, Azhar B. 2016. Effects of monoculture and polyculture farming in oil palm smallholdings on terrestrial arthropod diversity. *JAPE. 19*(2):415-421.1
[13] Lucey JM, Hill JK. 2012. Spillover of insects from rain forest into adjacent oil palm plantations. *Biotropica*. **44**: 368–377
[14] BirdLife International. 2010. Birds are very useful indicators for other kinds of biodiversity [internet]. [downloaded 15 Maret 16]. Available at http://www.biodiversityinfo.org/casestudy.php?r=introduction&id=61
[15] Koh LP. 2008. Can palm oil plantations be made more hospitable for forest butterflies and birds?. *Appl Ecol*. **45**: 1002-1009
[16] Bibby CJ, Burgess ND, Hill DA, Mustoe SH. 2000. *Bird Census Techniques Second edition*. London (UK): Academic Press.
[17] Arshad. MI, Zakaria M, Sajap AS and Ismail A (2000). Food and feeding habits of Red Junglefowl. Pakistan Journal of *Biological Science* **3**:1024-1026
[18] Heru SB. 2000. Large scale use of barn owl (*Tyto alba*) for controlling rat population in palm oil plantations in Riau, Sumatera *in* E. Pushparajah, editor. *Plantation Tree Crops in the New Millennium: the Way Ahead*; 2001 Des 1- 3; Kuala Lumpur, Malaysia. Kuala Lumpur (M): The Incorporated Society of Planters. p 529-544
[19] MacKinnon J, Philipps K. 2010. *Field Guide for Bird in Sumatera, Java, Bali and Kalimantan*. Jakarta (ID): Puslitbang Biologi-LIPI
[20] Magurran AE 2004 *Measuring Biological Diversity* (New Jersey: Blackwell Science Ltd)
[21] Santosa Y, Sunkar A, Purnamasari I, Yohana. 2017. *Sejarah Perkembangan Status, Penggunaan Lahan, dan Keanekaragaman Hayati Kebun Kelapa Sawit Indonesia*. Laporan Hasil Penelitian. Badan Pengelola Dana Perkebunan Kelapa Sawit & Lembaga Penelitian dan Pengabdian Kepada Masyarakat. Bogor
[22] Azhar B, Chong LP, Zakaria M, Hassan N, Arif M. 2014. Effects of monoculture and polyculture practices in oil palm smallholdings on tropical farmland birds. *Basic and Applied Ecol.* **15**: 336–346
[23] Erniwati, Zuhud EAM, Santosa Y, Anas I. 2016. The value of secondary forest patches for birds conservation in palm oil landscapes of Riau, Sumatera. *Biodiversitas* **17**(2): 791−798. https://doi.org/10.13057/biodiv/d170256
[24] Kwatrina RT, Santosa Y, M Bismarck, Santoso N, 2018. Ecological Impacts of Oil-Palm Plantation on Butterfly and Bird Species Diversity. Jurnal Manajemen Hutan Tropika Vol. **24**, (1): April 2018 EISSN: 2089-2063 DOI: 10.7226/jtfm.24.1.23
[25] Maas B, Tscharntke T, Saleh S, Dwi PD, Clough Y. 2015. Avian species identity drives predation success in tropical cacao agroforestry. *Appl Ecol*. **52:** 735-743

IOP Conf. Series: Earth and Environmental Science **336** (2019) 012030 doi:10.1088/1755-1315/336/1/012030

Impact of oil palm plantation on herpetofauna species diversity at PT Waimusi Agroindah South Sumatera

Sani Sari Sri Rejeki[1]*, Yanto Santosa[1]

[1]Wildlife Ecology and Management Division, Department of Forest Resources Conservation and Ecotourism, Faculty of Forestry, Agricultural Institute of Bogor, Bogor, West Java, Indonesia

*E-mail: sanirejeki@gmail.com

Abstract. Oil palm plantation playing an important role in Indonesia's economic development, whereas generates the largest foreign exchange and boosts communities' economic sector, also absorbs the workforce. However, there are some accusation to oil palm plantation that it caused biodiversity decreased in the minimum of 60%. The study was done to identify the impact of oil palm plantations on herpetofauna diversity, community similarity and percentage loss or gain of community. The observation was done using Visual Encounter Survey (VES) method in the 6 land cover types which classified into 2 land areas. Rubber plantation and shrubs were considered as the land cover before the oil palm plantation establishment. While, young, intermediate, old oil palm plantation and high conservation value forest were considered as the land covers after the plantation establishment. The study was conducted in May 2017 in the South Sumatera Province. The data analysis showed that old palm plantation had a higher species number (19 species) compared to rubber plantation (11 species) and shrubs land cover (3 species). However, rubber plantation and old oil palm plantation had the highest species composition similarity (IS= 0.62). While the lowest happened in both shrubs-old oil palm plantation and shrubs-conservation area (IS= 0.19). The transformation land cover into oil palm plantations led to the gain of herpetofauna species as many as 8 species for rubber plantation and 17 species for shrub, while on the other hand only cause loss of 1 species from shrub and no one species loss for rubber plantation.

1. Introduction

Oil palm plantation playing an important role in Indonesia's economic development, whereas generates the largest foreign exchange and boosts communities' economic sector, also absorbs workforce [1]. Besides having a positive impact on the economy, the expansion of oil palm plantations also has a negative impact on ecological aspects. Up to 2017, the area of oil palm plantations in Indonesia reached 12.31 million ha [2]. This makes many negative allegations about oil palm plantations such as increasing carbon emissions, climate and deforestation, thus disrupting environmental balance [3].

One animal that is susceptible to environmental changes is herpetofauna, especially amphibians. Because herpetofauna is poikilotherm animals whose body temperature is strongly influenced by the environment [4]. Also, amphibians also have semi-permeable skin and 2 phases of the life cycle [5]. Herpetofauna has an important role in human life such as predators of insect species that can become disease vectors, Environmental bioindicators and climate change [6], food and export commodities [7]

In Indonesia, herpetofauna is currently not widely known, both regarding taxonomy, biological and ecological characteristics [4]. Similar to herpetofauna studies on oil palm plantations which are also still rare. This makes it important to research the impact of oil palm plantations on the diversity of herpetofauna species. The purpose of this study was to identify the impact of oil palm plantations on herpetofauna diversity, community similarity and percentage loss or gain of community.

2. Materials and Methods

The impact of oil palm plantation on species diversity and composition is the difference between species diversity and composition in the baseline (before the existence of the oil palm plantation land cover) and after the oil palm plantation. In determining the type of land cover before plantations established, Landsat imagery used 2 years before the plantations established. Based on the results of the interpretation, it was found that the dominating area was rubber plantation and shrubs.

2.1. Study Area

Data collection was conducted out in May 2017 at PT Waimusi Agroindah, South Sumatra province. There are 2 categories of land cover. The first is the land before oil palm plantations established (rubber plantation and shrubs), the second is land cover after of oil palm plantation established which are young growth (2012), intermediate growth (2006), old growth (2001) and High Conservation Value area.

2.2. Study Method

The observation of herpetofauna used VES (Visual Encounter Survey) method. Starting from 7.00 p.m-9.00 p.m with 3 repetitions. The observation was made simultaneously on 6 land covers.

2.3. Data Analysis

To find out the impact of oil palm plantation on herpetofauna species diversity, the calculation of some species, evenness index and species richness index are calculated as follows [8]:

- Species richness index (Dmg)

  Margalef Index (species richness) used to determine the species richness of each species in the community that found in the field, the formula is:

  $$Dmg = S\text{-}1 / \ln N$$

  Information:

Dmg	= species richness index
S	= number of types
N	= total number of individuals of all species

- Evenness index (E)

  Evenness Index has a function to determine the evenness of each type in every community.

  $$E = H' / \ln(S)$$

  Information:

E	= evenness index (a value between 0-10)
H '	= diversity
ln	= natural logarithm
S	= Number of types

Whereas to determine the impact of oil palm plantation on species composition, a number of herpetofauna diversity loss and gain in PT Waimusi Agroindah, the following formula is used:

- Similarity index

  Similarity index is used to determine the similarity between the observation location based on the type herpetofauna found.

  $$IS = 2c\frac{1}{a+b}$$

Information:

IS	= similarity index
a	= the number of species that are only found in community A
b	= the number of species that are only found in community B
c	= the number of species found in community A and B

3. Results and Discussion

There were 20 species of herpetofauna (12 amphibians and 8 reptiles) with a total of 126 individuals which were successful at the observation site. Table 1 presents a list of the species of herpetofauna found before and after the oil palm plantation established.

Table 1. Variation of herpetofauna in the different type of land cover

No	Scientific Name	Family	Before		After				
			RP	SH	YG	IG	OG	HCV	OPP
1	*Duttaphrynus melanostictus*	Bufonidae	0	0	0	0	4	0	4
2	*Hylarana erythraea*	Ranidae	4	0	1	3	5	5	14
3	*Pulchrana glandulosa*	Ranidae	0	0	1	0	5	0	6
4	*Amnirana nicobariensis*	Ranidae	1	0	0	0	3	0	3
5	*Fejervarya cancrivora*	Dicroglossidae	3	3	10	2	4	8	24
6	*Fejervarya limnocharis*	Dicroglossidae	2	3	9	3	3	6	21
7	*Pulchrana baramica*	Ranidae	0	0	0	3	1	0	4
8	*Chalcorana chalconota*	Ranidae	0	0	0	0	2	0	2
9	*Polypedates leucomystax*	Rhacophoridae	0	0	1	0	3	2	6
10	*Ingerophrynus biporcatus*	Bufonidae	1	0	0	4	1	0	5
11	*Limnonectes paramcrodon*	Dicroglossidae	1	0	0	0	0	1	1
12	*Leptobrachium hasseltii*	Megophryidae	0	0	0	2	0	0	2
13	*Naja sumatrana*	Elipidae	0	0	0	0	1	0	1
14	*Dendrelaphis pictus*	Colubridae	1	0	0	0	2	0	2
15	*Hemidactylus frenatus*	Gekkonidae	1	0	1	0	0	0	1
16	*Eutropis multifasciata*	Scinidae	2	0	0	0	1	1	2
17	*Takydromus sexlineatus*	Lacertidae	1	0	0	0	0	1	1
18	*Bungarus candidus*	Elipidae	1	0	0	0	1	0	1
19	*Bungarus pasciatus*	Elipidae	1	0	0	0	1	0	1
20	*Xenochropis vitattus*	Natricidae	0	1	0	0	0	0	0
	Number of Individuals		**18**	**7**	**23**	**17**	**37**	**24**	**101**
	Number of Species		**11**	**3**	**6**	**6**	**15**	**7**	**19**

Where: RP=Rubber plantation; SH=Shrub; YG=Young growth oil palm; IG=Intermediate growth oil palm; OG=Old growth oil palm; HCV=High Conservation Value area; OPP=Oil palm plantation

The number of species in the rubber plantation area (11 species) was higher compared to the number of species in the OPP, except for the old oil palm area (15 species). Similar to the research of [8] in one of the oil palm plantations in North Sumatra Province which also found that the RP area had a greater number of species than the SH area, except the OG area. However, other research that was done in West Kalimantan found that the RP had the lowest number of species than OPP [9].

Several factors influence the difference in the number of species. Canopy cover is one of the most influential factors in the presence of herpetofauna at PT Waimusi Agroindah (WMA). This can be seen in figure 1, where the area of SH that has no canopy cover has the lowest number of species and species richness index. It the same case with YG and IG areas which have a lower number and species richness index than the RP area. Based on the results of [10] research, herpetofauna was not found in habitats that have open canopy cover, even though they have adequate sources of food and water.

Unlike the HCV area, even though it has a denser canopy cover with more diverse vegetation, the HCV area has a lower number of species compared to the RP area. This is because the HCV area has many predators for herpetofauna. The results of the study showed that in the HCV area there were 4 species of mammals and 33 species of birds. This amount is the highest number among 5 other land cover. One species of mammal that dominates in the HCV area is long-tailed macaque or *Macaca fascicularis* (28 individuals). The results of the study of *M. fascicularis* in Papua New Guinea by [11] found that the presence of *M. fascicularis* had a very large negative effect on the abundance of herpetofauna. [12] explained that although the main feed of *M. fascicularis* is in the form of fruits, *M. fascicularis* is also an opportunistic omnivore animal which means it can eat various types of food if there is an opportunity. Meanwhile, bird species that dominate in the HCV area include Indus *Haliastur, Acridotheres javanicus, Alcedo coerulescens, Alcedo meninting, Pelargopsis capensis, Ardea cinerea* and *Amaurornis phoenicurus*. [13] state that the main predators of amphibians are birds. According to [14] family of Accipitridae is a carnivorous animal that likes to eat small birds, snakes and fish [15] also explained that eagle feed is snakes, frogs, lizards and small mammals.

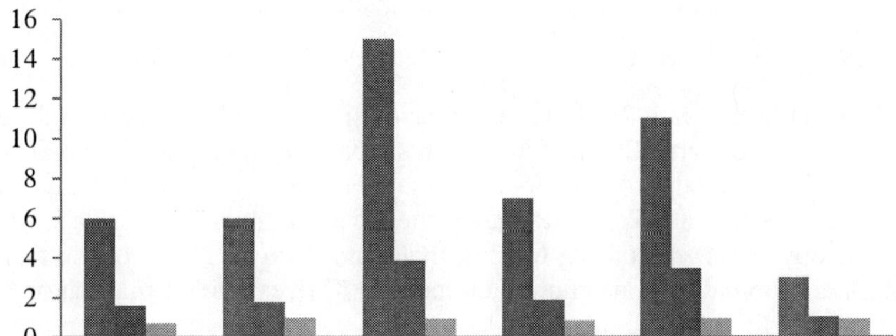

	YG	IG	OG	HCV	RP	SH
■S	6	6	15	7	11	3
■Dmg	1,6	1,77	3,88	1,89	3,46	1,03
■E	0,71	0,98	0,94	0,84	0,94	0,91

Figure 1. Comparison of the number of species, species richness and evenness index at type of land cover

Based on Figure 1, it can be seen that the area of OG area has the highest number of species among the 5 other types of land cover. Besides the denser canopy cover factors, the OG area has more water sources than the RP area. Also, the water in the RP area is not permanent and dependent on rainfall, while the water source in the OG area is a canal where the water is permanent and does not depend on rainfall. Research [16] found that the number of amphibians in habitats whose water volume depends on the weather is lowest than habitats whose water sources do not depend on the weather. Amphibians are very dependent on the presence of water sources because amphibians have two life cycles where one cycle requires water as a breeding medium or habitat [17]. Besides being a breeding ground, water sources are also used as a place to protect themselves from drought. Herpetofauna is a species of poikilotherm whose body temperature follows the environment temperature [18].

In addition to having the highest number of species, the OG area has the highest species richness than the baseline area (RP and SH). According to [8] species richness index is used to determine species richness each community encountered. Unlike the case with the species evenness index, the area with the highest evenness index value is the IG area, while the lowest evenness value is found in the YG area. The species evenness index is used to describe the stability of a community [19].

When compared as a whole, the OPP area has a higher number of species and species richness index than RP and SH areas (figure 2). Factors that cause a high number of species and types of wealth in the OPP area because in the OPP area have a lot of water sources and dense canopy cover. The RP area has a lower number of species than the OPP area because in the RP area there are fewer water sources than the OPP area. Water is needed by herpetofauna, especially amphibians to reproduce [20].

	S	DMG	E
OPP	19	3,9	0,79
RP	11	3,46	0,94
SH	3	1,03	0,91

Figure 2. The impact of oil palm plantation on herpetofauna species diversity

Similarly, the SH area also has a lower number of species and species wealth than the OPP area. There are only 3 species of herpetofauna found in this area. [21] in oil palm plantations in Riau Province and [22] in Central Kalimantan Province found the same thing, namely the SH area had a lower number of species than the OPP area. Like an RP area, the lower number of species in the SH area is also influenced by canopy cover. According to [23], that canopy cover is one of the factors that affect the species richness and abundance of amphibian species. [24] explained that canopy cover affects the temperature and humidity of a habitat.

	YG	IG	OG	HCV	OPP
RP	0,35	0,53	0,62	0,67	0,62
SH	0,44	0,44	0,22	0,24	0,19

Figure 3. Comparison of Sorensen similarity index between each oil palm plantation land cover with the baseline

Based on figure 3, it can be seen the value of similarity between the areas before and after the oil palm plantations established. Overall, the value of species similarity in PT WMA tends to be influenced by canopy cover. This can be seen from the similarity value in the OPP area with the RP area which has the highest value in the HCV area and the lowest value in the YG area. It is the same as the comparison of species similarity in the OPP area with shrubs. The highest similarity value was found in the area of YG-SH and IG-SH which had more open canopy cover than the OG and HCV areas. [21] states that habitats that have dense canopy cover have a higher reptile species richness than habitats that have open canopy cover.

The results of the analysis of loss and gain species found in figure 4 show changes in rubber plantations to oil palm plantations resulted in 64% of species gain and did not result in species loss. If compared to each land cover, the area that has the highest number of species loss is the IG area with a value of 64%. This is due to differences in canopy cover and ground cover. Besides having canopy cover which is more open than rubber plantations, the ground cover floor in the IG area is the least vegetation than the other areas. Both of these factors greatly influence the presence of herpetofauna because these factors affect the humidity in the place. Meanwhile, the area with the lowest number of loss species is the OG area (27%).

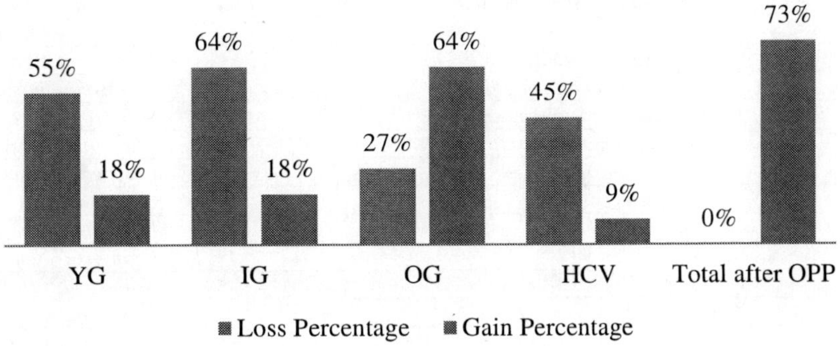

Figure 4. Comparison of herpetofauna species loss and gain percentage from a rubber plantation

Species loss due to land conversion from shrubs to oil palm plantations is 33%, and species gain is 533% (Figure 5). The difference is caused because of significant differences in habitat conditions between OPP and SH area. The OPP area has a dense canopy cover, while the SH area has no canopy cover. The highest species gain is found in OG areas which have a gain of up to 400%. Losses species in each type have the same value of 33%. The only species found in the SH area and not found in the RP area or OPP is *Xenochropis vittatus*.

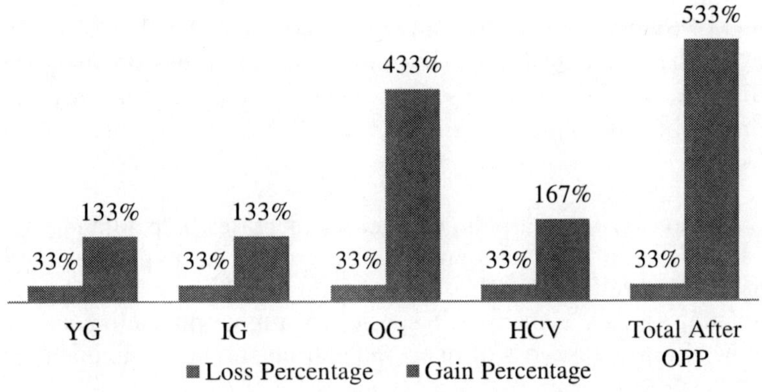

Figure 5. Comparison of herpetofauna species loss and gain percentage from shrub

In Table 2, it can be seen that there are 8 species of herpetofauna that are increasing and there are no species of herpetofauna lost due to conversion from RP to OPP area. Unlike the case with the conversion from SH to OPP area. There are one species of reptile loss due, but there are 17 species of herpetofauna that have increased due to this conversion. This is because the OPP area has canopy cover and ground cover is denser and has enough water sources. [25] which states that many herpetofauna is found in canopy cover conditions which are denser because they cause the habitat to become moist.

Species that can be found in the RP and SH areas are those that have high adaptation, such as green paddy frog or *Hylarana erythraea.* [26] study was conducted in oil palm plantations and found *H. erythraea* become the dominant species and can be found in all lines of observation. It was also stated by [27] that *H. erythraea* has a high tolerance for interference and has a wide distribution on the Indonesian islands.

Table 2. List of species of herpetofauna that are lost and gained

Rubber Plantation		Shrubs	
Loss	**Gain**	**Loss**	**Gain**
-	*Duttaphrynus melanostictus*	*Xenochropis vitattus*	*Duttaphrynus melanostictus*
	Pulchrana glandulosa		*Hylarana erythraea*
	Pulchrana baramica		*Pulchrana glandulosa*
	Chalcorana chalconota		*Amnirana nicobariensis*
	Polypedates leucomystax		*Pulchrana baramica*
	Leptobrachium hasseltii		*Chalcorana chalconota*
	Naja sumatrana		*Polypedates leucomystax*
	Bungarus candidus		*Ingerophrynus biporcatus*
			Limnonectes paramcrodon
			Leptobrachium hasseltii
			Naja sumatrana
			Dendrelaphis pictus
			Hemidactylus frenatus
			Eutropis multifasciata
			Takydromus sexlineatus
			Bungarus candidus
			Bungarus pasciatus

In addition, *Fejervarya cancrivora* and *Fejervarya limnocharis* are also species of amphibians that have high adaptability. This can be proven by the 2 amphibians that can be found in all observation areas, including the Shrub (SH) area which does not have canopy cover. As well as the research of [28] who said that *F.limnocharis* has a high adaptation because it can be found in all seasons. [29] stated that this frog could live in extreme environmental conditions.

One species only found in the SH area is *Xenocrophis vittatus*. This snake can be found in several places on the island of Sumatra, Java [30] and the eastern part of Sulawesi [31]. [32] states that *X. Vittatus* or striped keelback is a snake that lives in water and preys on fish. This is following the condition of the SH area, which is home to several species of fish such as *Clarias nieuhofii, Mystus sp, Channa striata* and *Channa pleurophthalma.*

4. Conclusion

The transformation land cover into oil palm plantations increases herpetofauna species diversity. The existence of oil palm plantations has an impact on differences in herpetofauna species composition. (Rubber Plantation=38%; shrub=81%). The transformation land cover into oil palm plantations led to the gain of herpetofauna species as many as 8 species for rubber plantation and 17 species for shrub, while on the other hand only cause loss of one species from shrub and no one species loss for rubber plantation.

Acknowledgments

Thanks to PT WAIMUSI AGROINDAH for funding this research, providing the permission and facilities provided during the study period.

References

[1] Purba JHV and Sipayung T. 2017. Perkebunan kelapa sawit Indonesia dalam perspektif pembangunan berkelanjutan. Academic Forum on Sustainability I, Jakarta

[2] Direktorat jendral perkebunan. 2016. *Statistik Perkebunan Indonesia 2015-2017*. Jakarta (ID):

Kementerian Pertanian Republik Indonesia

[3] Ulloa JG, Sloan S, Pacheco P, Ghazaoul J, Koh LP. 2012. Lowering environmental costs of oil-palm expansion in Colombia. *Conservation Letters* 366-357

[4] Juniarmi R, Nurdin J, Zakaria IJ. 2014. Population density and lizard distribution (*Mabuya multifasciata*) of small island in Padang city. *Jurnal Biologi Universitas Andalan*. **3**(1): 51-56

[5] Simon E, Puky M, Braun M, Tothmerez B. 2011. Frogs and toads as biological indicator in environmental assessment. Chapther7: Biology ecology and uses. New York (US): Nova Science Publisher

[6] Stebbins RC and Cohen NW. 1997. *A Natural History of Amphibian*. New Jersey (US): Princeton University

[7] Kusrini MD. 2003. Predicting the Impact of the Frog Leg Trade in Indonesia: An Ecological View of the Indonesia Frog Leg Trade, Emphasizing Javanese Edible Frog Species. Bogor (ID): Bogor Agricultural University

[8] Magurran AE. 1988. *Ecological Diversity and Its Measurement*. New Jersey (US): Princeton University

[9] Santosa Y, Sunkar A, Hilwan I, Ginoga LN. 2018. Pendugaan Besaran Perolehan dan Kehilangan Keanekargaman Hayati Akibat Perkebunan Kelapa Sawit. Bogor (ID): Bogor Agricultural University

[10] Muslim T, Sari UK. 2016. Keanekaragaman herpetofauna di lahan reklamasi tambang batubara PT Singlurus Pratama Kalimantan Timur. Seminar Nasional Biologi

[11] Kemp NJ and Burnett JB. 2003. Final Report: A Biodiversity Risk Assessment and Recommendations for Risk Management of Longtailed Macaques (*Macaca fascicularis*) in New Guinea. Washington (US) : Indo-Pacific Conservation Alliance

[12] Poirier FE and Smith EO (1974). The Crab-eating Macaque (*Macaca fascicularis*) of Angaur Island, Palau, Micronesia. *Folia Primatology* **22**: 258-306

[13] Duellman WE dan Trueb L. 1994. *Biology of Amphibian*. London (UK): Johns Hopkins University

[14] Soendjoto MA, Riefani MK, Triwibowo D, Anshari MN, Metasari D. 2015. Fauna in the reclamation area of PT Adaro Indonesia, South Kalimantan, after revegetated less than two years. Seminar Nasional Konservasi dan Pemanfaatan Sumber Daya Alam, Universitas Sebelas Maret, Surakarta, Indonesia

[15] Aryanti NA, Maarif S, Prabowo A. 2017. Status konservasi konservasi jenis burung di kawasan lereng Gunung Argopuro, Probolinggo. Prosiding Seminar Nasional III, Universitas Muhammdiyah Malang

[16] Putra K, Rizaldi, Tjong DH. 2012. Komunitas anura (amphibia) pada tiga tipe habitat perairan di kawasan hutan harapan Jambi. *Jurnal Biologi Universitas Andalas*. **1**(2) : 158-165

[17] Hanifa BF, Ismi N, Setyobudi W, Utami B. 2016. Kajian keanekaragaman dan kemelimpahan ordo anura sebagai indikator lingkungan pada tempat wisata di Karasidenan Kediri. Seminar Nasional Pendidikan dan Saintek 2016 2557-53

[18] Ariza YS, Dewi BS dan Darmawan A. 2014. Keanekaragaman jenis amfibi (ordo anura) pada beberapa tipe habitat di youth camp desa hurun Kecamatan Padang Cermin Kabupaten Pesawaran. *Jurnal Sylva Lestari*.**2** (2): 21-30

[19] Arista A, Winarno GD dan Hilmanto R. 2017. Keanekaragaman jenis amfibi untuk mendukung kegiatan ekowisata di desa Braja Harjosari Kabupaten Lampung Timur. *Biosfera* **34** (3): 103 -109

[20] Iskandar DT. 1998. Amfibi Jawa dan Bali LIPI-Seri Panduan Lapang Bogor. Bogor (ID): Puslitbang Biologi Lipi

[21] Santosa Y, Dalimunthe Y, Purnama Sari I. 2017. Impact of oil palm plantation on mammal and herpetofauna diversity: a case study in South Sumatra Province. *International Journal Of Environmental And Ecological Engineering*. **11** (9)

[22] Santosa Y dan Rejeki SSS. Variation of herpetofauna in different landcover types in PT ASMR

oil palm plantation Central Kalimantan Province. AIP Conference Proceeding 2019 060003

[23] Wanger TC, Iskandar DT, Motzke I, Brook BW, Sodhi NS, Clough Y, Tscharntke T. 2009. Effect of land use change on community composition of tropical amphibians and reptile in Sulawesi, Indonesia. *Conservation Biology*. **24** (1): 795-802

[24] Putra PT. 2011. Evaluasi kenyamaan pada beberpa taman kota di Jakarta Pusat Bogor(ID): Institut Pertanian Bogor

[25] Halliday t and Adler k. 2000. The encyclopedia of reptiles and amphibians. Fact on file inc. New York

[26] Ghafariansyah FM. 2017. Komunitas amfibi di perkebunan kelapa sawit, Kabupaten Kotawaringin Barat, Provinsi Kalimantan Tengah. Bogor (ID): Bogor Agricultural University.

[27] Inger RF and Stuebing RB (2005). *A Field Guide To The Frogs Of Borneo*. Kota Kinabalu (ID): Natural History Publications

[28] Shahriza S and Ibrahim J. 2014. A preliminary checklist of amphibians of Uli Paip Recreational Forest, Kedah, Malaysia. *Journal Of Species List And Distribution*. **10** (2): 253-259

[29] Almeria ML and Nuneza OM. 2013. Diet of seven anuran species (Amphibia: Anura) in Agusan Marsh, Mindanao, Philippines. *Animal Biology and Animal Husbandry*. **5**(1): 116-126

[30] Pallewatta N, Reaser JK, Gutierrez AT. 2003. Prevention and management of invasive alien species: proceedings of a workshop on forging corporation throughout south and southeast Asia. Global invasive species programme cape town, South Africa

[31] Pauwels OSG. Wallach V. David Patrick. 2008. Global diversity of snakes (serpentes; reptilia) in freshwater. *Hydrobiologia*. **595**; 599-605

[32] Wicaksono A, Madang K, Dayat E. 2015. Identification of types of snake in the village Muktijaya Muara Telang District Banyuasin and its contribution to biology in high school. Universitas Sriwijaya

IOP Conf. Series: Earth and Environmental Science **336** (2019) 012031 doi:10.1088/1755-1315/336/1/012031

Impact of oil palm plantations on herpetofauna species diversity in KGP and CNG, West Kalimantan

Putri Rahmadiyanti[1]*, Yanto Santosa[2]

[1]Magister Student of Tropical Biodiversity Conservation, Faculty of Forestry, Bogor Agricultural University

[2]Department of Forest Resources Conservation and Ecotourism, Faculty of Forestry, Bogor Agricultural University

*E-mail: prahma189@gmail.com

Abstract. Palm oil (*Elaeis guineensis* Jacq) is one of the fastest growing in the tropics so that the development of oil palm plantation is the one threats of the decrease of biodiversity including herpetofauna. That is why this research becomes important to study the impact of oil palm plantation on diversity and composition of herpetofauna. This research was conducted in July 2018 in Ketapang District, West Kalimantan Province in two oil palm plantation companies (KGP and CNG). Data collecting were using *Visual Encounter Survey* (VES) method combined with Strip Transect Method, in five lands covers (young-aged, middle-aged, old-aged oil palm plantation, high conservation value, and shrubs area) taken simultaneously with three repetitions. The results of this study showed that the establishment of oil palm plantations has impacted species diversity and species composition of herpetofauna. This research showed that the species number of herpetofauna in oil palm plantations is higher than the baseline area. The result of this study also showed that the species diversity and richness of herpetofauna in the high conservation value area is higher compared to another land cover. However, the overall results showed that oil palm plantation has positive impacts to increased and loss of herpetofauna species, there are losing one species and gaining seven species in CNG oil palm plantation, and losing two species and gaining fifteen species in KGP oil palm plantation.

1. Introduction

Land cover change is used to explain the occurrence of loss of diversity because it causes a shift in the composition of the community where some species can disappear (Species Loss) while some other species increase (Species Gain). In 1950 Indonesia's forest cover reached 145 million ha, but there has been massive deforestation since the 1970s caused the land cover has decreased to 20 million ha from 1985 to 1997. The biggest changes in land cover occurred in Sumatera, Kalimantan and Sulawesi due to the clearing of agricultural land and large-scale development [1]. One of the major changes in land cover is oil palm plantations. Based on statistical data from the Indonesian Plantation of Palm Oil Commodities [2], up to now the potential for industry and expansion of the area of Indonesian oil palm plantations have continued to increase according to concession status from 1970 to 2016. The total area of oil palm plantations in Indonesia in the last 10 years as of 2006 was 6.594.914 ha and in 2016 increased to an area of 11.914.499 ha. The data showed that the increasing number of area of oil palm plantations is in line with the increasing demand for palm oil production in Indonesia.

The development of oil palm plantations has a positive impact on social and economic because it provides greater income to cultivated farmers [3]. Also, according to the data [4] the development of oil palm plantations on the islands of Sumatera, Kalimantan, and Sulawesi attracts not only economic growth in the centers of oil palm plantations but also attracts economic growth in Java, Bali and others. However, the palm oil plantation sector is often indicated as a cause of a decline in biodiversity and environmental damage. Every research of [5]; [6]; [7], said that oil palm plantations are globally poor habitat for the lives of several taxa so that the types that normally inhabit forests are reduced due to changes in land cover to oil palm. The same study also mentions that the rapid development of oil palm plantations in the tropics, such as those in Malaysia and Indonesia, which do a lot of oil palm cultivation, causes a decrease in diversity due to habitat fragmentation [8]. Research related to herpetofauna is still rare, so it is important to prove whether oil palm plantations are the trigger for the loss of diversity in herpetofauna species or, on the contrary, increase the type of herpetofauna. The research aims to determine the impact of oil palm plantations on the diversity of herpetofauna species, to know the impact of oil palm plantations on the composition of herpetofauna species and to know the amount of loss and acquisition caused by oil palm plantations in two areas of oil palm plantations.

2. Materials and methods

2.1. Time and location

The research was conducted in two types of areas: (1) baseline area (shrubs area), and (2) oil palm plantation areas where categories become young-aged oil palm plantations area, middle-aged oil palm plantations area, old-aged oil palm plantations area, and high conservation value area. The research were conducted in two companies of oil palm plantations i.e. KGP and CNG in Ketapang District, West Kalimantan Province, on July 2018.

2.2. Data collection

Data collection method for identifying land cover types before being oil palm plantation area is using Landsat imagery interpretation from one year before. Herpetofauna diversity was conducted using *Visual Encounter Survey (VES)* method combined with Strip Transect Method [9] and time search method for 2 hours starting from 7:00 p.m. to 9:00 p.m. with three repetitions simultaneously in all research locations. The research also observed the condition of the location, such as canopy cover, the condition of the ground and undergrowth vegetation, availability of water source as well as temperature and relatively humid.

2.3. Time and location

Impacts of oil palm plantations on herpetofauna diversity were determined by comparing the current biodiversity found on each plantation site, with those found on the adjacent land cover types that resembled the previous land cover types. To determining the herpetofauna diversity were used species richness according to species richness index (Margalef index) and Evenness index. To compare the similarity of baseline area and oil palm plantations area, Sorensen index of similarity was used. Data analysis using the formula as follows:

a) **Species Number**
The number of types found on the observation in each type of land covers.

b) **Species Richness Index (D_{mg})** Margalef Index (species richness) used to serves to determine the species richness of each species in each community encountered. The following formula is used [10].

$$Dmg = \frac{(S-1)}{\ln(N)}$$

Note:
D_{mg} : margalef diversity

S : number of species observed
N : number of individuals
ln : natural logarithm

c) **Evenness Index (E)**

Evenness Index has a function to determine the evenness of each type in every community [10].

$$E = \frac{H'}{\ln(S)}$$

Note:
E : evenness index
H': diversity
ln : natural logarithm
S : number of species observed

For comparing the similarity of the age of oil palm plantations, it is used the index of similarity of Sorensen [10] with formula:

$$IS = \frac{2C}{(A+B)}$$

Note:
IS : Sorensen Index
A : the number of species that only found in community A
B : the number of species that only found in community B
C : the number species which is found in both two communities

3. Results and discussion

3.1. Comparison of herpetofauna diversity in baseline and oil palm plantation area

Based on result of the observation of herpetofauna species in CNG oil palm plantations, in all research area i.e the baseline area (the shrubs area with vegetations), and oil palm plantations areas were categories the young-aged oil palm plantations (2011 years old), middle-aged oil palm plantations (2008 years old), old-aged oil palm plantations (2007 years old) and high conservation value (HCV) area, has found fifteen species (eleven amphibians and four reptiles) from seven families, there are Dicroglossidae family (four species), Ranidae family (five species), Rhacophoridae family (two species), Colubridae family (one species), Gekkonidae family (one species), Pythonidae family (one species) and Viperidae family (one species). The type and distribution of all discovered herpetofauna in CNG oil palm plantations, from observation sites, can be seen in table 1.

Table 1. Type and distribution of the discovered herpetofauna in CNG oil palm plantations

Family	Species	Oil Palm Plantations				Baseline Area
		Young-aged Oil Palm	**Middle-aged Oil Palm**	**Old-aged Oil Palm**	**HCV Area**	**Shrubs**
Dicroglossidae	*Fejervarya limnocharis*	3	0	0	1	3
Dicroglossidae	*Limnonectes paramacrodon*	1	0	6	2	1
Dicroglossidae	*Fejervarya cancrivora*	0	0	1	0	0
Dicroglossidae	Limnonectes sp	0	0	0	1	1
Ranidae	*Hylarana erythraea*	1	0	2	5	0
Ranidae	*Pulcharana baramica*	4	2	0	2	4
Ranidae	*Amnirana nicobariensis*	0	0	2	0	0
Ranidae	*Chalcorana raniceps*	0	0	1	2	0
Ranidae	*Pulcharana glandulosa*	0	0	0	1	0
Rhacophoridae	*Polypedates macrotis*	1	0	1	0	1
Rhacophoridae	*Polypedates leucomystax*	0	2	1	0	1
Colubridae	*Dendrelaphis pictus*	0	0	1	0	0

Family	Species	Oil Palm Plantations				Baseline Area
		Young-aged Oil Palm	Middle-aged Oil Palm	Old-aged Oil Palm	HCV Area	Shrubs
Gekkonidae	*Hemidactylus platyurus*	0	0	0	0	1
Pythonidae	*Malayopython reticulatus*	1	0	0	0	0
Viperidae	*Trimeresurus borneensis*	1	1	0	1	1
Individual Number		**12**	**5**	**15**	**15**	**13**
Species Number		**7**	**3**	**8**	**8**	**8**

Based on the data in table 1, the type that is often found is from the family Dicroglossidae and Ranidae in each land cover. *Limnonectes paramacrodon* of the family Dicroglossidae is one of the species that found in almost every land cover research area, except for middle-aged oil palm plantations. Referring to the research of [11] states that *Limnonectes paramacrodon* is classified as the most commonly found individual in the region that has flowing water. The same study was also found in [12] that the genus Limnonectes was found in many riverbank areas. This correlates with the results of the study, all types of *Limnonectes paramacrodon* were found in areas that have water sources except for middle-aged oil palm plantations. The reason is that the location of the oil palm plantation is close to the employee housing and does not have a flow of water at this location, so it is difficult to find *Limnonectes paramacrodon*. The diversity of frog species is less found in disturbed habitats such as gardens and residential areas [13]. Similar to the Ranidae family of the *Hylarana erythraea* species, it is often found in the research area except for middle-aged oil palm plantations and shrub areas. According to [14] *Hylarana erythraea* like waterlogged places and can be found in fragmented habitats such as oil palm plantations. So that this species is mostly found in the research area (young-aged, old-aged oil palm plantations and HCV areas) besides due to water factors, these frogs can adapt to disturbed habitats.

While, based on these results in KGP oil palm plantations, the baseline area used as the shrubs area with not canopy cover, and for observation area in oil palm plantations used the young-aged oil palm plantations (2010 years old), middle-aged oil palm plantations (2008 years old), old-aged oil palm plantations (2007 years old) and high conservation value (HCV) area. The observations on all research area has found seventeen species (twelve amphibians and five reptiles) from ten families, they are Bufonidae family (one species), Dicroglossidae family (four species), Megophryidae family (one species), Ranidae family (four species), Rhacophoridae family (two species), Elapidae family (one species), Gekkonidae family (one species), Homalopsidae family (one species), Scincidae family (one species) and Viperidae family (one species). The type and distribution of all discovered herpetofauna in KGP oil palm plantations, from observation sites, can be seen in table 2.

Table 2. Type and distribution of the discovered herpetofauna in KGP oil palm plantations

Family	Species	Oil Palm Plantations				Baseline Area
		Young-aged Oil Palm	Middle-aged Oil Palm	Old-aged Oil Palm	HCV Area	Shrubs
Bufonidae	*Pseudobufo subasfer*	0	0	0	1	0
Dicroglossidae	*Fejervarya cancrivora*	0	1	0	0	0
Dicroglossidae	*Fejervarya limnocharis*	0	0	0	0	1
Dicroglossidae	*Limnonectes paramacrodon*	0	0	0	4	0
Dicroglossidae	Limnonectes sp	0	2	0	1	0
Megophryidae	*Leptobrachium abbotti*	0	0	0	0	1
Ranidae	*Chalcorana raniceps*	0	1	1	4	0
Ranidae	*Hylarana erythraea*	4	1	0	1	0
Ranidae	*Pulcharana baramica*	0	0	3	0	0
Ranidae	*Pulcharana glandulosa*	4	0	2	0	0
Rhachoporidae	*Polypedates leucomystax*	0	1	0	1	0

Family	Species	Oil Palm Plantations				Baseline Area
		Young-aged Oil Palm	Middle-aged Oil Palm	Old-aged Oil Palm	HCV Area	Shrubs
Rhachoporidae	*Polypedates macrotis*	0	0	0	1	0
Elapidae	*Naja sumatrana*	1	0	0	0	0
Gekkonidae	*Hemidactylus platyurus*	0	1	0	0	0
Homalopsidae	*Homalopsis buccata*	0	0	0	1	0
Scincidae	*Eutropis rudis*	1	1	0	0	0
Viperidae	*Trimeresurus borneensis*	0	0	0	1	0
Individual Number		**10**	**8**	**6**	**15**	**2**
Species Number		**4**	**7**	**3**	**9**	**2**

Based on the data in table 2, the most common types were from the family Ranidae of the *Pulcharana glandulosa* and *Chalcorana raniceps*. *Pulcharana glandulosa* species was found in young-aged oil palm plantations and old-aged oil palm plantations. Both locations have water sources in the form of rivers and small ditches. According to [15] *Pulcharana glandulosa* species like lowland and swamp habitats, which are drained with calm water. Based on the findings of the study, this species was not found in other locations such as HCV areas and shrubs. This species is dominantly found in oil palm plantation habitat. So that it can be indicated that this type is easy to adapt to damaged habitat such as in oil palm plantations. Referring to the study [16], it was stated that the family Ranidae of the genus Hylarana became the character of degraded habitat. Other species of the Ranidae family found in various land coverings namely *Chalcorana raniceps* were found in middle-aged, old-aged oil palm plantations and HCV areas. This type is commonly found in HCV areas, with habitat conditions varying at the level of vegetation. *Chalcorana raniceps* is a species that inhabits extensive primary and secondary lowland forest habitats [17]. This species has close relatives with the types of *Hylarana chalconota* which are spread in Sumatera, Java, Bali, Kalimantan and Sulawesi [18].

In CNG oil palm plantations, the highest number of species and richness index was found in the baseline area (S= 8; Dmg= 2.73). While the lowest total number species and richness index was found in the middle-aged oil palm plantation (S= 3; Dmg= 1.24). Meanwhile, the evenness index for all lands cover varies between 0.87 to 0.96, indicating no dominant species on all lands covers. While in KGP oil palm plantations, the highest number of species and richness index was found in the HCV area (S= 9; Dmg= 2.95). While the lowest total number species and richness index were found in the baseline area (S= 2; Dmg= 1.44). Meanwhile, the evenness index for all lands cover varies between 0.86 to 1. The total number of species, species richness index and evenness index in all study area can be seen in figure 1.

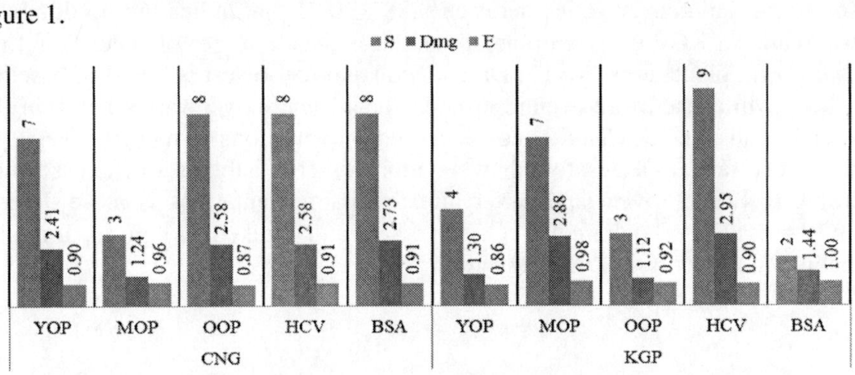

Notes: YOP: Young-aged oil palm, MOP: Middle-aged oil palm, OOP: Old-aged oil palm, HCV: High conservation value area, BSA: Baseline area

Figure 1. Comparison of the total number of species, species richness, and evenness index

In both oil palm plantations, the high number of species is directly proportional to the species richness index (Dmg). In CNG oil palm plantations the highest number of species and wealth was found in the baseline area (eight species), and the lowest was found in middle-aged oil palm cover (three species). The high number of species and species richness found in the baseline area is caused by differences in habitat conditions, where dense canopy cover with forest floor conditions is covered by understorey. The structure and composition of canopy cover influence the presence of herpetofauna, based on research [19] many amphibians occupy available habitat for feed resources, suitable microclimate and the existence of shelter and breeding. Whereas the low number of species and species richness found in middle-aged oil palm plantations is due to habitat conditions having fairly open canopy cover with clean garden floors and no water sources found in this area, these conditions do not support the existence of herpetofauna. This is consistent with the research of [20] that herpetofauna tends to be found more in watery places because according to its microhabitat conditions, herpetofauna likes moist places. Also, this location is adjacent to employee housing, so that the higher level of disturbance causes a low number of species found because herpetofauna especially amphibians are very sensitive to environmental changes [21].

While the oil palm plantation KGP the highest number of species and wealth was found in the HCV area (nine species), and the lowest was found in the baseline area (two species), and the evenness index in both oil palm plantations varied between 0.86 to 1 indicating that no species dominated the entire plantation palm oil. According to [22] the greatest evenness index value is 1 (maximum evenness index). Based on figure 1, the high number of species and species richness found in the HCV area is because this location has a tight canopy cover condition which is dominated by the tree level, then there are several river streams around this area. A similar study also states that significant differences in species richness and composition between forests, not forests, and plantations can be seen from the compilers of habitat structures and the characteristics of microclimates. Forested areas such as HCV area have a more diverse structure of terrestrial and arboreal habitat diversity [14]. Whereas the low number of species and species richness found in the baseline due to very open habitat conditions, the ground floor is dominated by reeds. Also, the location of the water source was not found and was very dry, so the level of encounter with herpetofauna was very low. This indicates that herpetofauna, especially amphibians, like high humidity sites to protect the body from drought [18].

3.2. Impacts of oil palm plantation on herpetofauna diversity

Comparison between the baseline and oil palm plantations (figure 2) showed that in CNG oil palm plantations, the highest number of species and richness index was found in the palm plantations (S= 14; Dmg= 3.38), and the lowest is found in the baseline area (S= 8; Dmg= 2.73). Meanwhile, the score evenness index for all land cover varies between 0.88 to 0.91 that indicating no dominant species on both locations. While In KGP oil palm plantations the highest number of species and richness index was found the oil palm plantations (S= 15; Dmg= 3.82) and the lowest is found in baseline area (S= 2; Dmg= 1.44). Meanwhile, the score evenness index for all land cover varies between 0.91 to 1. The high addition or low loss in each land cover is caused by variations in habitat diversity which affect the variety of herpetofauna species found, were more diverse habitats to increase the diversity of species according to [23]. Proven land cover after oil palm plantations is more diverse due to the presence of HCV areas and oil palm cover. On the other hand, land cover before palm oil is only in the form of one habitat in the form of scrub vegetation.

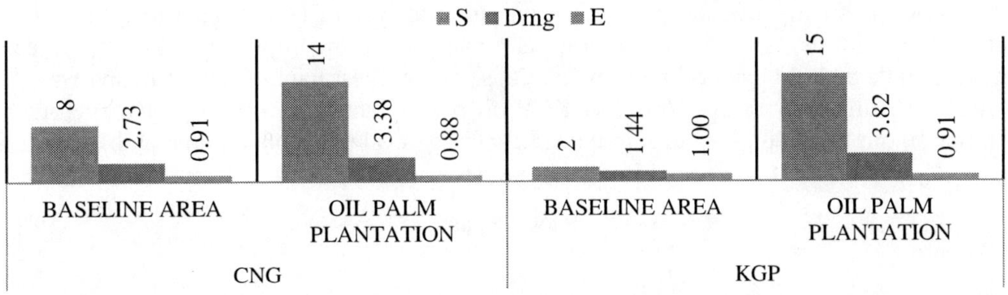

Figure 2. Comparison of total number of species, species richness and evenness index in baseline and oil palm plantations area

In overall results in this research, the conversion from shrubs area to oil palm plantation has led to an increase in species and species loss. In CNG oil palm plantations there is an increase of seven species they are *Fejervarya cancrivora, Hylarana erythraea, Amnirana nicobariensis, Chalcorana raniceps, Pulcharana glandulosa, Dendrelaphis pictus*, and *Malayopython reticulatus*, and losing one species is *Hemidactylus platyurus*. Meanwhile, KGP oil palm plantations there was an increase of fifteen species they are *Pseudobufo subasper, Fejervarya cancrivora, Limnonectes paramacrodon,* Limnonectes sp, *Chalcorana raniceps, Hylarana erythraea, Pulcharana baramica, Pulcharana glandulosa, Polypedates leucomystax, Polypedates macrotis, Naja sumatrana, Hemidactylus platyurus, Homalopsis buccata, Eutropis rudis*, and *Trimeresurus borneensis*, and losing two species is *Fejervarya limnocharis* and *Leptobrachium abbotti*. The total loss or gain of herpetofauna species in each oil palm plantation can be seen in table 3 and 4:

Table 3. Total number of loss and gain species (baseline-oil palm plantations) in the CNG oil palm plantation

No	Species Loss	No	Spescies Gain
1	*Hemidactylus platyurus*	1	*Fejervarya cancrivora*
		2	*Hylarana erythraea*
		3	*Amnirana nicobariensis*
		4	*Chalcorana raniceps*
		5	*Pulcharana glandulosa*
		6	*Dendrelaphis pictus*
		7	*Malayopython reticulatus*

Table 4. Total number of loss and gain species (baseline-oil palm plantations) in the KGP oil palm plantation

No	Species Loss	No	Species Gain
1	*Fejervarya limnocharis*	1	*Pseudobufo subasper*
2	*Leptobrachium abbotti*	2	*Fejervarya cancrivora*
		3	*Limnonectes paramacrodon*
		4	Limnonectes sp
		5	*Chalcorana raniceps*
		6	*Hylarana erythraea*
		7	*Pulcharana baramica*
		8	*Pulcharana glandulosa,*
		9	*Polypedates leucomystax*
		10	*Polypedates macrotis*
		11	*Naja sumatrana*
		12	*Hemidactylus platyurus*
		13	*Homalopsis buccata,*
		14	*Eutropis rudis*
		15	*Trimeresurus borneensis*

The loss or gain of herpetofauna species in detail to each type of land cover can be seen in (figure 3). Based on the result in research, In CNG oil palm plantations, the highest loss of herpetofauna was found in the middle and old aged oil palm plantations. The highest gain of herpetofauna was found in the old-aged oil palm plantations. While, in KGP oil palm plantations the loss of herpetofauna was found similar among the young-aged, middle-aged, old-aged oil palm plantations and HCV area. The highest gain of herpetofauna was found in the HCV area.

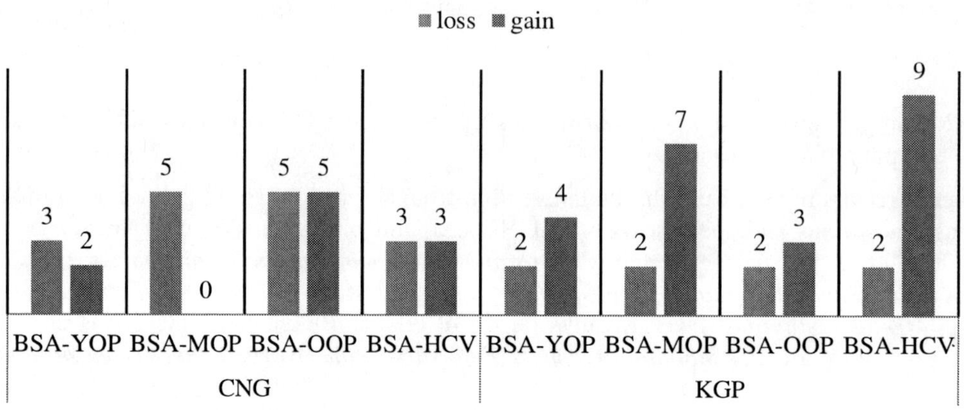

Notes: YOP: Young-aged oil palm, MOP: Middle-aged oil palm, OOP: Old-aged oil palm, HCV: High conservation value area, BSA: Baseline area

Figure 3. The total number of loss and gain species for all lands covers

Based on the above data on CNG oil palm plantations, the loss of herpetofauna species occurs in each land cover, the highest loss is in middle-aged and old-aged oil palm plantation five species. Meanwhile, the conversion of shrubs to oil palm plantations has brought new species of herpetofauna, but no additional species has occurred in middle-aged oil palm plantations. Whereas in oil palm plantations KGP, the loss of herpetofauna was found similar to two species of young-aged, medium-aged, old-aged and HCV area. Meanwhile, the conversion of shrubs to oil palm plantations has brought new species of herpetofauna to all land cover, the highest increase being found in the HCV area. Judging from the addition of species from the two companies, it was shown that HCV land cover gave an important role to species addition, such as in KGP oil palm plantations. HCV land cover carries an increase of nine species, this is because HCV has a more diverse land cover, there are many trees, close to water sources and soil floors overgrew with dense vegetation. So the HCV area can be said as a habitat that supports the existence of wildlife because of the availability of physical and biotic factors that support the survival of herpetofauna [24].

3.3. Impacts of oil palm plantation on species composition

The similarity degree of discovered species of herpetofauna based on Sorensens Index can bee seen in table 5. This value shows the similarity of the baseline area with each land cover in two oil palm plantation companies.

Table 5. Impact of oil palm plantations to the composition of herpetofauna species diversity

Oil Palm Plantation	BSA-YOP	BSA-MOP	BSA-OOP	BSA-HCV	Total (after becoming oil palm)
CNG	0.67	0.55	0.38	0.63	0.64
KGP	0.00	0.00	0.00	0.00	0.00

Notes: YOP: Young-aged oil palm, MOP: Middle-aged oil palm, OOP: Old-aged oil palm, HCV: High conservation value area, BSA: Baseline area

The comparison of Sorensen Index from all oil palm plantation land cover with before palm oil cover (baseline) shows that in oil palm plantations CNG the cover of young-aged oil palm land has the highest similarity with the baseline land cover of 0.67. According to [25] which states that community similarity can be formed by several environmental factors such as the distance between adjacent habitats and the same composition of vegetation. The high similarity of these types can be caused by having similar characteristics of the habitat. Young-age palm cover tends to be almost the same because of the close closure of the canopy, the ground floor is covered with understorey and litter, and the number of herpetofauna species in both land cover is almost the same whereas the smallest similarity value is indicated by the cover of old oil palm land of 0.38, which means that 0.62 shows the difference between the old palm oil variety and the baseline area. Meanwhile, if the similarity value is calculated as a whole on CNG oil palm plantations with a baseline the similarity value is 0.64, which means that 36% indicates different types of both. While in the oil palm plantation company KGP there is no similarity in the type of land in all fields, this indicates that no similar species is found in all land cover. The reason is that the baseline land cover (shrubs) has different habitat characteristics with young, medium, old, and HCV oil palm, is the location is open because there is no canopy cover, dominated by Imperata, there is no water source around the location, so the soil is relatively dry. So that none of the same types is found in this location.

4. Conclusion

The study concluded that the highest herpetofaunas species diversity was found in oil palm plantations area than in the baseline land cover (shrubs). Oil palm plantation has positive impacts on increased or gain and loss of herpetofauna species in both oil palm plantations, where there is losing one species and gaining seven species in CNG oil palm plantation, and lossing two species and gaining fifteen species in KGP oil palm plantation.

Acknowledgments

Further thanks to BPDP (Badan Pengelola Dana Perkebunan Kelapa Sawit) for funding this study. Thanks are also due to PT. KGP and CNG for providing the permissions and facilities provided during the study period.

References

[1] [FWI]. Forest Watch Indonesia. 2001. Keadaan Hutan Indonesia. Bogor (ID): Forest Watch Indonesia dan Washington DC Global Forest Watch

[2] [Ditjenbun]. Direktorat Jendral Perkebunan. 2016. Statistika Perkebunan Indonesia 2015-2017. Jakarta (ID): Direktorat Jendral Perkebunan

[3] Castiblanco C, Etter A, Ramirez A. 2015. Impacts of oil palm expansion in Colombia: What do socioe conomic indicators show?. Land Use Policy **44**(1):31-43

[4] [GAPKI] Gabungan Pengusaha Kelapa Sawit Indonesia. 2013. Indonesia dan perkebunan kelapa sawit dalam isu lingkungan global [Internet]. [Download 20 September 2018]. https://www.acedemia.edu/7536010/INDONESIA_DAN_PERKEBUNAN_SAWIT_DALAM_ISU_LINGKUNGAN_GLOBAL.pdf

[5] Srinivas A, Koh LP. 2016. Oil palm expansion drives avifaunal decline in the Pucallapa region of Peruvian Amazonia. Global Ecology and Conservation **7**(1):183-200

[6] Correa F, Juen L, Rodrigues LC, Silva-Filho HF, Santos-Costa MC. 2015. Effects of oil palm plantations on anuran diversity in the eastern Amazon. Animal Biology **55**(1):321-335

[7] Kissinger, Pitri RMN, Hamdhani. 2016. Perubahan komposisi jenis vegetasi dan burung setelah penanaman kelapa sawit di lahan kering Pelaihari Kalimantan Selatan. Enviro Scinteae **12**(1):28-34

[8] Vargas LEP, Laurence WF, Clements GR, Edwards W. 2015. The impacts of oil palm agriculture on Colombia's biodiversity: what we know and still need to know. Tropical Conservation Science **8**(3):828-845

[9] Heyer WR, Donnelly MA, McDiarmid, Hayek LC, Foster MS. 1994. Measuring and Monitoring Biological Diversity: Standared Methodes for Amphibians. Washington (US):

Pricenton University Press
[10] Magurran A E. 1988. Ecological Diversity and Its Measurement. New Jersey (US): Princeton University Press
[11] Putra K, Rizaldi, Tjong DH. 2012. Komunitas anura (amphibia) pada tiga tipe habitat perairan di Kawasan Hutan Harapan Jambi. J.Bio.UA **1**(2):156-165
[12] Wanda IF, Novarino W, Tjong DH. 2012. Jenis-jenis anura di Hutan Harapan Jambi. J.Bio.UA **1**(2): 99-107
[13] Ul-Hasanah AU. 2006. Amphibian diversity in Bukit Barisan Selatan National Park, Lampung Bengkulu [Skripsi]. Institut Pertanian Bogor
[14] Gillespie GR, Ahmad E, Elahan B, Evans A, Ancrenaz M, Goossens B, Scroggie MP. 2012. Conservation of amphibians in Borneo: Relative value of secondary tropical forest and non-forest habitats. Biological Conservation **152**:136–144
[15] Zainudin R, MD Zain BM, Ahmad N, M Nor S. 2017. Microhabitat partitioning of closely related Serawak (Malaysian Borneo) frogs species previously assigned to the genus hylarana (amphibia:anura). Turkish Journal of Zoology **41**:876-891
[16] Zainudin R. 2002. Frog diversity at Sedilu Peat Swamp Forest Reserve. Sarawak Mal Nat J **56**:217-223
[17] Inger RF, Stuebing RB. 2005. A Field Guide to the Frogs of Borneo. Kota Kinabalu (MY): Natural History Publications.
[18] Iskandar DT. 1998. Amfibi Jawa dan Bali. Bogor (ID): Pustlitbang LIPI
[19] Beljai M, worabi MS. 2018. Struktur dan komposisi vegetasi serta keanekaragaman jenis amfibi di hutan Pegunungan Arfak, Papua Barat. Proc. SEM NAS MASY BIODIV INDON **4**(1):1-12
[20] Muslim T, Sari UK, Yassir I. 2017. The dominant species of herpetofauna in the coal mining area at East Kalimantan Indonesia. Biodiversitas **18**(2):773-779
[21] Yudha DS, Eprilurahman R, Trijoko, Alawi MF, Tarekat AA. 2014. Keanekaragaman jenis katak dan kodok (ordo) anura di sepanjang sungai opak Provinsi Daerah Istimewa Yogyakarta. Jurnal Biologi **18**(2):52-59
[22] Santosa Y, Ramadhan EP, dan Rahma DA. 2008. Studi keanekaragaman mamalia pada beberapa tipe habitat di stasiun penelitian Pondok Ambung Taman Nasioal Tanjung Puting kalimantan Tengah. Media konservasi **13**(3):1-7
[23] Kusrini MD. 2008. Pedoman Penelitian dan Survei Amfibi di Alam. Bogor (ID): Fakultas Kehutanan Institut Pertanian Bogor
[24] Alikodra HS. 2002. Pengelolaan Satwaliar Jilid 1. Bogor (ID): Yayasan Penerbit Fakultas Kehutanan Institut Pertanian Bogor
[25] Kendeigh SC. 1980. Ecology with Special Reference to Animal and Men. New Delhi (IN): Prentice Hall of India Private limited

Impact of oil palm plantation on the butterfly diversity: a Case study in KGP & CNG, Ketapang, West Kalimantan

A R Mutmainnah[1], Y Santosa[1]

[1]Wildlife Ecology and Management, Forest Resource Conservation and Ecotourism Department, Faculty of Forestry, Bogor Agricultural University, IPB Dramaga Campus, Bogor 16680, West Java, Indonesia

*E-mail: rezki_007@apps.ipnb.ac.id

Abstract. Oil palm plantation in Indonesia is rapidly growing, encouraging the expansion of oil palm plantation area. The expansion estimated may decrease the biodiversity including the butterfly diversity. The research aimed to see the impact of oil palm plantation on the butterfly diversity that occurs in KGP and CNG oil palm plantation. The research was conducted in July 2018 by strip transect method. Impact of the establishment oil palm plantation on butterfly diversity was determined by comparing the current butterflies found on oil palm plantations area with the butterflies found on the baseline (shrub area). The result showed an increase in butterfly diversity and species richness index in KGP and CNG oil palm plantation. Total number of species and richness index in both oil palm plantations area were higher than the baseline area. The establishment of oil palm plantation in KGP oil palm plantation result gaining 22 species and losing 2 species, while in CNG oil palm plantation result is gaining 31 species and losing 5 species.

1. Introduction

Expansion of oil palm plantation in Indonesia according to data Directorate General of Plantations 2016 has increased significantly by approximately 10 million hectares in the last 20 years. The expansion is encouraged by the high economic value of palm oil [1]. Extensive growth has implications for increasing outcome that support the national economy and as a people's livelihood [2]. The expansion was obtained from conversion of various types of land cover. However, the controversy over changing land cover into oil palm plantations was already familiar and often debated. The issue of land conversion was often accused by several experts as one of the main factors causing deforestation and decreasing biodiversity [2]; [3]; [4]. [5] stated that primary and secondary forest converted into oil palm plantation lowered the butterfly species richness by around 79-83%.

The paradigm change towards the sustainable oil palm plantations can be achieved by considering aspects of economic, social, cultural and environmental policies, especially the ecological aspects that are the main of international attention. Consideration of the policy aspects related to the economy, socio-culture, and environment which are the most important parts of the HCV Classification (High Conservation Values) and Roundtable for Sustainable Palm Oil (RSPO) are certainly worth noting towards the management of sustainable oil palm plantations [6]. Butterflies are part of the biodiversity that must be preserved from extinction and species diversity decline. Butterflies have important values, such as ecological values (pollinators), endemism, conservation, education, culture, aesthetics, and economics [7]. Therefore, research about the impact of oil palm plantations on biodiversity including butterfly is important to carry out. The aim of this research is to (1) Identify the impact of oil palm plantation on a total number of butterfly, and (2) Identify the impact of oil palm plantation on species composition on the butterfly.

2. Materials and methods

2.1. Time and Location

The study was conducted in Ketapang, West Kalimantan Province, in July 2018. There were 2 types of land covers: (1) baseline area (Shrubs area); and (2) Oil palm plantation area (young-aged oil palm plantations, middle-aged oil palm plantations, old-aged oil palm plantations and high conservation value area).

2.2. Data Collection

Historical data of land cover before the establishment of the plantations were collected by literature study, interview with manager and local communities around the plantation and interpretation of satellite imagery one year or 2 years before establishment. Based on the satellite imagery interpretation, the baseline (a type of land cover that dominates the area before the development of oil palm plantations) namely shrubs area. The observation was conducted using transect method from 09.00-11.00 with 3 repetitions simultaneously in all research location. Observations were carried out at a constant speed by using a butterfly net to catch butterflies directly and also helped by a butterfly trap which was installed at ± 1m height, there are 6 traps on each observation track. The conditions that were also observed such as the condition of undergrowth vegetation, canopy cover, the presence of water sources, temperature and humidity.

2.3. Data Analysis

Data analysis that used to find out the variety of species diversity in various types of land cover is the number of species, species richness index (Margalef index) and Evenness index and species similarity index.

2.3.1. Number of Species (S).

The number of species is the number of all types of butterfly found in PKWE plantation.

2.3.2. *Species Richness Index (Dmg).*

Species Richness Index serves to show the number of species in a certain type of habitat using Margalef index [10], with the formula as follows:

$$Dmg = \frac{(S-1)}{\text{In N}} \quad (1)$$

Information:
Dmg : Species Richness Index
S : Number of species
N : Total number of individuals of all types

2.3.3. Evenness Index (E)

$$E = \frac{H'}{\text{In}(S)} \quad (2)$$

Information:
E : Evenness index
H' : Shannon-Wiener index (diversity index)
S : Number of species

2.3.4. Sorensen's Species/Community Similarity Index.

To find out the impact of oil palm plantation on the composition, the number of loss and gain of butterfly species, Sorensen's species/community similarity index was applied [8] as follows:

$$IS = \frac{2c}{a+b} \quad (3)$$

Information:

a : Number of species found in community a

b : Number of species found in community b

c : Number of species found in community a and b

3. Results and discussion

3.1. Comparison of Butterfly Diversity in Baseline and Oil Palm Plantations Area

In the KGP oil palm plantation, this research found 32 species of butterfly from 198 individuals and 4 families, namely Pieride (4 species), Nymphalidae (21 species), Lycanidae (5 species), and Hesperidae (2 species). Data about butterfly species found in KGP oil palm plantation as presented in table 1:

Table 1. Butterfly species in various types of land covers in KGP oil palm plantation.

No	Species	Family	YA	MA	OA	HCV	BA	Σ
1	Amathusia phidippus	Nymphalidae	7	26	9	20	0	62
2	Elymnias hypermnestra	Nymphalidae	9	3	8	1	1	22
3	Ypthima baldus	Nymphalidae	1	4	9	0	0	14
4	Zizina otis	Lycanidae	9	0	0	1	2	12
5	Jamides pura	Lycanidae	8	1	0	2	0	11
6	Eurema hecabe	Pieridae	1	1	1	2	3	8
7	Orsotriaena medus	Nymphalidae	4	1	0	0	3	8
8	Neptis hylas	Nymphalidae	2	2	0	1	1	6
9	Danaus melanippus	Nymphalidae	2	0	0	1	2	5
10	Telicota augias	Hesperidae	4	0	0	1	0	5
11	Other (22 species)		16	7	6	10	6	45
	Total number of species		19	14	9	15	10	
	Total number of individuals		63	45	33	39	18	198

Notes : YA = Young-aged; MA = Middle-aged; OA = Old-aged; HCVA = High Conservation Value Area; BA = Baseline Area.

While in the CNG oil palm plantation, this research found 41 species of butterfly from 193 individuals and 5 families, namely Papilionidae (3 species), Pieridae (6 species), Nymphalidae (26 species), Lycanidae (4 species), Riodinidae (1 species), and Hesperidae (1 species). Data about butterfly species found in CNG oil palm plantation as presented in table 2:

Table 2. Butterfly species in various types of land covers in CNG oil palm plantation.

No	Species	Family	YA	MA	OA	HCV	BA	Σ
1	Amathusia phidippus	Nymphalidae	2	7	7	8	2	26
2	Elymnias hypermnestra	Nymphalidae	6	2	7	3	5	23
3	Eurema hecabe	Pieridae	1	4	1	13	3	22
4	Appias olferna	Pieridae	4	0	1	5	0	10
5	Eurema sari	Pieridae	1	3	2	3	0	9
6	Junonia atlites	Nymphalidae	6	2	1	0	0	9
7	Neptis hylas	Nymphalidae	1	1	2	2	2	8
8	Tanaecia pelea	Nymphalidae	0	0	0	8	0	8
9	Ypthima philomela	Nymphalidae	0	8	0	0	0	8

No	Species	Family	YA	MA	OA	HCV	BA	Σ
10	Ideopsis vulgaris	Nymphalidae	1	1	1	2	0	5
11	Others (31 species)		15	13	5	22	10	65
	Total number of species		20	16	12	10	20	
	Total number of individuals		37	41	27	66	22	193

Notes : YA = Young-aged; MA = Middle-aged; OA = Old-aged; HCVA = High Conservation Value Area; BA = Baseline Area.

The dominance of the Nymphalidae family can be seen on both plantations, in KGP found 73.74% of the family Nymphalidae, whereas in CNG found 63.21% of the family Nymphalidae. This is because this family has the largest members and the widest distribution. [9] stated that Nymphalidae is a family of butterflies that has the highest number of species and cosmopolitan properties, the distribution of families spread in many regions of the world, has a high ability to survive in various types of habitats because it is polyphagic. Polifag is a group of animals that live and eat on various types or many types of plants from various families [10]. The number of family members that produce from this family is also getting bigger, and the nature of the polyps due to this family is not only in one feed plant which varies more in various environmental conditions. According to [11], host plants of the family Nymphalidae are Annonaceae, Asteraceae, Moraceae, Rubiaceae and Anacardiaceae. In addition, reference [12] stated that the food of butterfly from the family of Nymphalidae was Asteraceae, Melastomataceae, Solanaceae, and Poaceae.

A total number of species on both plantations, KGP oil palm plantation and CNG oil palm plantation implicate to the species richness index value. Comparison of total number of species, species richness index value and evenness index value on both plantations as presented in the following figure 1:

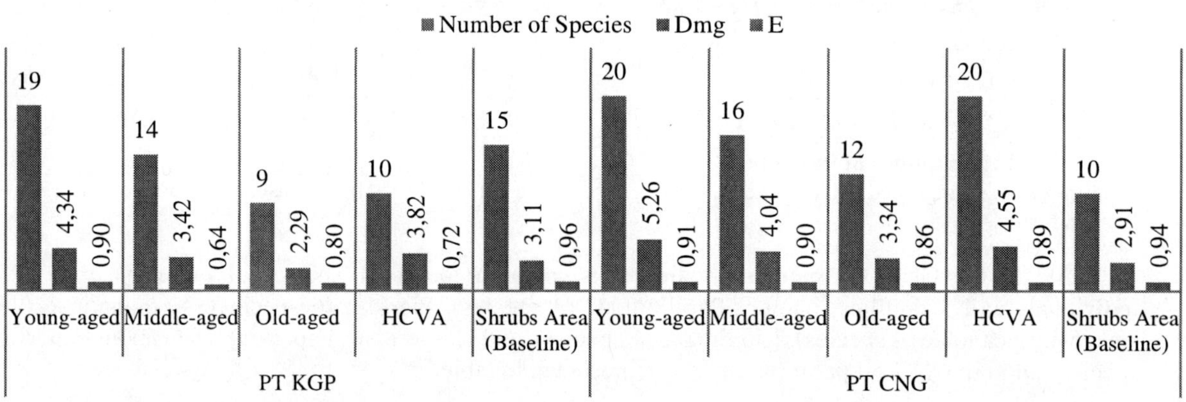

Figure 1. Comparison of total number of species, species richness and evenness index

In KGP oil palm plantation, the highest number of species and richness index values were found in the Young-aged oil palm (S=19 and Dmg=4.34), this result were similar as at the CNG oil palm plantation that the highest number of species and species richness index was in the young-aged oil palm (S = 20 and Dmg = 5.26). Even though, if we look at the number of species in HCVA in CNG plantations it is also 20 species but the value of species richness in young-aged oil palm higher than HCVA area. The highest number of species and species richness in the young-aged oil palm is because this area has water source from the river located in the north of the young-aged oil palm area, as it is known that butterflies like areas that have water sources [13]. In addition to water sources, the presence of butterflies is also affected by canopy cover [14]. The young-aged palm oil has more open canopy cover compared to other oil palm canopy cover. Although the shrubs area in the KGP plantations has more open than the area of the young oil palm because reeds dominate it, but this area has no water source and has very dry soil conditions, so the number of butterflies was found less.

Meanwhile, the lowest number of species and species richness in KGP plantations are found in old oil palm areas (S = 9 and Dmg = 2.29). These results are different from CNG plantations, namely the area with the lowest number of species and species richness value was in the area of scrub (S = 10 and Dmg = 2.91) which is the original land cover. This difference can be due to differences in habitat conditions in each location. The old oil palm area on the KPG plantation has tight canopy cover conditions while the butterflies like the open area so that the butterflies found in this area are less than another land cover in the KPG plantation. Whereas on CNG plantations, the area of shrubs has the lowest number of species and value of species richness because no water source was found in the area and the canopy cover was very tight with rocky soil conditions. Habitat area of CNG bushes is not suitable as a butterfly habitat. Optimal habitat characteristics for butterflies is sufficient light factors, clean or unpolluted air and water as material needed for the humidity of the environment where butterflies live [15].

Generally, the evenness value of both plantations shows a variety of values with the lowest evenness value is 0.64. Evenness values of these types can be classified into evenly categories because the values obtained are high or close to the maximum evenness value (E = 1). The type evenness value can be used to see the distribution of the number of individual species in each type of land cover, the high evenness value means that no particular species dominates a particular habitat.

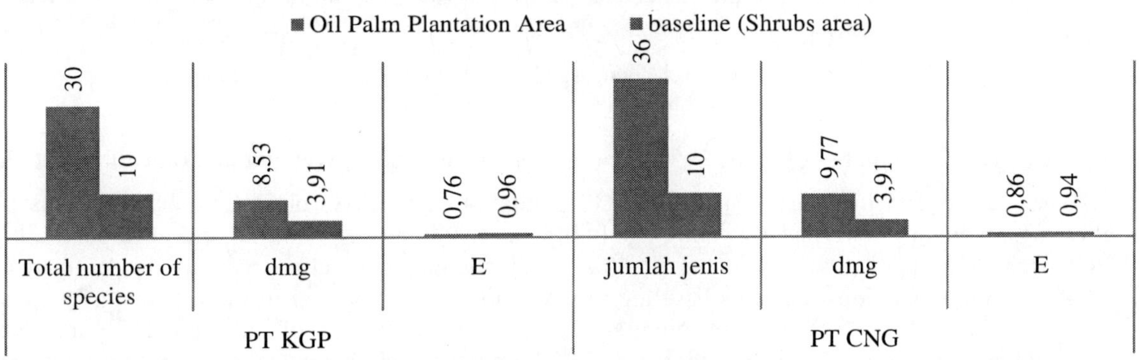

Figure 2. Comparison of total number of species, species richness and evenness index in baseline and oil palm plantations area

The establishment of oil palm plantations requires a land conversion that is the impact on changes its biodiversity [16]. The results of the comparison between the baseline data and the land area in both of plantations showed the same results as the number of species in the plantation area was higher than the baseline area. Likewise, the richness values of butterflies in the plantation area higher than the baseline area. Comparison of the total number of species, species richness index and Evenness index in baseline and both of plantation presented on (figure 2). In the KGP population, there were 30 species of butterflies were found with a value of Dmg = 8.53 at the plantation area while the number of butterflies found at the baseline (Shrubs area) was 10 species with the Richness value or Dmg = 3.91. In the CNG plantations found 36 species of butterflies with Dmg = 9.77 in the plantation area, while the number of butterflies found at the baseline (Shrubs area) was 10 species with the richness value Dmg = 3.91. Evenness values for both oil palm plantations showed that the lowest value was 0.76 which indicated that no butterfly species dominated in both oil palm plantations. According to [17] the high evenness index indicates that locations have similar habitat or even abundance of individuals.

Overall, changes the land cover from shrubs to oil palm plantations in KGP plantations (into types of young oil palm cover, medium palm oil, palm oil and HCV areas) show a loss of 2 butterfly species (*Junonia orithya* and *Moduza procris*), gain of 22 species (*Amathusia phidippus, Rapala scintilla, Catopsilia pyranthe, Catopsilia Scylla, Cethosia hypsea, Euploea mulciber, Eurema sari, Ideopsis vulgaris, Jamides pura, Junonia atlites, Junonia hedonia, Leptosia nina, Mycalesis anapita, Mycalesis fuscum, Mycalesis horsfieldi, Pandita sinope, Parantica agloides, Pelopidas agna, Tanaecia godartii, Telicota augias, Ypthima baldus, Ypthima philomela*) and 8 species survive (*Danaus melanippus,*

Elymnias hypermnestra, Eurema hecabe, Neptis hylas, Orsotriaena medus, Spindasis lohita, Ypthima pandocus, Zizina otis. Similarly, changes the land from shrubs to oil palm plantations on CNG plantations (into types of young oil palm cover, medium oil palm, oil palm and HCV areas) show that 5 species losses (*Neocheritra amrita, Pandita sinope, Papilio memnon, Troides helena, Zizina otis*), 31 species gain (*Acraea violae, Appias lybithea, Appias olferna, Catopsilia phyranthe, Catopsilia schylla, Cethosia hypsea, Cupha erymnathis, Danaus melanippus, Drupadia ravindra, Elymnias kamara, Euploea mulciber, Eurema sari, Faunis stomphax, Hympolimnas bolina, Ideopsis juventa, Ideopsis vulgaris, Junonia atlites, Leptosia nina, Melantis leda, Mycalesis fuscum, Mycalesis horsfieldi*) and 5 species survive (*Amathusia phidippus, Everes lacturnus, Elymnias hypermnestra, Eurema hecabe, Neptis hylas*.

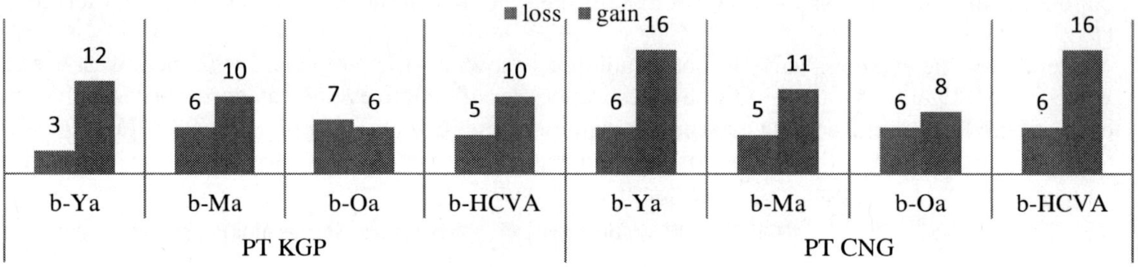

Notes : b = Baseline; Ya = Young-aged; Ma = Middle-aged; Oa = Old-aged; HCVA = High Conservation Value Area;

Figure 3. The total number of butterfly lost and gained due to land conversion into oil palm plantations

Base on the data in figure 3, both of KGP plantations and CNG plantations have various land coverings, including coverings by young-aged oil palm, medium-aged palm oil, old-aged palm oil and HCV areas. Both plantations also come from the same land cover (baseline), namely the scrub area. In KGP oil palm plantations, changes baseline to plantation areas have led to the lost butterfly species lower than gained butterfly species in young-aged oil palm, medium-aged oil palm, and HCV area, while in the old-aged oil palm has the lost butterfly species higher than gained butterfly species. This is because the old palm area has the least number of species and has a closed canopy cover which prevents sunlight from entering while the butterfly likes an open place [18].

In CNG oil palm plantations, all types of land cover after becoming plantations area (young-aged oil palm, medium-aged oil palm, old-aged oil palm, and HCV area) has gained butterfly species higher than lost butterfly species. This shows that the habitat formed from new land cover after the conversion of land from shrubs to plantation areas on CNG plantations still has sufficient ability to support the presence of butterflies including having the availability of living resources such as feed, host plants, shelter and breeding enough for various types of butterflies found.

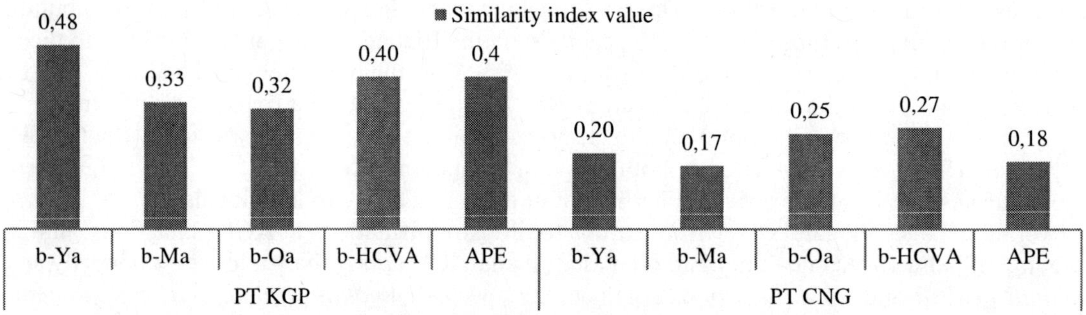

Notes : b = Baseline; Ya = Young-aged; Ma = Middle-aged; Oa = Old-aged; HCVA = High Conservation Value Area; APE = After Plantation Establishment

Figure 4. Species similarity index among Oil Palm Plantation and Baseline Area

Generally, the results of analysis similarity value on both plantations with baselines are quite low, both the plantation area as a management unit or based on the type of land cover that is value below 0.5. Low species similarity values indicate that most species of butterflies found in the plantation area were not found at the baseline (bush area) and vice versa. The similarity between the 2 locations is influenced by the habitat characteristics of each location. According to [19], the low similarity coefficient causes species that are found to be different or almost not in the same composition at these 2 different locations, this indicates that the greater the similarity of species, the greater the similarity of habitat characteristics or vice versa. This is also in line with [20] statement that different habitats have low similarities in species.

4. Conclusion

The highest butterfly diversity was found in oil palm plantations area, both on KGP oil palm plantation and CNG oil palm plantation. The total number of species found in baseline and oil palm plantations area in KGP oil palm plantation was 8 species on each. Conversion of Shrubs area to oil palm plantation has resulted in the gain of 22 species of butterfly and loss 2 species of butterfly. The total number of species found in baseline and oil palm plantations area in CNG oil palm plantation were 5 species on each. Conversion of Shrubs area to oil palm plantation has resulted in the gain of 31 species of butterfly and loss 5 species of butterfly.

Acknowledgments

Further thanks to BPDPKS (Badan Pengelola Dana Perkebunan Kelapa Sawit) for funding this study. Thanks are also due to KGP and CNG oil palm plantation for providing the permissions and facilities provided during the study period.

References

[1] [Ditjenbun] Direktorat Jendral Perkebunan. 2016. *Statistika Perkebunan Indonesia 2015-2017*. Jakarta (ID): Direktorat Jendral Perkebunan

[2] Sheil D, Casson A, Meijaard E, Noordwijk MV, Gaskell J, Groves JS, Wertz K, Kanninen M. 2009. *The Impacts and Opportunities of Oil Palm in Southeast Asia*. Bogor (ID): CIFOR

[3] Danielsen F, Beukema H, Burgess ND, Parish F, Bruhl CA, Donald PF, Murdiyarso D, Phalan B, Reijnders L, Struebig M, Fitzherbert EB. 2009. Biofuel plantations on forested lands: double jeopardy for biodiversity and climate. *Conserv Biol*. **23**(2): 348–358

[4] Obidzinski K, Andriani R, Komarudin H, Andrianto A. 2012. Environmental and social impacts of oil palm plantations and their implications for biofuel dumberproduction in Indonesia. *Ecology and Society*. 17(1):25

[5] Hamer KC, Newton RJ, Edwards FA, Benedick S, Bottrell SH, Edwards DP. 2015. Impacts of selective logging on insectivorous birds in Borneo: the importance of trophic position, body size and foraging height. *Biol Conserv*. **188**:82–88

[6] Brown E, Dudley N, Lindhe A, Muhtaman DR., Stewart C, dan Synnott T (ed.). 2013. *Common guidance for the identification of High Conservation Values*. HCV Resource Network

[7] Lestari DF, Putri RDA, Ridwan M, Purwaningsih AD. 2015. Keanekaragaman kupu-kupu (Insekta: Lepidoptera) di Wana Wisata Alas Bromo, BKPH Lawu Utara, Karanganyar, Jawa Tengah. *Prom Sem Nas Masy Biovid Indon*. **1**(6): 1284-1288

[8] Magurran AE. 1988. *Ecological Diversity and Its Measurement*. New Jersey (US): Princeton University Press

[9] Tabadepu H, Damayanti B, Bandung S. 2008. Butterfly record from salak mountain, Indonesia. *J.Entomol Indonesia*. **5**(1):10-16

[10] Sreekumar, P.G dan M. Balakrishnan. (2001). Habitat dan altitude preferences of butterflies in Aralam Wildlife Sanctuary, Kerala. *Tropical Ecology*. **42**(2):277-281

[11] Dendang B. 2009. Keragaman kupu-kupu di Resort Selabintana Taman Nasional Gunung Gede Pangrango, Jawa Barat. *Jurnal Penelitian Hutan dan Konservasi Alam*. **4**(1): 25-36

[12] Wahyuni I. 2016. Keanekaragaman kupu-kupu pada berbagai tipe tutupan lahan di perkebunan kelapa sawit PT Adimulia Agrolestari, Kampar, Riau. [skripsi]. Bogor (ID) : Institut Pertanian Bogor

[13] Mustari AH, Pramana Y, dan Nurlinda R. 2013. Keanekaragaman kupu-kupu di Taman Nasional Bantimrung Bulusaraung. *Media Konser*. **18**(2):63-68

[14] Koh KP,Sodhi NS. 2004. Importance of reverse, fragments and parks for butterfly conservation in a tropical urban landscape. *Ecological Applications*. **14**(6):1695-1708

[15] Aidid L. 1991. Studi Penangkaran Kupu-Kupu di Taman Nasional Bantimurung Kabupaten Maros, Propinsi Sulawesi Selatan. [skripsi]. Bogor (ID): Institut Pertanian Bogor

[16] Muin, A., 2013. Pengusahaan perkebunan kelapa sawit berwawasan konservasi. [Disertasi]. Bogor (ID): Institut Pertanian Bogor

[17] Priyono B dan Abdullah M. 2013. Keanearagaman jenis kupu-kupu di taman kehati Unnes. *Biosaintiifika*. **5**(2):100-105

[18] Spitzer K, Jaros J, Leps J. 1997. Effect of Small-Scale Disturbance On Butterfly Communities Of An Indochinese Montane Rainforest. *Biol Conserv*. **80**(1):9-15

[19] Meidilaga. 2013. Keanekaragaman Kupu-kupu. Desain Penangkaran dan Pengembangannya sebagai Objek Wisata Hutan Pendidikan Gunung Walat Sukabumi [skripsi]. Bogor (ID): Institut Pertanian Bogor

[20] Koneril L, Saroyo. 2012. Distribusi dan keanekaragaman kupu-kupu (lepidoptera) di gunung manado tua, kawasan taman nasional laut bunaken, Sulawesi Utara. *Jurnal Bumi Lestari*. **12**(2):357-365

IOP Conf. Series: Earth and Environmental Science **336** (2019) 012033 doi:10.1088/1755-1315/336/1/012033

Impact of oil palm plantation on species diversity of tropical vegetation

Iwan Hilwan[1] and Yanto Santosa[2]

[1] Department of Silviculture, Faculty of Forestry, Bogor Agricultural University
[2] Department of Conservation of Forest Resources and Ecotourism, Faculty of Forestry, Bogor Agricultural University

*E-mail: ihilwan@yahoo.co.id

Abstract. The area size of oil palm plantation in Indonesia increases significantly every year. The great expansion of oil palm plantation stimulates the emergence of negative accusation that oil palm plantation development caused deforestation, reduction of biodiversity, and environmental degradation. This research aimed to analyze the impact of oil palm plantation development on species diversity of tropical vegetation. This research was conducted on 5 types of land cover in BPME oil palm plantation (young growth oil palm, medium growth oil palm, and old growth oil palm, bush land, and coconut plantation) by single plot method. Research results show that , the change of bush land to oil palm plantation caused biodiversity loss as many as 7 species (38.89%), biodiversity gain as many as 44 species (244.44%) and 18 species being found both in bush land and oil palm plantation, while 3 species of epiphyte moss were found too. On the other hand, change from coconut plantation to become oil palm plantation caused biodiversity loss as many as 6 species (60%), biodiversity gain as many as 52 species (520%) and 10 species being found both in coconut plantation and oil palm plantation, while 3 species of epiphyte moss were also found.

1. Background

Indonesia has succeeded in becoming the largest oil palm producer in the world with significant contribution toward national economic development. According to Directorate General of Plantation, Ministry of Agriculture [1], such contribution was among others in the form of state foreign exchange earning as large as around Rp 287 trillion in the year 2017, employment opportunity for more than 5.5 million people in on-farm sector (plantation) and in other multiplier effects sector. Area size of oil palm plantation in Indonesia exhibited significant increase from year to year, and in the year 2017 has reached area size of 14.03 million hectare. Such high rate expansion of oil palm plantation, stimulate the emergence of negative accusation toward oil palm management, namely accusation of oil palm plantation as the cause of biodiversity reduction, deforestation and environmental degradation.

One of the negative accusation toward the development of oil palm plantation as the cause of biodiversity reduction, especially for vegetation, should be proven in the field through research. Therefore, the objectives of this research were: (1) making inventory of plant species diversity on land before being converted to oil palm plantation (which in this research was in the form of coconut plantation and bush) and on young growth oil palm plantation, moderate age oil palm plantation and old growth oil palm plantation, and (2) estimating the magnitude/ value of loss / gain of plant species diversity resulting from oil palm plantation development. The desired output of this research is data and information on the magnitude of loss/gain of species diversity due to oil palm plantation development.

Data and information obtained from this research will serve as one important source of information concerning the impact of oil palm plantation development on loss/gain of plant species diversity. These data will also serve as consideration for developing oil palm plantation which supports biodiversity conservation. Particularly for companies, data, especially those which are related with biodiversity, could enrich the collection of data which are needed for certification process by ISPO (*Indonesian Sustainable Palm Oil*) or RSPO (*Roundtable on Sustainable Palm Oil*). Besides that, the research results could also be used by the Indonesian government or private institutions to give clarification or explanation in responding to the negative accusation, and for designing the best efforts to handle the issue of biodiversity reduction

2. Materials and methods

This research was conducted in the area of BPME plantation, regency of Indragiri Hilir, the province of Riau, at 5 kinds of land cover, namely: (1) young age oil palm plantation (4 years old), (2) moderate age oil palm plantation (6 years old), (3) old age oil palm plantation (9 years old), (4) bush land area, and (5) coconut plantation. Bush land and oil palm plantation were land cover type before conversion to oil palm plantation.

On the 5 land cover types, there were construction of single compartment measuring 40 m x 40 m for each land cover type. The area of 40 m x 40 m is optimal sample plot for measuring plant species diversity in low land tropical forest [2]. Afterwards, in every single compartment, there was construction of 20 subcompartments measuring 2m x 2m for analysis of undergrowth vegetation and shrubs. The collected data were in the form of plant species and their abundance.

In data analysis, several equations were used, namely:

a. Species Richness Index of Margalef :

$$Dmg = (S-1)/\ln N$$

b. Species Diversity Index of Shannon-Wienner :

$$H' = - \sum pi. \ln pi$$

c. Species Similarity Index of Sorensen :

$$IS = 2\,C / (A + B)$$

d. Percentage of species biodiversity loss = $\frac{\text{Number of lost species}}{\text{Number of similar species}} \times 100\ \%$

e. Percentage of Biodiversity Gain = $\frac{\text{Number of Species Gain}}{\text{Number of similar species}} \times 100\ \%$

3. Results and discussion

3.1. Number of species and diversity

Table 1 shows that the number of species in the initial land cover was as many as 35 species comprising SRB (25 species) and CCP (16 species). In oil palm plantation land cover, there were found 64 plant species which could be detailed as follows: in young oil palm plantation (YPO) there were 23 species, in moderate age oil palm plantation (MPO) 46 species, and in old age oil palm plantation (OPO) 34 species. Therefore, number of plant species in land cover in the form of oil palm plantation was greater than those in initial land cover in the form of SRB and CCP. In terms of their habitus, undergrowth vegetation in the form of herbs and shrubs possessed a far greater number of species as compared with those of tree habitus. The number of plant species in the oil palm plantation, as many as 64 species was a little bit higher than that in study results of reference [3] in oil palm plantation in Sumatera, namely as many as 56 species. On the other hand, reference [2], reported that undergrowth vegetation in Dharmasraya area, West Sumatera, comprised only 39 species.

Table 1. Number of plant species in each land cover type

Land Cover		**Species Number**					
		Herb	**Seedling**	**Sapling**	**Poles**	**Tree**	**Number of Spec.**
Previously	SRB	24	1	1	0	0	25
	CCP	11	2	3	0	1	16
				Total			35
Oil Palm Plantation	YPO	22	0	0	0		23
	MPO	45	1	2	0		46
	OPO	33	1	0	0		34
				Total			64

Notes: SRB = Shrub, CCP = Coconut Plantation, YPO = Young Growth Oil Palm, MPO = Medium Growth Oil Palm, OPO = Old Growth Oil Palm

Undergrowth species with the highest abundance was pakis harupat or *Nephrolepis biserrata* (jalak harupat) and *Asystasia gangetica*. Species *N. biserrata* was so abundant because this species was left to grow uncontrolled (not being eradicated) by the plantation manager. The purpose of this policy was for maintaining soil surface humidity in order to keep it to remain high. According to [4], the presence of *N. biserrata* as land cover crop reduce water deficit by 51.53% at the condition of maximum water deficit in oil palm plantation land. Besides that, another contribution of *N. biserrata* was as the contributor of nutrients as large as 15.7 tons of dry matter/ha/year or 7.7 ton C/ha/year. Species *A. gangetica,* which is like *N. biserrata,* is categorized as undergrowth vegetation which can grow under the shade of oil palm. [3] found the species *A. ganggetica* to be very dominant, with highest IVI in 10 years old oil palm plantation in Muaro Jambi. This made the species *A. ganggetica,* which was also categorized as oil palm weeds, to be very difficult to eradicate. Other species which was also categorized as having high population were *Peperomia pellucida, Digitaria adscendens,* and *Stenochlaena palustris* (pakis kelakai).

Table 1 also shows that the number of plant species varies greatly between land covers. This phenomenon also resulted in varying values of Shannon-Wienner Species Diversity Index (H') and Margalef Species Diversity Index (Dmg) (figure 1). The number of species will linearly affect the value of H' and Dmg. Area of MPO with highest S (46 species) also possessed the highest value of H' and Dmg (Dmg = 5.78 and H' = 2.31). On the other hand, in area of CCP with smallest S (18), the value of Dmg was also smallest (2.07). The low species richness in CCP was supposed possibly due to considerably high disturbance by human activities. In figure 1, it can be seen the number of species (S) and species richness (Dmg) at various land covers.

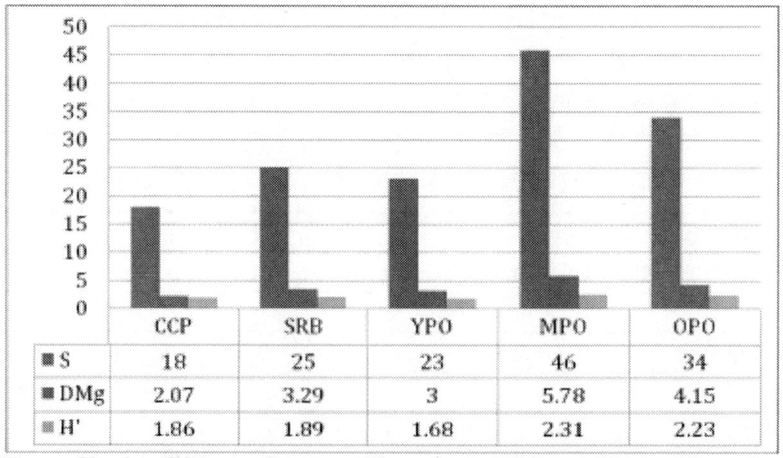

Notes: SRB = Shrub, CCP = Coconut Plantation, YPO = Young Growth Oil Palm, MPO = Medium Growth Oil Palm, OPO = Old Growth Oil Palm, S = Number of Species, Dmg = Richness Index of Margalef, H' = Diversity Index of Shannon-Wienner

Figure 1. Number of species and level of plant species diversity at each type of land cover.

It can be seen, in figure 1, the increase of number of plant species (S), Shannon-Wiener Species Diversity Indexes (H') and Margalef Species Diversity Index (Dmg) due to change from bush land and coconut plantation to oil palm plantation. In young oil palm plantation (YPO), such increase of plant species diversity had not occurred, but this diversity increased in medium age oil palm plantation (MPO) or old age oil palm plantation (OPO). Therefore, in terms of plant species diversity, conversion from bush land to oil palm plantation was very positive.

3.2. Community similarity
Levels of plant community similarity between land cover types were shown by Sorensen Similarity Index (IS). The IS measure similarity of species composition between 2 types of land cover being compared. Based on measurements, there were no 2 types of land cover which possessed similarity of species composition which composed them. This appears from IS values which are all less than 0.75. The largest value of IS, namely 0.60 occurred between moderate age oil palm plantation (MPO) and old age oil palm plantation (OPO). Complete values of IS can be seen in table 2.

Table 2. Indexes of Similarity between 2 communities or land cover types

Land cover	YPO	MPO	OPO	CCP	SRB
YPO	1	0.38	0.56	0.17	0.54
MPO		1	0.60	0.20	0.25
OPO			1	0.27	0.42
CCP				1	0.33
SRB					1

Notes: SRB = Shrub, CCP = Coconut Plantation, YPO = Young Growth Oil Palm, MPO = Medium Growth Oil Palm, OPO = Old Growth Oil Palm.

3.3. Biodiversity loss and gain
In the change from bush land to oil palm plantation there was loss of plant species, while at the same time, there was gain of plant species, while also there were plant species which were found in both types of land cover being compared (table 3). In medium age oil palm plantation (MPO), number of plant species obtained (*gain*) was categorized as high, namely as many as 34 species or as large as

261.54 %, and the lowest was found in young age oil palm plantation (YPO) which was only 10 species or as large as 76.92%. On the whole, within the area of oil palm plantation (POP), there were 44 plant species being obtained or as large as 244.44%, with only 7 plant species being lost , or as large as 38.89 %.

Table 3. Loss and gain of species biodiversity between bush land and oil palm plantation.

SRB	YPO	MPO	OPO	POP	Loss	Gain	Similar	% Loss	% Gain
25	23				12	10	13	92.31	76.92
25		47			12	34	13	92.31	261.54
25			36		11	22	14	78.57	157.14
25				59	7	44	18	38.89	244.44

Notes: SRB = Shrub, YPO = Young Growth Oil Palm, MPO = Medium Growth Oil Palm, OPO = Old Growth Oil Palm, POP = Palm Oil Plantation

The change from coconut plantation to oil palm plantation also exhibited plant species loss, and gain in plant species, as well as plant species which were found in both types of land cover being compared (table 4). Number of plant species obtained (*gain*) was categorized as high in medium age oil palm plantation (MPO), namely as many as 38 species or 422.22%, while the lowest was found in young age oil palm plantation (YPO) namely as many as 15 species or as large as 187.50%. On the whole, within the area of oil palm plantation (POP), there were 52 plant species being obtained or as large as 520.00%, with only 6 plant species being lost or as large as 60 %.

Number of obtained plant species (*gain*) due to change from coconut plantation to oil palm plantation (52 species) was greater than that of change from bush land to oil palm plantation (44 species).

Table 4. Loss and gain of species diversity between coconut plantation and oil palm plantation

CCP	YPO	MPO	OPO	POP	Loss	Gain	Similar	% Loss	% Gain
16	23				8	15	8	100.00	187.50
16		47			7	38	9	77.78	422.22
16			36		7	27	9	77.78	300.00
16				62	6	52	10	60.00	520.00

Notes: CCP = Coconut Plantation, YPO = Young Growth Oil Palm, MPO = Medium Growth Oil Palm, OPO = Old Growth Oil Palm

4. Conclusion

Change of land cover from bush land and coconut plantation to oil palm plantation had a positive impact, namely increase in plant biodiversity. This was indicated by the following phenomena:

1. Number of plant species in oil palm plantation was 2.5 and 4 times greater than in bush land and than that of coconut plantation respectively.
2. Change of land cover from bush land to oil palm plantation obtained a gain of 44 species of plants (244.44%) and loss of 7 species (38.89%).
3. Change of land cover from coconut plantation to oil palm plantation obtained a gain of 52 species of plants (520%) and loss of 6 species (60 %).

References

[1] Direktur Jenderal Kementerian Pertanian. 2018. Industri Kelapa Sawit Berkontribusi Besar terhadap Ekonomi

[2] Kusuma S. 2007. Penentuan Bentuk dan Luas Plot Contoh Optimal Pengukuran Keanekaragaman Spesies Tumbuhan pada Hutan Hujan Dataran Rendah: Studi Kasus di TN Kutai. [Tesis] Sekolah Pascasarjana IPB

IOP Conf. Series: Earth and Environmental Science **336** (2019) 012033 doi:10.1088/1755-1315/336/1/012033

[3] Hutwan S, Novianti S, Adriani A. 2017. Analisis Indeks Kepekaan Ekologi terhadap Hijauan Pakan di Bawah Perkebunan Kelapa Sawit. Jurnal Ilmu-Ilmu Peternakan Vol.**20** No.1: 25-32. eISSN: 2528 0805 pISSN: 1410 7791

[4] Purwantari ND. 2016. Sumberdaya Genetik Tanaman Pakan Ternak Toleran Naungan. WARTAZOA Vol. **26** No.2 Th 20016 Hal. 051-056

[5] Rahmadayanti R, Nursyahra, Rizki. 2016. Komposisi Vegetasi Dasar pada Perkebunan Kelapa Sawir di Trimulya, Kecamatan Timpeh, Kabupaten Dhramasraya. STKIP PGRI Sumatera Barat

[6] Ariyanti M. 2016. Peranan Tanaman Penutup Tanah *Nephrolepis biserrata* pada Teknik Konservasi Tanah dan Air terhadap Neraca Air di Perkebunan Kelapa Sawit. [Thesis] Sekolah Pascasarjana IPB

AUTHOR INDEX